Gas Globes Amoco to Mobil & Affiliates

Scott Benjamin & Wayne Henderson

Schiffer Publishing Ltd
4880 Lower Valley Road, Atglen, PA 19310 USA

This book is dedicated to our families and all our friends everywhere.

Title page photos:
ST: SOCONY MOTOR GASOLINE (porcelain lens), 1915-1926, 14" METAL SPC, $2000-2500. *Courtesy of PCM Archives.*

2ND: GILMORE RED LION W/TETRAETHYL, 1933-1942, 15" METAL BODY GLOBE, $8000-10000. *Courtesy of PCM Archives.*

3RD: WHITE EAGLE GASOLINE, 1918-1920, ONE PIECE ETCHED GLOBE, $7500-10000. *Courtesy of PCM Archives.*

Library of Congress Catalog Card Number: 98-88859

Designed by "Sue"
Type set in Americana XBd BT/Zurich BT

ISBN: 0-7643-0668-5
Printed in China
1 2 3 4

Published by Schiffer Publishing Ltd.
4880 Lower Valley Road
Atglen, PA 19310
Phone: (610) 593-1777; Fax: (610) 593-2002
E-mail: Schifferbk@aol.com
Please visit our web site catalog at
www.schifferbooks.com

This book may be purchased from the publisher.
Include $3.95 for shipping.
Please try your bookstore first.
We are interested in hearing from authors
with book ideas on related subjects.
You may write for a free catalog.

In Europe, Schiffer books are distributed by
Bushwood Books
6 Marksbury Rd.
Kew Gardens
Surrey TW9 4JF England
Phone: 44 (0)181 392-8585; Fax: 44 (0)181 392-9876
E-mail: Bushwd@aol.com

Contents

Acknowledgments ... 4
Cross Reference to Affiliated Gas Companies ... 5
Chapter One. How to Use Your New Gasoline Globe Book ... 7
Chapter Two. Collecting Gasoline Globes ... 12
Chapter Three. The Different Types of Globes ... 13
Chapter Four. Theory of Pricing Gasoline Globes ... 18
Chapter Five. Gasoline Globes as Investments ... 19
Chapter Six. Reproduction Gasoline Globes ... 20
Chapter Seven. Packing Gasoline Globes for Shipment ... 26
Chapter Eight. Restoring Gasoline Globes ... 27
Chapter Nine. Overview of the Petroleum Collectibles Hobby ... 29
Chapter Ten. Major Oil Company Globes: A-M (listings begin on page 129) ... 31
- Amoco & Pan-Am ... 31
- Asamera & Affiliates ... 33
- Ashland & Affiliates ... 33
- Atlantic & Affiliates ... 45
- Champlin & Affiliates ... 48
- Chevron & Affiliates ... 49
- Cities Service & Affiliates ... 50
- Clark & Affiliates ... 57
- Colonial & Affiliates ... 57
- Conoco & Affiliates ... 58
- Co-Operative Associations ... 60
- Crown & Affiliates ... 67
- Crystal & Affiliates ... 67
- Derby & Affiliates ... 68
- Dixie Distributors ... 69
- DX & Affiliates ... 71
- Erickson & Affiliates ... 75
- Exxon & Affiliates ... 75
- Fina & Affiliates ... 84
- Flying A/Getty ... 86
- Gulf & Affiliates ... 92
- Hess & Affiliates ... 96
- Hudson & Affiliates ... 97
- Husky & Affiliates ... 99
- Independent Oil Men of American (IOMA) ... 103
- Kendall & Affiliates ... 104
- Kerr McGee & Affiliates ... 107
- Lion Oil Company ... 108
- Mapco & Affiliates ... 108
- Marathon & Affiliates ... 116
- Martin & Affiliates ... 116
- Midwest & Affiliates ... 118
- Mobil & Affiliates ... 120

Glossary ... 202
Index ... 204

Acknowledgments

The following people have made significant contributions to this new globe book in terms of supplying numerous photos, consultation or general help. Their efforts are greatly appreciated. It is impossible to list everyone and we apologize if your name is not listed.

Pat Anderton
Preston Aust
Bruce Bednar
Calloway Oil Company
Don Cooper
Jim Covers
Tom Dooley
Lonnie Hop
George Kraus
Don Loughren
Jim Massen
Bob Mewes
Kyle Moore
Rick Pease
Pergl Gas Pump Globes
Pat Quail
Ed Schwartz
Bob Scism, Jr.
Larry Starkweather
Jim Welty
Walt Wimer, Jr.
R.V. Witherspoon

Countless others have contributed photos and with their help only then could we compile such a large volume. The readers of *Petroleum Collectibles Monthly* sent many photos and their help is greatly appreciated as well.

And, last but not least I'd like to thank my wife, Jeannette Benjamin, who helped compile, type and edit this book as well as photograph globes and my children, Zachary, Breanna and Jaren for not breaking any globes before we could photograph them!

—Scott Benjamin

In addition to those listed above, to whom I also offer a heartfelt thanks, I would like to thank my wife, Joanne Crowley, and my children, Jason and Chris Henderson, and Lora, John and Worth Crowley for without their patience and co-operation, and their interest, this book would not have been possible. Joanne, who worked on this book during the period of our engagement, helped immensely with typing and editing, and encouragement, in the finishing days of the book. Finally I would like to thank my parents, Carl and Jackie Henderson. Without them, and a little wooden Gulf sign they carved for their toddler son, none of this would have been possible.

— Wayne Henderson

LION PETROLEUM PRODUCTS, 1930-1946, 12" METAL, $1800-2500. *Courtesy of PCM Archives.*

SOCONY, an affiliate of Mobil. See more in this line on page 120.

Cross Reference to Affiliated Gas Companies

The individual companies listed were purchased by the major oil companies (A-M) and are found under the listings for the major oil companies in Chapter 10. The major oil companies are italicized in this cross reference, following the name of the individual companies.

A Gas Company, *IOMA*
A.R. Newcombe Oil Company, *Cities Service*
Aetna Oil, *Ashland*
Alert Oil Company, *Exxon*
Amerada Hess, *Hess*
American Liberty, *Fina*
American Oil Company, *Amoco & Pan-Am*
American Oil Company, *IOMA*
American Petrofina, *Fina*
Andy-Dandy Oil Company, *Asamera*
Anglo-American Oil Company, *Gulf*
Arctic Oil Company, *Gulf*
Asamera Oil Company, *Asamera*
Ashland Oil and Refining, *Ashland*
Ashland, *Ashland*
Associated Oil Company, *Flying A/Getty*
Atlantic Refining Company, *Atlantic*
Aurora Gasoline Company, *Marathon*
Bareco, *DX*
Barkhausen Oil Company, *IOMA*
Barnsdall Refining Company, *DX*
Bartles-McGuire Oil Company, *Mobil*
Beebee Oil Company, *IOMA*
Billups Eastern Petroleum, *Hess*
Blake Oil Company, *IOMA*
Bonded, *Marathon*
Brandt Oil Company, *IOMA*
Brilliant Bronze, *Exxon*
British American Oil Company, Ltd., *Gulf*
Buford Oil Company, *Dixie*
Bulk Petroleum Company, *Gulf*
CSC Oil Company, *IOMA*
California Company, Rocky Mtn. Marketing, *Chevron*
California Oil Company, Northeast Marketing, *Chevron*
Capitol Oil Company, *Amoco & Pan-Am*
Capitol Oil Company, *Mapco*
Carpenter Oil Company, *Gulf*
Carter Hall Company, *Cities Service*
Carter Oil Company, *Exxon*
Central Co-Operative Wholesale, *Co-Operative Associations*
Champlain Oil Company, Ltd., *Exxon*
Champlin Oil and Refining, *Champlin*
Cheker Oil Company, *Marathon*
Cincinnati Oil, *Ashland*
Citgo, *Cities Service*
Cities Service Oil Company, *Cities Service*
Cities Service Oils Once Always, *Cities Service*
Clark Oil & Refining Corporation, *Clark*
Clipper Oil Company, *Mobil*
Co-Operative GLF, *Co-Operative Associations*
Coast Oil Company, *Kerr McGee*
Col-Tex, *Fina*
Colonial Beacon Oil Company, *Exxon*
Colonial Oil Company, *Fina*
Colonial Oil Industries, *Colonial*
Colonial/Progressive, *Marathon*
Consolidated, *Marathon*
Consumers Oil Company, *IOMA*
Continental Oil Company, *Conoco*
Cosden Petroleum, *Fina*
Crew-Levick Company, *Cities Service*
Crown Central Petroleum, *Crown*
Crystal Oil & Land Company, *Crystal*
Crystal Petroleum Company, *Crystal*
Dance, *Ashland*
Deep Rock Oil Company, *Kerr McGee*
Delta Oil Company, *Mapco*
Derby Oil Company, *Derby*
Direct Service Oil Company, *Conoco*
Dixie Distributors, Inc., *Dixie*
Dobson Brothers Oil Company, *IOMA*
Douglas Oil Company, *Conoco*
Duro Oil Company, *Marathon*
Eason Oil Company, *Champlin*
Eastern Petroleum Corporation, *Crystal*
Elwood Oil Company, *Cities Service*
Enco, *Exxon*
Energy Oil Company, *Crown*
Erickson Oil Company, *Erickson*
Erickson Oil Products, *Erickson*
Exxon, *Exxon*
F.P. Kendall & Company, *Conoco*
Falcon, *Co-Operative Associations*
Farmer Educational & Co-Operative State Union, *Co-Operative Associations*
Farmers Co-Operative Exchange, *Co-Operative Associations*
Farmers Gas & Oil Company of Michigan, *Co-Operative Associations*
Farmers Union Co-Operative Organization of Iowa, *Co-Operative Associations*
Farmland Industries, *Co-Operative Associations*
Farmland Union Central Exchange, *Co-Operative Associations*
Fina, *Fina*
Fisca Oil, *Hudson*
Freedom Oil Works, *Ashland*
Freedom- Valvoline, *Ashland*
Frei-Ulrich Oil Company, *IOMA*
Frontier Oil, *Husky*
Frontier Refining, *Ashland*
Gas & Oil Products, Ltd., *Gulf*
Gasamat, *Asamera*
Gaseteria Oil Company, *Exxon*
General Petroleum Corporation, *Mobil*
Gerkins Oil Company, *IOMA*
Gibson Brothers Oil Company, *IOMA*
Gilmore Oil Company, *Mobil*
Golden Rule Oil Company, *Champlin*
Golden Rule, *Kerr McGee*
Growmark Industries, *Co-Operative Associations*
Gulf Oil Company, *Gulf*
Guthrie Refining Company, *Fina*
H. Earl Clark, Inc., *Husky*
H.K. Stahl & Company, *Midwest*
H.S. Goodell & Sons, *IOMA*
Harbor, *Champlin*
Hawkeye Oil Company, *DX*
Hawkeye Oil Company, *IOMA*
Hess Oil & Chemical Company, *Hess*
Hi-Way Refineries, Ltd., *Gulf*
High Hat, *IOMA*
High Octane Terminal Company, *Crystal*
Highway Oil Company, *Hudson*
Home Oil Company, *IOMA*
Home Oil Company, Ltd., *Exxon*
Hood Oil Company, *Ashland*
Hoosier Oil Company, *IOMA*
Hoosier Petroleum Corporation, *Exxon*
Hornet Oil Company, *Hudson*
Hudson Oil Company, *Hudson*
Hudson Oil Company, Kansas City, *Hudson*
Humble Oil and Refining, *Exxon*
Husky Oil Company, *Husky*
Husky-Hi Power, Inc., *Husky*
Imperial Oil Limited, *Exxon*
Imperial Refineries, *Marathon*
Independent Oil Company, *IOMA*
Independent Oil Company of Pennsylvania, *Mobil*
Independent Oil Works, *Ashland*
Independent Oil Men of American, *IOMA*
Iowa Farm Service Company, *Co-Operative Associations*
J.A. Hogshire & Son, *Marathon*
Jenny Manufacturing Company, *Cities Service*
Johnson Oil Company, *Marathon*
Johnson Oil Refining Company, *Exxon*
Jones Brothers Oil Company, *IOMA*
Kansas Farmers Union, *Co-Operative Associations*
Kendall Oil, *Husky*
Kermac, *Kerr McGee*
Kerr McGee Oil Industries, *Kerr McGee*
Kerr McGee, *Kerr McGee*
Kimmel Oil Company, *IOMA*
Kriegler, *Conoco*
L.L. Coryell & Sons, *Gulf*
Landmark Co-Operatives, *Co-Operative Associations*
Life Oil Company, *Erickson*

Lincoln Oil Company, *Cities Service*
Lincoln Oil Refining Company, *Marathon*
Lindsay-McMillan Company, *Cities Service*
Lion, *Lion*
Louisiana Oil Refining Company, *Cities Service*
Lubrite Refining Company, *Mobil*
Magnolia Petroleum, *Mobil*
Major's, *Hudson*
Malco Refining, *Conoco*
Mapco, *Mapco*
Marathon Oil Corporation, *Marathon*
Martin Oil Company, *Martin*
Martin Oil Service, *Marathon*
Martin Oil Service, *Martin*
Maryland Co-Operatives, *Co-Operative Associations*
Matheson Oil Company, *IOMA*
McCormick Oil Company, *IOMA*
McQuarter Oil Company, *IOMA*
Meadville Corporation, *Hess*
Mercury Discount Stations, Inc., *Crystal*
Metro, *Mobil*
Mid-Continent Petroleum, *DX*
Midland Cooperatives, *Co-Operative Associations*
Midwest Oil Company, *Midwest*
Mileage Mart, *Kerr McGee*
Milton Oil Company, *Cities Service*
Milton-Savex, *Cities Service*
Minnesota Farm Bureau, *Co-Operative Associations*
Missouri Farm Bureau, *Co-Operative Associations*
Missouri Farmers Association, *Co-Operative Associations*
Mobil Gas/Socony-Vacuum, *Mobil*
Mobil, *Mobil*
Mohawk Oil Company, *Flying A/Getty*
Monarch Manufacturing Company, *DX*
Musco Oil Company, *IOMA*
Mutual Oil Company, *Kerr McGee*
Naph-Sol Refining Company, *Marathon*
National Oil Company, *Conoco*
National Refining, *Ashland*
New Mexico Oil Company, *Conoco*
Northland Gasoline, *IOMA*
Northwest Oil Company, *Flying A/Getty*
Northwest Refining Company, *Exxon*
Northwestern Refining Company, *Erickson*
Ohio Oil Company, *Marathon*
Oklahoma Oil Company, *Exxon*
Old Dutch Refining Company, *Marathon*
Olentangy Oil Company, *Exxon*
Olixir, *Hess*
O'Neils Paint & Hardware Company, *IOMA*
Onyx, *Fina*
Oriental Refining Company, *Asamera*
Osceola Refining Company, *Marathon*
Oskey Bros. Petroleum, *Ashland*
Oval E, *Exxon*
Owens Oil Company, *Clark*
Pan-Am Southern Corporation, *Amoco & Pan-Am*
Panhandle Oil Company, *Fina*
Pate Oil Company, *Exxon*
Pathfinder Petroleum, *Flying A/Getty*
Payless, *Ashland*
Payless-Dance, *Ashland*
PDQ, *Chevron*
Pendergast Oil Company, *IOMA*
Penn Valley Oil Company, *IOMA*
Peoples Oil Company, *Kerr McGee*
Pepper, *Ashland*
Petersons Oil Company, *IOMA*
Petro-Atlas, *Fina*
Petroleum Marketing Company, *Fina*
Pettico Oil Company, *IOMA*
Pioneer, *Cities Service*
Platolene, *Crystal*
Powerine Oil Company, *Exxon*
Premier, *DX*
Pure Pep, *IOMA*
Puritan Oil Company, Ltd., *Gulf*
Pyroil Chemical Company, *Ashland*
Quality Gasoline Company, *Exxon*
Quincy, *IOMA*
Red "C" Oil Company, *Atlantic*
Red Ace Petroleum, *Mapco*
Red Fox Oil Company, *Marathon*
Red Head Oil Company, *Ashland*
Republic Oil Company, *Marathon*
Road Chief, *Marathon*
Rock Island Refining, *Marathon*
Royal & Royal, *IOMA*
Royalite Oil Company, Ltd., *Gulf*
Ryan's, *Exxon*
S.M. Oil Company, *IOMA*
Seaside Oil Company, *Flying A/Getty*
Senco, Inc., *Exxon*
Shoemaker Coal Company, *Cities Service*
Signal, *Chevron*
Silver King, *Hudson*
Smith Oil Company, *Gulf*
Southern Farmers Association, *Co-Operative Associations*
Southern Independent Oil & Refining Company, *Mapco*
Southern Oil Company, *Exxon*
Southern Oil Company, *Hess*
Southern Oil of New York, *Ashland*
Southern States Co-Operatives, Inc., *Co-Operative Associations*
Southport Oil Corporation, *Kerr McGee*
Sovereign Service, *Derby*
Spar, *Hess*
Sparky, *Kerr McGee*
Standard Oil Company of California, *Chevron*
Standard Oil Company of Louisiana, *Exxon*
Standard Oil Company of New Jersey, *Exxon*
Standard Oil Company of New York, *Mobil*
Standard Stations, Inc. & Standard Oil/Chevron Dealers, *Chevron*
Stephen, *Marathon*
Stone Oil Company, *Crystal*
Sun Glo, *Marathon*
Sunray Oil, *DX*
Sunset Oil Company, *Gulf*
SuperAmerica, *Ashland*
Superior Oil Company, *Gulf*
Swanson Oil Corporation, *IOMA*
Tan-Kar Stations, *Kerr McGee*
Target-Western Oil Company, *Clark*
Tennessee Co-Operatives, *Co-Operative Associations*
Texas Pacific Coal & Oil, *Exxon*
Thrift, *Crown*
Tidal Refining Company, *Flying A/Getty*
Tidewater Associated Oil Company, *Flying A/Getty*
Tidewater Oil Company, *Flying A/Getty*
Tower Oil Company, *Marathon*
Trackside Service Stations, *Kerr McGee*
Traffic Oil Company, *IOMA*
Transcontinental Oil Company, *Marathon*
Tresler Oil, *Ashland*
Tri Star, *IOMA*
Tri-State, *Ashland*
Triangle Refineries, *Kerr McGee*
Tulsa Oil Company, *Crystal*
Tulsa Oil Company, *Marathon*
UFA Central Co-Operatives Association, Ltd., *Co-Operative Associations*
Union Oil Company, *Ashland*
Union Oil Company, *Co-Operative Associations*
United Co-Operatives, Inc., *Co-Operative Associations*
United Oil Company, *Kerr McGee*
United, *Marathon*
Vacuum Oil Company, *Mobil*
Valley Oil Company, *Cities Service*
Valley Oil Company, *Conoco*
Valvoline Oil Company, *Ashland*
Vapor Gasoline, *IOMA*
Vaught Oil Company, *Kerr McGee*
Vortex Oil Company, *Exxon*
Wadhams Oil Corporation, *Mobil*
Wallace & Walco, *Kerr McGee*
Waverly Oil Works, *Husky*
Weeks Oil Company, *IOMA*
Western Oil & Fuel, *Conoco*
Western Oil & Fuel, *Husky*
Western States Petroleum, *Husky*
Western, *Conoco*
Westland Oil Company, *IOMA*
Weyanewga Oil Company, *IOMA*
White Eagle Oil & Refining, *Mobil*
White Rose/Mid-West Refining, *Ashland*
White Star Oil Company of Canada, *Mobil*
White Star Refining Company, *Mobil*
Wides Oil Company, *Fina*
Wilshire Oil Company, *Gulf*
Winona Oil Company, *Cities Service*
Wood Oil Company, *Exxon*
Yale Oil, *Exxon*
Zenith, *Conoco*

GILMORE "ROAR WITH GILMORE", 1925-1942, 15" METAL BODY GLOBE, $7000-9000. *Courtesy of PCM Archives.*

Chapter 1. How to Use Your New Gasoline Globe Book

The globe listings on the following pages are divided into the major oil companies (A-M) and their affiliates in alphabetical order.

When looking up a globe, use the cross-reference to affiliated oil companies or the index. If your globe does not appear in the Major Oil Companies and their affiliates listing, then it may be an affiliate of one of the Major Oil Companies (N-Z), an independent, or either a Foreign Oil Company or a generic globe — all found in the second volume of this series. The possibility does also exist that your globe is not listed in either book! Although we have added over one thousand new listings we know there are many more globes left out that have yet to be discovered. Please contact us if your globe does not appear in these books:
Scott Benjamin & Wayne Henderson
P.O. Box 556
LaGrange, OH 44050-0556
440/355-6608

Writing this book has proved to be a lesson in history for all of us. Pure Oil Co. bought out Hi Speed Oil. For example, your Hi Speed globe will be found in the cross-reference section directing you to Pure Oil Co. So, you will need to look in the "P" section of the Major Oil Co. headings to find Pure Oil.

Independents are listed alphabetically, so, if it is not a Major Oil Company, look in the Independent section. Canadian/Foreign globes are listed accordingly.

Please study the next section thoroughly for understanding each globe listing.

Here is a typical *example* and its breakdown:

Item Description	Rarity	Years Used	Body Type	Body Style	Value
Cities Service Oils Once Always	6	1914-1918	OPE	W	$2500-3800

Cities Service Oils Once Always = name of globe or name on globe
6 = rarity guide listing (see below)
1914-1918 = approximate years the particular item was used
OPE = one-piece etched
W = wide body (see section describing different styles)
$2,500-$3,800 = wholesale to retail price value

Example #2:

Item Description	Rarity	Years Used	Body Type	Body Style	Value
Cities Service Oils Once Always	3	1918-1936	16.5" metal	H/L black	$400-650

Here we are describing a Cities Service Oils Once Always 16-1/2" metal used from 1918-1936, rarity 3, which will be found on a High or Low Profile black body with a wholesale value of $400 and a retail value of $650.

Below is the information you'll need to decipher the new codes, the new rarity guide and other information.

Rarity Guide - Please read as this has been expanded since the first book was written:
(1) A very common globe - readily available in collector circles.
(2) A common globe - *fairly* available among most collectors.
(3) Available - these globes, though available, are a little more scarce.
(4) Rare - few known to exist. Around but much harder to find.
(5) Extremely rare - less than 10 known; rarely found or offered for sale.
(6) Unique - one of a kind. If a 6 appears, there is only one known. If another 6 is found, that globe then becomes a 5.
(7) Known only in old photographs. If a 7 appears, then as of this writing one does not exist! A discovered 7 will then become a 6!
(8) An unfortunate category. If an "8" appears then the only known example of this globe was broken and discarded! This has happened to a few globes and we needed a special category for it. If another 8 is found then that globe will become a 6. Let's hope that happens.
(9) Strong evidence suggests this globe exists. If a globe appears as a 9 then we have official drawings made by a company, a known mate to that globe - regular/ethyl, or a sketch on a road map, etc. During the writing of this book *many* 9s became 6s or were discovered in a collection. You may have such a globe. If so, let us know and when this book is updated we will categorize it properly.

These additional categories will immediately tell the rarity of a globe. It will sort out things like why a Sinco one-piece globe is never seen—because it is a 7.

The prices in this book are listed as a range of wholesale to retail. Ranges are left between the two for condition, negotiation, etc. A price of $1,500-$2,200 simply means that $1,500 is a wholesale value and that $2,200 is the top retail value.

A simple review of some different types of globes and how they will appear is as follows:

1: ONE-PIECE ETCHED, BAKED, CAST, PAINTED: Solid one-piece body wherein the design is directly applied, either by etching and paint, baked-on paint, or a painted cast (raised) design.

2: 12, 14, 15 or 16.5" METAL: Metal globe body, of any style, designed to hold a lens of the size listed. The body style would be indicated as either H or L or H/L indicating whether the metal body used is a High Profile or Low Profile style or both. Any special metal bodies are described in the text. Other sizes exist as well.

3: 12.5, 13.5 or 15" GLASS: Glass globe body, solid or hull type, designed to hold a lens of the size listed, with two mounting screws positioned at three and nine o'clock. Many varieties and wide or narrow widths exist. The 15" inserts, however, have no notches at three and nine o'clock. Balcrank bodies have notches at 4, 8 and 12 o'clock.

4: 13.5" CAPCO: Any plastic body globe designed to hold a 13.5" lens with notches at three and nine o'clock.

5: GILL, RIPPLE GILL, 14" GILL: Glass body globe, hollow in the center, designed to hold a lens, with an external ring closed at the bottom with a bolt and nut. Ripple Gill bodies are cast in clear glass, with a textured finish and a clear or baked-on color finish designed to diffuse light. Lenses for Gill bodies have no notches.

6: OVAL GLASS: Oval glass body designed to hold an oval glass lens, with screws positioned at three and nine o'clock.

7: OVAL CAPCO: Oval plastic body with removable lenses. Lenses do not have any notches.

8: HULL: A hollowed out glass body. Specified as different from a GLASS BODY only when necessary, as in a Fleet-Wing red Hull glass body.

9: BANDED GLASS: Similar to a Hull glass body but wider. A flat ring with small ridge holds the inserts in place with four screws holding it to the body. Lenses do not have notches like Gill lenses.

10: NEON METAL: Metal band body, extremely narrow and approximately 18" in diameter that is designed to hold lens as well as a ring of neon tubing inside the frame that lights up the lens, particularly through clear bands that appeared around the edge of most lenses designed for these globes.

11. Any specialty bodies/shapes, such as two piece glass or plastic. Anything unusual is described in the globe text.

Note that some one-piece and metal frame globes had 7" bases. Typically Standard Oil used 7" base globes but occasionally other companies did as well. In our opinion, pricing these and desirability is the same.

Other codes used are as follows:

FOR THE 3-PIECE GLASS GLOBES: N = Narrow *Glass* body; W = Wide *Glass* body.

FOR METAL GLOBES: L or LP = Low Profile *Metal* Body; H or HP = High Profile *Metal* Body. If we know the original color for the metal body it would be listed as "H red" or "L blue", etc.

FOR 1-PIECE GLOBES: N = Narrow One-Piece Body; RW = Rounded Wide One-Piece Body; W = Wide One-Piece Body.

Photos of these examples appear under the Types of Bodies section and should be studied.

FOR ONE-PIECE GLOBES: OPB means one-piece baked, OPE means one-piece etched, OPC means one-piece cast, OPP means one-piece with painted-on, not baked-on details.

GLOBE BODY STYLES, A Pictorial Overview with Values:

PRICES ARE FOR BODIES ONLY

All prices are for mint original globe bodies only. Dates are approximate introduction date for this body style. Globes with copper or steel bases are usually older and were originally a higher priced globe. They are more desirable among collectors.

Narrow 13.5" Glass Body - Capco Model 16-NR, 1928. *Courtesy of PCM Archives.* $100-150

Wide 13.5" Glass Body - Capco Model 16-R, 1928.
Courtesy of PCM Archives. $100-150

15" Glass Body, 1930. *Courtesy of PCM Archives.* $250-400

Banded Body, 1930. *Courtesy of PCM Archives.* $150-250

Early Hull Crimped Base, 1930. *Courtesy of PCM Archives.*
$150-200

Left: Metal Body - Low Profile, 1914. $150-175
Right: Metal Body - High Profile, 1914. $150-200
Courtesy of PCM Archives.

Jeweled Metal Body, 1920.
Courtesy of PCM Archives.
$1000-1600

White Porcelain Metal Body - High Profile, 1920s. *Courtesy of PCM Archives.* $200-300

Above right: 14" Gill Body, 1930.
Courtesy of PCM Archives.
$175-275

Right: Gill Body, 1930. *Courtesy of PCM Archives.* $150-250

Metal Body with Neon Tube, 1930. *Courtesy of PCM Archives.* NL (**N**o **L**isting)

Left: One Piece Wide.
Center: One Piece Rounded Wide.
Right: One Piece Narrow.
Courtesy of PCM Archives.

WIDE 13.5" GLASS BODY (2) 1928 $100-$150
White glass body approximately 8" wide, designed to hold 13.5" glass lenses with two machine screws.
NARROW 13.5" GLASS BODY (2) 1928 $100-$150
White glass body approximately 5" wide, designed to hold 13.5" glass lenses with two machine screws.
12.5" GLASS BODY (4) 1935 $100-$175
White glass body approximately 6" wide, designed to hold 12.5" glass lenses with two machine screws.
BANDED BODY (4) 1930 $150-$250
White glass body designed to hold approximately 13 5/8" diameter lenses with an internal ring attached to the body by four screws.
BALCRANK 3-SCREW BODY (4) 1928 $175-$300
White glass body designed to hold 13.5" glass lenses with three machine screws positioned at 12, 4 and 8 o'clock.
15" GLASS BODY (4) 1930 $250-$400
Special glass body designed to hold 15" diameter glass lenses.
NOTE: Globe bodies with copper or metal bases are older and more desired among collectors.

CLOVER BODY (4) 1932 $250-$450
White glass body designed to hold Cities Service clover shaped glass lenses.
CLOVER BODY FOR ROUND INSERTS (4) 1932 $300-$500
White glass clover shaped hull body designed to hold a special 14" diameter lens with two machine screws.
GLASS OVAL BODY (4) 1930 $200-$300
White glass body designed to hold oval shaped glass lenses with two machine screws.
CAPCOLITE 216 BODY (1) 1932 $25-$50
Two piece white plastic body, introduced in 1933, designed to hold notched glass lenses by sandwiching them between the two plastic halves.
CAPCOLITE 218 OVAL BODY (2) 1946 $50-$75
Same as CAPCOLITE 216, above, designed to hold oval lenses without notches.
RED CAPCO (3) 1932 $50-$75
Same as CAPCOLITE 216, above, cast in red plastic.
BLUE CAPCO (4) 1932 $75-$125
Same as CAPCOLITE 216, above, cast in blue plastic.
GREEN CAPCO (4) 1932 $50-$100
Same as CAPCOLITE 216, above, cast in green plastic.
YELLOW CAPCO (4) 1932 $50-$100
Same as CAPCOLITE 216, above, cast in yellow plastic.
ORANGE CAPCO (4) 1932 $75-$125
Same as CAPCOLITE 216, above, cast in orange plastic.
NOTE: There are early plastic bodies dating 1932/1933 that only say "CAPCOLITE" at the bottom. Each plastic halve is different where they come together at the top. Also, some are very wide.

HULL 13.5" GLASS BODY (3) 1928 $150-$200
White narrow glass body with hollowed out area behind globe faces.
YELLOW HULL BODY (6) 1930 $900-$1400
RED HULL BODY (4) 1930 $800-$1200
Same as HULL, above, cast in red painted glass.
ORANGE HULL BODY (6) 1930 $1000-$1500
Same as HULL, above, cast in orange painted glass.
METAL BODY LOW PROFILE (2) 1914 $150-$175
Rolled metal globe body with the band formed in one piece, therby appearing nearly flat when looking at the lenses.
METAL BODY HIGH PROFILE (2) 1914 $150-$200
Stamped metal globe body formed as two halves and assembles back to back with a high seam in the center, appearing much larger than the lens it holds.
JEWELED METAL BODY (4) 1920 $1000-$1600
Metal body with holes around the circumference, fitted with one inch glass jewels which light up.
GILL BODY (3) 1930 $150-$250
White glass hollow body designed to hold a 13 5/8" +/- lens with an external band that is joined at the bottom by a single machine screw.
14" GILL BODY (4) 1930 $175-$275
Same as GILL, above, for larger diameter lenses. Usually wider than a standard Gill body.
CLEAR RIPPLE GILL (3) 1930 $700-$900
Gill body, same as above, cast in clear glass with a textured surface designed to diffuse light.
RED RIPPLE GILL (3) 1930 $800-$1100
Clear ripple Gill body with a baked on red finish inside.
BLUE RIPPLE GILL (4) 1930 $1200-$1600
Clear ripple Gill body with a baked on blue finish inside.
WHITE RIPPLE GILL (3) 1930 $800-$1000
Clear ripple Gill body with a baked on white finish inside.
GREEN RIPPLE GILL (4) 1930 $1500-$2000
Clear ripple Gill body with a baked on green finish inside.
ORANGE RIPPLE GILL (3) 1930 $1000-$1350
Clear ripple Gill body with a baked on orange finish inside.
YELLOW RIPPLE GILL (3) 1930 $900-$1200
Clear ripple Gill body with a baked on yellow finish inside.
BROWN RIPPLE GILL (6) 1930 NO LISTING
Clear ripple Gill body with a baked on brown finish inside.
TEAL RIPPLE GILL (6) 1930 NO LISTING
Clear ripple Gill body with a baked on teal finish inside.

Chapter 2. Collecting Gasoline Globes

Collecting gasoline globes has changed significantly since our first book on globes which was published in 1993.

Most globes have increased in value over the years. Some a lot and some very little.

The hobby seems to have shifted towards collecting the rarer and prettier globes, especially among the more experienced collectors. However many of the new collectors who can afford it will start off their collecting by just wanting the best. Some collectors go after numbers of globes. This, of course, has driven prices up on the better globes. In turn, collectors today often specialize in collecting certain types of globes. The most popular trend is to collect globes from your state or region. Many collectors will specialize in just Gulf or Sinclair or Texaco, etc. And, there are many among us, including myself, who just like all those globes and have little control at all in collecting!

The best way to collect globes and really get somewhere doing it is a multi-step approach. The first rule is to advertise a lot. Let people know what you want. Next, "network" with other collectors. This step is a must for the serious collector. Most collectors buy, sell and trade, so it is to your great advantage to befriend the ones who do. Next, one should attend the gas and oil shows and learn as much as you can about globes. This includes buying books and asking a lot of questions.

Since our first globe book, countless globes have turned up, still at gas stations, still at bulk plants, in attics, sheds, etc., etc. And though globes do get scarcer as time goes on, globes will again be found in the future at these places.

Globes have always been a good investment and I feel will continue this trend. New collectors come into the hobby every day and as long as that continues so will globes' collectibility. We will go into great detail about the different globes and variations in later chapters.

In the meantime, remember that gasoline globes are exciting, fun to collect, a great investment and may be found at your neighbor's garage sale!

Chapter 3. The Different Types of Globes

As stated in our first book, there are four basic globe types or categories. To succeed in this hobby, you must understand the differences and the significance of each.

Plastic-Body, Glass Insert: These bodies have 13.5" glass inserts and were produced from about 1932 to the present. They were most common in the 1950s.

Glass-Body, Glass Insert: These have 13.5" glass inserts, were produced from the late 1920s through the early 1960s and were most common in the mid-1930s and 1940s.

Metal-Body, Glass Insert: The usual size for this body style's inserts is 15 or 16.5". They were produced from the mid-teens through mid-1930s, though some were made as late as 1960. They were typically used in the 1920s and early 1930s.

One-Piece, No Insert: Produced and used primarily between 1912-1931, these include etched, raised letter, baked-on, fired-on, painted on and chimney cap globes.

There are several variations of each of the four main categories. Most globes have a standard 6" diameter base, but some are found with 7" as well as 4" bases.

Plastic-Body, Glass Insert Globes

Typically, a glass insert fits on each side of a two-piece plastic body. The insert diameter is 13.5", with notches at nine o'clock and three o'clock. These notches are not as deep as the 13.5" inserts for glass-body globes and the inserts themselves are thinner than most for glass.

If you took a circle and cut it in half, this is about what the notches look like for plastic-body globes (a semi-circle).

The notches on inserts for glass bodies are elongated semi-circles. They are much deeper or longer. Lenses for plastic globes are usually thinner, and those for glass bodies are generally thicker. However, we have seen thin lenses with deep notches. Collectors say, "These go on plastic, right?" Not really. Deep-notched inserts were made for glass bodies, whether they are thin or thick glass. The reason for this was basically to allow for a better fit on the glass body. The inserts we have seen that are thin glass with deep notches were too old to be on a plastic globe for that particular company. By the way, any 13.5" insert without notches must go on a Gill or banded-glass body.

People outside the hobby and novice collectors sometimes relate original plastic globes with reproductions. I hear this a lot. They will look at an original plastic globe and say, "Hey, that's a repro." This is because the first repro globes were plastics and today most repros are still plastic. So, they assume all plastics are repros. Repro plastics date back over 25 years now and many new collectors didn't know most companies originally switched to plastic globes in the 1950s. Some as early as the 1930s!

The oldest plastic globes were used by Phillips and Conoco and date back to 1932. Again, these globes had the typical plastic frames and glass inserts. The only big difference is that early plastic frames went together slightly different at the top and said "Capcolite" on the underside of the globe body near the base. The newer plastics—not repros—said "Capcolite 216" which was embossed in this area. The "216" was a model number. The first plastics were model #116, but I've never seen this number on an early plastic. *Oval* plastics were model #218. In the early 1980s some repro plastics were made without the original mold and said nothing underneath this area. Later the original mold was found and today most all plastic repro bodies again say Capcolite 216.

Capco or plastic globes are some of the most popular today. Their graphics offer collectors some of the best logos such as Hudson with the truck, Sunray with the sunset, Texas Rose with a rose, Dance with the ballerina, Covey with race cars, etc. Occasionally I will hear a collector saying, "I don't collect plastics." To each his own, but they are missing some of the best graphics the globe hobby has to offer. Ever see a Salyer's globe?

Plastic bodies are usually white, but yellow, red, green, orange, tan/cream, and blue originals exist. "CAPCO" or "Capcolite" is often inscribed on the underside of one of the two pieces. As mentioned, CAPCO stood for Cincinnati Advertising Products Company, from Cincinnati, Ohio. They manufactured many types of globes. These bodies were patented in about 1932. So some, particularly Phillips and Conoco, can be quite old.

One screw on the top and two through the base hold the globe together. To assemble or disassemble the globe, point the base at three o'clock, loosen the screws completely, and the upper half body comes off. Then, carefully remove each insert separately.

Some older plastic bodies from the early 1930s have a ridge only on one half where they come together at the top. These date from the early to mid-1930s. Inserts for glass bodies don't fit well into a plastic body. They are too thick, and don't line up correctly. We've seen some fit adequately, but you risk chipping the edge of any lens that fits too tight.

Phillips was a big user of the wide plastic bodies. Wide plastic bodies can also come in a variety of colors and are sought after by many collectors. Some colors were red (the most common), green, yellow, tan, orange and blue. For some reason many of the plastic framed globes from 1949-1951 have the date stenciled into the underside of the bottom of the base. Dixie used yellow bodies, Fleetwing used red, Conoco used green, Sunray used red and ange and Carter used blue. Several other companies used colored plastics, but again, these globes are quite scarce. Colored plastics demand prices of $60-$125—just for the plastic frame.

All plastic frame inserts are the common 13-1/2" size, except for the ovals, of course. Plastic oval inserts are unique in that they are not notched. The inserts are glass, of course. The plastic frame goes together in two halves, just like the round version. Few companies used oval globes and these make quite an attraction to any collection. I've never seen a colored oval body but it wouldn't surprise me if someone found one some day. Acquiring more than a dozen ovals in a collection is quite a task.

One-piece or two-piece plastic globes are quite interesting to look at. Some are large and boxy looking things, some are shield shaped, like the Phillips and some are just strange overall. Co-Op used an unusual shield shaped two-piece plastic and a two-piece in the shape of a figure eight on it's side which is held together by an aluminum band.

Many of these all plastic globes came much later in the history of globes. Years used were the 1950s, 1960s and 1970s. By the way, the Phillips shield shaped globes, Phillips 66 and Phillips 66 Flite-Fuel had *plastic inserts*, not glass. Sometimes Champlin and Harbor used plastic inserts in plastic frames. An Esso plastic insert was recently discovered.

Plastic framed globes offer an affordable entry level into the world of collecting gasoline globes. Many plastics can be picked up in the $100 to $250 range. The popular Texaco star, the Sinclair Dino globes, Phillips, Conoco, Standard Oil, Cities Service and hundreds of companies all used plastic framed globes that can be purchased in this price range. Graphic or picturesque plastic framed globes can run in the $500 to $1,500 range and higher. So when collecting gasoline globes, I wouldn't pass up the plastics. You'd be missing some of the best and most interesting globes available.

Glass-Body, Glass-Insert Globes

This category has many exceptions and variations. Insert diameter is typically 13.5" as previously mentioned and notches are deeper at both three and nine o'clock. The inserts are also normally a bit thicker than those for plastic bodies. Lenses with no notches are for Gill, banded, or 15" glass bodies, which will be described shortly.

The Gulf Oil Company and several others used 12.5" glass inserts that fit only on smaller, wide-glass bodies. Other than Gulf globes, not many of these ever turn up.

Some companies used inserts with diameters of 13-3/8, 13-5/8, 14, 14-1/4, 14-7/8 and 15". There probably were other sizes as well. Bodies for these inserts are difficult to find. Most of these odd sizes go on various Gill-type bodies. A Gill globe is basically a narrow glass body that is hollow in the center. The notchless inserts are held on by a narrow metal ring that attaches to the body, with one screw at the bottom of the ring. These bodies were made by the Gill Glass Company of Pennsylvania, hence the name.

Gill also manufactured ripple-textured globes, so named for their ripple-type surface on the outside of the glass. The inside glass was often painted and then baked on so that when the globe lights up, it looks quite beautiful. Red is the most common color found today. Other colors include clear, white, orange, yellow, blue, teal, green and brown. The latter three are the rarest. There are also different shades of the green, red and yellow. The older versions of these globes have a screw-on base that is usually made of copper. Some were also made of steel and aluminum. In fact, any screw-on base on any type of glass body signifies an older glass body, from around the early 1930s.

The 14-7/8 and 15" inserts were used by only a couple of companies and attach to a large, solid (in the center) glass body. A ring similar to a Gill body holds the inserts onto the body—talk about a large and heavy globe! These globes are extremely rare. The inserts will fit a metal frame 15" globe and when one can't find one of these unusual glass bodies, metal will have to do. Kanotex, Pennzoil and a couple other companies used these large bodies.

Hull bodes, made by Hull Glass Company, are hollow, like the Gills except that they hold notched-type inserts. Some Hull bodies have unusual shaped bases and some are inscribed "Hull" toward the bottom of the globe. Many Hull inserts and bodies are dated! There are many other versions of glass bodies that hold inserts, too many to mention them all. The two most common types are the narrow-glass and the wide-glass bodies. Both are solid pieces, the narrow one is approximately 5" wide (front to back) and the wide version around 7 to 8" wide. These bodies hold most insert types. Again, many variations exist with different widths, different bases, styles, and so on. Many companies were making globes in those days, thus the reason for so many different types. It is not so important to be familiar with all these versions. However, it is important to be able to differentiate between a Gill, Hull, ripple and regular glass globe.

On rare occasion, some inserts will have notches at twelve, four and eight o'clock. These types went on the more unusual bodies with holes that lined up in the same respective positions. Marland Oil, among others, sometimes used these globe types; Cincinnati Balcrank was the manufacturer. We call these Balcrank globes. Another version that we recently found is an unusual glass body with notches at 3 o'clock, 6 o'clock and 9 o'clock. It was found with Texaco inserts.

Another unusual style of globe body was used by D-X, Empire Refining and a few other companies and is asked about by many collectors. We call this a banded-glass body or globe. The globe body is hollow like the Hull body, but with a recessed metal rim that holds each insert on with four screws from the inside. It's also like a Gill, whereas the inserts are not notched. Some lenses are interchangeable with Gills. The globe manufacturer is unknown.

Gill bodies, including ripples, are sought after more because of their rarity and, in the case of ripples, for their beauty. Some prefer the Hull body to the regular glass globe while others opt for the narrow glass over the wide glass and vice versa.

Inserts are much more common than bodies, so collectors will often come across inserts without bodies. If you have inserts without notches, you will need to find a Gill body. Notched lenses give you more options. Certain gasoline companies used only wide-glass bodies or narrow-glass bodies. Although it would be impossible to list them all here, we'll mention a few. Oil companies that usually used the narrow-body glass globes included Sinclair, Texaco, Marathon, Fleet-Wing, Pennzoil and Cities Service. Companies that normally chose the wide-bodies included Sohio, Phillips, Skelly, Wolf's Head, Coreco, Ashland, and White Rose. Some companies used both. Amoco, Atlantic, Crown, Kanotex, Dixie, and Derby used a number of Gill bodies, while Kanotex, Bay, and Dixie globes are more often found on ripple bodies.

Colored-glass globes are very rare, much rarer than ripples. Fleet-Wing used red glass on occasion, and Freedom Oil used yellow glass. Browns Lubricants used orange glass.

There were a few other oddball glass globes manufactured. Waverly Oil Company and Fleet-wing used a two-piece glass globe that fastened in the center. Cities Service used clover-shaped globe bodies with clover-shaped inserts that fit like regular notched inserts. The most unusual one we've seen is a *clover-shaped* Cities Service Hull body that held *round* inserts.

Oval glass bodies exist today, but are rare. Only a few dozen companies used them, such as Cavalier, Hicksatomic and Pugh Peerless.

As mentioned, lenses made for glass bodies don't fit well into plastic bodies. The reverse is also true. Lenses made for plastic bodies are thinner glass and not deeply notched, so they don't fit well into glass bodies. However, certain narrow-glass bodies will hold inserts for plastic. Also, most notched 13.5" inserts will fit in a Gill body as well. The Gill band simply covers these notches. Some people take a pretty insert for plastic and put in a pretty Gill ripple body. This greatly increases the globe's value, but it is not original. For display purposes it is probably acceptable, but a potential buyer should be informed that this is not the way it originally came. It's like fitting your Model A with air conditioning—convenient, but not original.

Metal-Body, Glass-Insert Globes

Metal frame globes typically hold 15 or 16.5" glass inserts with a snap ring that holds the insert onto the body by pressure. The ring is split in one spot. These inserts are not notched, but sometimes have a small inverted cone-shaped filing on the inside bottom. The indentation fits over a seam on the inside of the metal globe body. But this is not all that important. If you line up the globe and look at it closely, the inserts often appear crooked.

There are many exceptions in this category, more for the type of inserts than the body styles.

Metal framed globes have a metal body or holder, with two inserts made of glass which attach to each side of the frame. Most metal framed globes, probably 98% are called either 15" or 16.5" metal, which describes the diameter of the inserts. Other sizes of metals, all rare, include 10-1/4, 11-1/2, 13-1/2, 13-3/4, 14, 18, 22" and other oddballs.

Just like glass body globes, some companies only used the high profiles while some only used the low profiles and then, some used both.

Here are some interesting facts about metal framed globes:

Pure Oil Company globes usually came on a high profile white porcelain body. I believe it is the only company using such unique metal holders.

Recently an orange porcelain body was found. The company using this is unknown. Expect that other colors will be discovered.

There are two basic body styles that fit both the 15 or 16.5" inserts, the high profile and the low profile bodies. It is important to distinguish between the two types. A picture of each is included in the Types of Globes section in this book. The advantages of high profile bodies are two-fold. One, they will hold all inserts except oddball types. Two, when looking at a high-profile body, you see the body around the lenses, which adds to the color and contrast of the globe. The low profile bodies do hold most inserts, but some fit tighter and some don't fit at all. When looking directly at a low-profile mounted globe you mainly see just the inserts. Again, as in the glass types, some companies used only the high profile, some only the low profile and some used both.

Texaco used a unique wide bodied metal frame which was made exclusively for their leaded glass inserts. These inserts measured 13-3/4" for the regular size leaded glass globes and about 22" for the large size leaded glass globes. Though other companies used leaded glass in buildings and door displays, Texaco leaded glass globes were unique in the oil industry.

Some early metal framed globes used small brass clips to hold the snap ring in as an additional precaution. Texaco leaded glass inserts are set in a metal holder and that whole assembly is screwed directly into the wide metal body.

The most interesting body type ever made was a rare 15" metal body with about twenty-four 1-1/4" "jewels" attached around the whole globe. Usually these jewels are red, blue/indigo and white. Dixie Distributors and Sterling used these on rare occasions and other companies used them as well. These globes are phenomenal when lit up!

All metal framed globes have 6" bases like most globes—with two exceptions. Mobil, Socony and affiliates used 7" bases on their metal frames and I have even seen a 4" base metal framed globe. The 4" base globe on metal is extremely rare and tells of a very early globe that adapted to an early teens pump.

Metal framed globes generally came after the one-piece globes—1920s, 1930s, however, several companies used them in the 1910s. On rare occasion some companies used the metals first, then the one-piece later. Skelly's first globe was a metal frame, their one-piece globes came later.

Texaco, White Star, Atlantic, and Socony used milk glass inserts. Socony and Texaco fired their logo onto the surface and White Star and Atlantic etched the surface, then painted on their logo. There are probably other companies that used milk glass inserts. However, except for Socony, these are all extremely rare.

Cast inserts are more common than milk glass overall, but are still very rare. Tydol, Socony, Atlantic, Cities Service, Hearn, Venus Oil Company, Pennzoil, and Valvoline, to name a few, used cast face inserts. These inserts are very thick, heavy glass with deep letters or logos molded into them. The letters were then painted or fired-on, or sometimes both techniques were used on the same insert.

We've been asked if the 16-1/2" metals are earlier than the 15" metals. The answer is no. The earliest metals were all 15" and 16-1/2" ones seem mostly to appear in the later 1920s and 1930s. Few companies actually used the larger 16-1/2" metals but, the familiar companies that did were

Atlantic, Mobil, Tyrol, Cities Service, Socony and the other Standard Oil companies.

Socony had many unique metal framed globes. In fact, they used more odd metals than any company. Socony had a unique globe with 14" metal inserts that were flat and porcelain. They also used a 15" convex porcelain as well as 13-1/2, 14, 15 and 16.5" milk glass inserts where the logo was fired on the outside of the insert. Socony also used a 15" and 16.5" cast face metal globe. Here the inserts were cast from a thick piece of glass, then painted later.

Gulf, Laurel Leaf, Pure, Texaco and a few others used 18" inserts on a very narrow metal body. Then, a neon light wrapped the inside of the body which showed through the inserts to light the globe. These globes are very rare with only a few existing. We have seen a photo of these being used on pumps, but arguably they could also have been used inside a window as a display. They were made my Neon Globe Company of Norfolk, Virginia.

ONE-PIECE GLOBES

One-piece globes have always intrigued us because they are the oldest globes. These globes were used from about 1912 to 1931. One-piece Crowns and Shells were made through the 1950s and though they are one-piece globes, they don't fall into the true rarer category of one-piece. Super Shells and lettered Crowns are true one-piece globes from the 1920s.

What is a one-piece globe? Just what it states. It is a gasoline globe made or blown into one piece of glass. No inserts or attachments adorn a one-piece globe, except for the chimney capped globes which we'll discuss in a minute.

In terms of how one-piece globes were made and decorated, there are four basic types we can identify—etched, baked-on/fired-on, raised/molded, and combinations thereof. Some painted only ones also exist.

In collecting one-piece globes one must know these types and which questions to ask when buying over the phone. A one-piece etched Sinclair H-C and a fired-on Sinclair H-C are identical in general appearance! Yet, one sells for $450 more than the other and is much rarer.

Etched one-piece globes were acid etched when made. The details, letters or backgrounds are *recessed* into the glass, depending on the logo on the globe. Etching can vary from 1/32 of an inch to nearly 1/4" deep. The etched portions are usually very rough and irregular and feel like a medium grit sandpaper. Unlike a fired-on one-piece, there is no smoothness to an etched one-piece gasoline globe around the colored areas. After such a globe was etched, those areas were spray painted with a glossy type paint. Globes with two or more colors were often painted the first color overall, then the second color in specific areas. This is why sometimes two colors can begin to show through as the globe fades. It is okay to repaint etched globes and does not hurt their value.

Etched globes were crudely made, and no two are alike. Runs of acid, spots, bubbles in the glass and irregular edges are typical of etched globes and make them very unique. Most chimney capped one-piece globes are etched, and date in the teens. These globes have a small vent hole in the top, about 4" diameter with a metal cap that covers this hole. This cap has small holes or vents in it to let gasoline fumes out that could have drifted upward from the internal workings of the pump.

Fired-on or baked-on globes are one-piece globes where the logo was applied to the surface and then baked/fired onto it in a kiln. The paint is nearly impossible to remove whereas etched globes exposed the paint to the elements with no way to keep the paint on, and so fired-on globes lasted much longer outdoors. Lit up the baked globes were a great improvement in appearance as well as durability. There is no roughness, gritty texture or whatever on a baked-on one-piece globe. As one runs his finger over the whole globe, he will feel the *slightest* bumps of the edges of the paint but overall the globe is very smooth. Most baked-on globes appeared in 1926 though we could argue their existence as early as the mid-teens.

If a company used etched and baked-on globes the etched ones will be older. Baked-on one-piece globes were often dated on the inside with an ink stamp. The earliest date I've seen is 1926. Nineteen hundred twenty-seven dates through 1929 dates are more common.

Repainting a worn baked-on globe destroys most of its value. This opens the door for reproductions and fantasy pieces. A suggestion would be to repaint one side and leave the other side to prove what was once on it, though it'll still have little value. Fired-on one-piece globes are a great addition to anyone's collection.

The third category of one-piece globes is the cast or molded globe. The logo or designs were part of the mold. Areas can be raised, recessed or layered and are generally smooth to the touch. Few companies used these cast globes. Texaco, Cities Service, Gulf and Tydol were the biggest users. Sinclair, Republic and the Columbus Oil Company used cast one-piece globes on rare occasions. These globes seem to date from the mid-1920s to 1931. The most common and familiar cast one-piece globes are the Texaco, Texaco Ethyl and Gulf globes. For some unknown reason most cast one-piece globes have a threaded copper base, but not all. Many of these globes are actually Hull glass globes, the same company that made the familiar Hull Glass three-piece globes, some were made by Gill Glass. Detailed areas of these globes were usually *fired-on*, rarely just painted. Larger areas were applied evenly with a screen or spray, and smaller areas, like the letters were often painted with a brush. Brush strokes can be seen on any one-piece Texaco or Texaco Ethyl on the black borders and letters. Many Texaco globes are dated '30, or '31 inside the base. Looking inside any cast one-piece you can also see the logo recessed into the glass! This is why we call it a cast or molded one-piece globe. These globes are okay to restore but never look as good as a mint original. Letters can stick up 1/8" to 1/4" above the surface. Again other areas may be recessed into the glass, but this is not etching.

Know what questions to ask when dealing over the phone. Let's say that you determine that a Texaco globe is a one-piece. Texaco globes can be etched or raised. Are the areas *raised* up or recessed by etching into the glass? The advanced collector knows a one-piece Texaco with a copper base can only be a raised letter one. So he can cut the questions short and asks, "Does this one-piece Texaco have a threaded base?" If the answer is yes, go no further, as you have a *raised* letter or cast one-piece Texaco.

The neatest cast globes are the rare Cities Service clover shaped globes. These do not have the threaded base. Yet their clover shape makes them very unique. Columbus Oil Company Silver Flash and Silver Flash Ethyl globes are octagon shaped cast globes and are quite unique. Tydol used a large cast one-piece globe with flat sides and a copper base. These are extremely rare and should not be confused with the 15" metal globes with cast *inserts*. The one-piece is much rarer and a whole different globe.

The last category is a catch basin for all the variations and combinations of one-piece globes. One such sub-category here is the etched one-piece globes with *painted* details. The Sterling Ethyl globe has an etched *background* but the letters and Ethyl logo were painted on top of that. It is impossible to restore one of these globes in worn condition. You have no distinct areas between the letters and background, so when it's worn, it's all gone. Try painting the details of an old Ethyl Corporation logo! In the Sterling Ethyl pictured in this book, only the rays and the *general background* of the Ethyl logo are etched, not the details. Penndrake, Sterling and Pennzoil are the few companies using these types of etched/painted one-piece globes. Since these companies base their operations in a similar region we'd probably find, with further research, a local glass company from back then that supplied these special globes.

I've seen a cast Good Gulf globe recently with an *etched* background to hold the paint—make the paint adhere easier. Remember, all cast globes are really a combining of a cast globe with a *baked-on* finish. This is an unusual one. A few early one-piece globes were painted on, no baking.

And last but not least, are figural globes—Crowns, Shells, Eagles or monument shaped globes like the Barnsdall, ROC, etc. Let's not forget the three sided cast Republic globe. Shell globes have *recessed* letters, *cast* with fired-on paint in most cases. Early Shells were etched. Also, the Musgo globes which are baked and hand painted! All these oddities I put in this last category.

As a final note, one-piece globes have different body styles too. Within this book the body styles used for a particular globe are in the listings. The three *basic* styles are wide body one-piece, rounded body one-piece and narrow body one-piece. Knowing these one-piece body styles will help protect us from reproductions. Refer to the section on Types of Globes for a photo showing the side view of the three basic styles. Take the rare one-piece Penndrake globe. It is only known to exist on a narrow body format. It is listed as Penndrake one-piece etched, narrow body style. We are familiar with most known one-piece globes and which styles were used. Musgos were only made on wide glass styles for instance. Unfortunately, some globes like the Sinclair H-C etched exist on all three body styles. But generally, this is not the case. This new globe book lists all variations of every known globe. Fifteen years ago when purchasing a White Plume one-piece etched globe—one of a kind—it came on the rounded one-piece style. Four years ago when a second one was found, it came on a narrow body style. It turned out to be a phony! One-piece globes are highly recommended to all collectors and are a great asset to any collection.

Lens Removal & Installation

To remove a snap ring on a metal body, you must get a sharp object under the ring without touching the insert. Then lift or pry up slowly and carefully. Keep your hands on the snap ring so it doesn't fly off. You really need three hands, so use a knee to hold back one side of the ring, then one hand on the other side, and one hand to pry or lift. Or, once you get one side of the snap ring up around about 1/4", put your hand under it and hold it tight. Move the ring inward so it becomes smaller and lift it around and out of the metal body. The ring then expands as the pressure is relieved, so don't let it fly! Rings can fly off and could possibly hit another globe which can be costly.

To install an insert, set it inside the globe body, put one side of the snap ring down, and bend the other side of the ring around and toward the center, carefully and slowly. It will snap, but won't break the insert.

Chapter 4. Theory of Pricing Gasoline Globes

The theory behind pricing gasoline globes which we wrote about in the first globe book has not changed at all. Of course, prices have escalated but the same facts and theories hold true. Rarity is still very second rate to beauty today and type of globes should fall into certain ranges. Beauty is still the key and probably will be forever.

Below is an updated version of the original theory behind pricing gasoline globes. We will discuss different pricing theories and how we arrived at the current prices in this book. If you read this you should better understand the task we had in front of us.

What's difficult is bearing the responsibility of changing the course of the world of gasoline prices for years to come. If prices wind up too low, everyone's investment seems worth a lot less. Too high and people will fuss about how they can't buy anything anymore. Certainly we can' t make everyone happy, yet we will do all we can to try.

Another problem is time. Since production on such a large volume can take eight to nine months, things we price today will be nearly a year old by the time the books gets printed, distributed and in your hands! So when you do have this new globe book at your fingertips, prices need to be current. People have asked, "How you gonna do that?" We all know some globes have increased little in several years and many have increased a lot. We take each globe and price it on a case by case basis. We culminate all the data we can find on what people have been paying—low prices paid, high prices paid, prices at the shows, dealers' prices, auction prices and plain old experience and start with that. Then we look at the market trends and decide whether this globe would be worth any more in November 1998 or January 1999 as compared to today's prices. This is all mixed together with gut instinct and presto!, out comes a price.

But seriously, then there is another problem. Years ago just about all of us could afford to collect any globes we wanted. Prices have forced most of us to specialize. We have Texaco collectors, Gulf collectors, Montana globe collectors, etc., etc., etc. These collectors often pay more for a globe than another person would. Some pay more than it is worth. Take for instance a Capco globe from Montana that says "Silver Gas". In our minds it is a $250 globe—nothing special really, but a nice globe. But that globe is worth a lot more to the Montana folks. Take it to the Des Moines show, run it in a catalogue auction, make a few calls and it'll bring $350-$500 to the right person. Similarly, if a wealthy person has to buy your property valued at $15,000 and offers you $25,000, it does not mean that the property is now worth $25,000. House values and property values are based on three comparison sales or "comps" as they call it.

Let us explain further. We hear people say things like, "That globe brought $3,500 at auction!" That does not determine value and if you think that establishes a price then you don't understand pricing. Two years ago hardly anyone had seen a Pure Ethyl 15" globe. It is extremely rare and is not to be confused with the Purol Ethyl globe. The first one sold to a Pure collector for around $1,500-$1,600. Then another one turned up. It sold for around $1,200. Then a third one sold for $1,000, etc. See what happens? This averages out to about a $1,250 value today, but it's not really a $1,500-$1,600 globe today. Some globes are so desirable that they *increase* in value each time one is sold. Welcome to the world of Gilmore globes!

Many globes have *never* sold. Price is established here by knowing the market, estimating a value and overall experience. Capco globes don't sell for $3,500 and any Indian globes with Indians pictured don't sell for under $1,000. Any detailed airplane globes are worth at least $2,500, etc. These are extreme examples of how one could figure out a price—starting with broad parameters and narrowing it down. And in the end, it is still an opinion. Yet, it's an opinion with some good judgment behind it.

As you can see, it is quite a task and a lot of factors need to be weighed carefully. We hope you understand pricing is a little more than what something sold for recently at auction. And again, we probably won't make everyone happy with the prices we've determined but, we gave it our best shot.

Chapter 5. Gasoline Globes as Investments

I have collected, bought and sold gasoline globes for nearly 25 years. Globes have always been a great investment. As we've discussed in this book, some globes increase a lot in value per year and some very little. It's easy to say the prettier the globe the quicker its value will increase. Collectors like the pretty or picturesque globes. Rare globes sometimes do well but they come in a distant second. Popular companies like Texaco, Standard, Gilmore, Sinclair, etc., well, their globes do well even if they are not so pretty. Unknown companies with plain logos are collectible, as are common globes, but they increase the least in value per year.

If you invest in a pretty globe that is also rare, then that may be a home run as we call it. These globes increase in value the fastest in most cases.

So, in investing in gasoline globes as opposed to general collecting, follow these rules:

Rule No. 1 - Buy the more colorful or picture type globes.

Rule No. 2- Buy the rarer type globes, especially ones with major brand names or more colorful logos.

Rule No. 3 - Avoid the very common globes, unknown plain looking globes, broken globes, etc.

I'm speaking here about investment only. We can appreciate the plain, unknown globes more than the colorful brands. Why? Because they are rare or unique, will never be reproduced and are just interesting overall. It is not wrong to collect these, it's just that these types appreciate slower than the prettier or known globes.

Avoid worn, damaged or problem type globes. Again, I'm speaking only for investment value. These sell slower and many people avoid them.

So, how do you know what to purchase as investments? Call people, read this book, ask questions, read Petroleum Collectibles Monthly or call us. Helpful starter ideas of what to collect could be the rarer Texaco globes, like the Texaco Diesels, Texaco Ethyls, Musgo globes, globes with Indians, airplanes, trucks, boats, animals, etc. All these globes appreciate in value more quickly than any other globes.

Again, many gasoline globes, assuming you didn't pay too much, will increase in value. Only experience and doing something will set you in the right directions.

Chapter 6. Reproduction Globes

Many new reproduction gasoline globes as well as fantasy pieces have been added to the original list of known reproduction globes since our first book. Luckily, most of these globes are now being dated, thanks to our efforts, the cooperation of people making these items, and to the hobby coming together overall.

We can't emphasize enough the importance in dating reproduction or fantasy globes. This includes the new colored ripple bodies being sold to hold Gill inserts. These ripple bodies are not being dated as of this writing. So here is what happens:
You see an auction in Texas selling five gasoline globes from an older person's long time collection. The globes look real and you certainly cannot tell anything from the flyer they sent. The auctioneer doesn't suspect these globes as phony—why should he? You buy these globes and suddenly everyone is in for a big surprise.

Auctioneers, in all due respect, know very little about gasoline globes. How many times will the above scenario happen over and over in various situations, all because of an undated gasoline globe? We get over 100 calls a year asking questions about undated phonies. We never get calls about dated repros!

Early reproductions and fantasy globes are now 25 years old! So again, that old saying, "Well it's been in the barn for 25 years," means nothing to smart collectors.

Today we can estimate that about 80% of all known reproduction globes are dated. And about 99% of all current reproductions being made are being marked as reproductions.

Most reproductions are the 13.5" inserts that will fit the glass or plastic frames. But in recent years, many 15" and one-piece globes have been made as well.

How does one spot a reproduction or fantasy piece? First, check for dates or the word "Reproduction" being marked at either the bottom of the insert, say at 6 o'clock or at 12 o'clock. Some may say "Licensed by Mobil Oil Corp." instead of a date. Or some may just say the word "reproduction". Of course, there are many that will have a date, for example, Sept. 1982 or Nov. 1995. Most of these dates appear at the bottom outside of an insert at around 6 o'clock. One must study and know cold the undated reproductions or carry this book around. There are enough undated repros and fantasy pieces to cause confusion in even the advanced collector.

The first repros or fantasy globes (which were not dated) were the car globes, such as "Cadillac Sales and Service," "Ford Sales and Service," and so on. About the same time came the generic "Gasoline" globe on a plastic body. Then, other brands and dealers got involved.

Again, study the list. Many repros imitate body styles or categories that never even existed. The first category, for instance, includes the 13.5" inserts for plastic or glass bodies. A number of companies never made globes in this category. Signal globes came only in 15" metal frames, so any Signal on a glass or plastic body is a reproduction. Musgos are one-piece globes only, so any Musgo on glass or plastic is a repro. Reading and studying the list will help with any questions you might have.

One item in particular makes it difficult for all collectors—the 13.5" repro metal body. Most metal globes with glass inserts are original and have 15 or 16.5" inserts. So, when you find one of these 13.5" repro inserts in a repro body and someone says, "It's in a metal frame...so it's original," it's not. This is where it gets confusing. Ninety-nine percent of all original metal frame globes have either 15 or 16.5" inserts, not 13.5". There are several metal frame 15" inserts being reproduced and none are dated, except a few Canadian types. Reference this list and know with whom you are dealing. Use the notes to help determine whether it is likely that the globe is original or not.

Get to know authentic globe colors. This can not be stressed enough. Repro colors are usually off a bit. Once you start seeing more and more originals, repros will stand out as phonies by their color alone.

It no longer holds true that all repros have a smooth backing. We've seen known originals from the teens and '20s with smooth backing.

We've seen a few repros where the paint on the back of the insert is not fired on or silk screened. A fingernail or edge of a coin will easily scrape the paint off the back of the insert. This has occurred on some 15" inserts for metal frame globes only.

Concerning the reproduced plastic, glass or metal bodies, we feel that as long as the seller tells you what you are buying, those bodies are not a big deal. There are never enough bodies to mount all the inserts around, so these repro bodies *are* needed. However, some collectors still prefer the original bodies.

The best advice we can offer is to use the provided list, remember the points made here, ask a lot of questions and you should do well in obtaining original globes.

List of Known Reproduction Gasoline Globes

The following reproductions are separated into the four major categories of gasoline globes: plastic body, glass body, metal body, and one-piece (as well as some subcategories).

Plastic or Glass Frame Gasoline Globes with 13.5" Glass Inserts

Although most of the repros in this first category are dated, not all are, so beware. See the list of known globes in this book for descriptions if not given here.

The following explains the terms used in the reproduction descriptions:

"Reproduction" or "Repro" - may be imprinted on base or on the lens itself indicating the item is a repro.

"dated" - an actual date appears on the repro such as 1981, 1994, etc.

"should be dated at bottom" - the *reproduction* only (not the original) *should* have an actual date such as 1981, 1994, etc.

"Licensed by Mobil Oil" (or any other company) - signifies a reproduction.

American Gas: Repros should be dated at bottom.

American Gasoline: With eagle; repros should be dated at bottom.

Amoco: Repros should be dated at bottom.

Ashland A Plus: Large "A" logo; 25 made by Ashland Oil in the 1980s; repros are not dated.

Ashland Flying Octanes: Blue "Flying Octanes", 25 made by Ashland Oil in the 1980s; not dated.

Associated Gasoline: Oil can with "More Miles to the Gallon"; all 13.5" inserts are repros; should be dated at bottom; existed as 15" inserts for metal frame only.

Atlantic: Rectangle logo; should be dated at bottom.

Atlantic Hi-ARC: Repros should be dated at bottom; most originals on Gill only.

Atlantic White Flash: All 13.5" notched lenses are repros; repros should be dated at bottom.

Bay: Shield shown; should be dated at bottom.

Beacon: Lighthouse shown; should be dated at bottom; originals very rare and of a different design.

Bengal Gasoline: All 13.5" notched lenses are repros; repros should be dated at bottom; originals in 15" only.

Buffalo Gasoline: Picture of buffalo; should be dated at bottom.

Chevron: Dated 1991.

Chevron: Originals rare; not dated (circa 1989).

Cities Service Green: Green letters with black outline; dated 1992; no original Cities Service globes have lettering outlined in black.

Cities Service Red: Red letters with black outline; dated 1992; no original Cities Service globes have lettering outlined in black.

Cities Service Yellow: Yellow letters w/black outline; dated 1992; no original Cities Service globes have lettering outlined in black.

Cities Service Koolmotor: Black, white and yellow; should be dated at bottom.

Clark: ***Note:*** Clark used globes into the 1980s and many late originals are of poor quality; we don't believe any repros exist, but possibly undated ones were made by a company in the late 1970s.

Conoco: Triangle; should be dated at bottom.

Conoco Ethyl: Repros should be dated at bottom.

Conoco Ethyl: All 13.5" notched lenses with yellow background are repros; repros should be dated at bottom.

Conoco Gasoline: All 13.5" inserts with Minuteman are repros; existed as 15" inserts for metal frame only.

Diamond D-X Lubricating Gas: Dated 1991; D-X Lubricating Gas; should be dated at bottom.

Dixie: With crossed flags; originals of this design say "Rebel"; all with "Dixie" are repros; these usually are not dated.

Doggone Gas: With dog; all are repros; not dated.

Esso: Should be dated at bottom.

Esso Extra: Block letters; should be dated at bottom.

Esso Script: Repros should be dated at bottom.

Ethyl: With Ethyl Corporation logo; should be dated at bottom.

Farmers Union Coop: Plow and shield; should be dated at bottom; originals are rare.

Fleet-Wing: Bird; should be dated at bottom.

Flying A Gasoline

Letter "A" with wings; should be dated at bottom.

Frontier Rarin' to Go: Black rearing horse; some not dated; early one with red band design; originals deeply notched for glass only; repros made to fit plastic body only. Black areas in reproductions have bubbles.

Gasoline: All 13.5" inserts are repros; red, orange or green backgrounds; originals only one-piece or metal frame.

General Petroleum: All 13.5" inserts are repros; circa 1989; not dated; originals 15" inserts for metal frame only.

Gilmore Blue Green: Lion shown; all 13.5" inserts are repros; some dated 1991; some have earlier date; existed as 15" inserts for metal frame only.

Gilmore, Roar With: Lion across top and checkered flag at bottom; all 13.5" inserts are repros; existed as 15" inserts for metal frame only; should be dated at bottom.

Golden West Oil Co.: Picture of sunset over water, mountains; originals very rare; should be dated at bottom.

Grizzly Dubbs Cracked Gasoline: Picture of standing bear; originals extremely rare; should be dated at bottom. Dubbs Cracked appears on left on 13.5" repro, right on original on Capco only.

Gulf: Should be dated 1982 at bottom.

Gulf: 12.5" inserts; not dated.

That Good Gulf Gasoline: All 13.5" inserts are repros; should be dated at bottom.

Hudson Gasoline: Pictures truck; repros should be dated at bottom.

Humble Gasoline: "Humble Oils" in center; should be dated at bottom.

Husky: Picture of husky dog; should be dated at bottom.

Imperial Refineries: Shield; should be dated at bottom. *Note:* Imperial used globes into the 1980s and many still exist today.

Indian Gas: With teepee; all are repros; should be dated at bottom; never existed as originals with this design.

Indian Gasoline: With designs; all are repros; never existed as originals with this design.

Johnson Gasoline: With hourglass; repros should be dated at bottom.

Liberty Gasoline: Picture of Statue of Liberty; should be dated at bottom; originals extremely rare and of a different design.

Lion Knix Knox: Picture of lion in center; most are dated at bottom.

Magnolia Gasoline: Flower in center; all 15" inserts are repros; should be dated at bottom; existed as 16.5" inserts for metal frame only.

Marathon Mile-maker: Running man in center; should be dated at bottom.

Mobil Premium: Flat shield logo; dated at bottom.

Mobil Regular: Flat shield logo; dated at bottom.

Mobilgas: Pegasus; no outer ring; should be dated at bottom.

Mobilgas: With blue outer ring; all are repros.

Mobilgas Special: Pegasus; should be dated at bottom.

Mohawk Gasoline: Large Indian head profile in center; all 13.5" inserts are repros; should be dated at bottom; existed as 15" inserts for metal frame only.

Mojave Gas: Indian; several colors; all are repros; not dated; never existed as originals; made exclusively for Disneyworld Florida.

Musgo Gasoline Michigan's Mile Maker: Indian in headdress; all 13.5" inserts are repros; originals were one-piece globes.

Mustang Gasolene: Horse pictured; all are repros; not dated; fantasy piece.

Olympic Quality Gasoline: Greyhound dog in center; should be dated at bottom.

Pennzoil: Liberty bell; should be dated at bottom.

Phillips 66: Shield; should be dated at bottom.

Phillips 66 Ethyl: Ethyl on shield; should be dated at bottom

Phillips 77 Aviation: Should be dated at bottom.

Phillips Unique: Should be dated at bottom.

Polly Gas: Shows bird; all 13.5" inserts are repros; should be dated at bottom; existed as 15" inserts for metal frame only.

Pure Firebird: Pictures bird; all 13.5" notched lenses are repros; repros should be dated at bottom.

Pure Ethyl: Repros should be dated at bottom.

Pure Products of the Pure Oil Company: Zigzag border; should be dated at bottom.

Purol Gasoline, The Pure Oil Company: Arrow through center; should be dated at bottom.

Rainbow Gasoline: Rainbow, pump, car shown; all 13.5" inserts are repros; should be dated at bottom; existed as 15" inserts for metal frame only.

Red Crown Gasoline: Small red crown in center; should be dated at bottom; original 13.5" inserts *extremely* rare.

Red Crown Gasoline: Smaller red crown in center; should be dated at bottom.

Red Crown Ethyl: Should be dated at bottom; no 13.5" originals known.

Red Indian: Repros should be dated at bottom. There are many undated ones as well. The only easy way to tell these is to count the feather tips. Start with the one pointing directly dead center to the "A" of the word "Indian" and go to the bottom ending with the feather tip pointing between the second "I" and "A" of "Indian". If you count four, it is original; five is a repro.

Richfield Ethyl: Shield with small Ethyl symbol; should be dated at bottom.

Richfield Ethyl Gasoline of Power: With bird; all 13.5" notched lenses are repros; repros should be dated at bottom.

Richfield Gasoline of Power: Should be dated; originals existed as 15" only.

Richfield Hi Octane: With bird; should be dated at bottom.

Road Runner Gasoline: Road runner pictured as on the oil can; circa 1991; all with this logo are repros; should be dated at bottom; originals are very rare.

Salyer's Gasoline: Woman riding goose; not dated; overall whitish backing on repros and the black is less distinct on repros; originals very rare.

Seaco Aviation: Airplane; not dated (circa 1989). No originals known.

Shell: Yellow shell pictured; all are repros; should be dated at bottom; originals exist as one-piece clam shell, one-piece round shape, 15" metal or 12-1/2" only.

Super Shell: All 13.5" lenses are repros; should be dated at bottom; originals exist as one-piece clam shaped or one-piece round shaped, or 15" metal only.

Signal: Stop light with "Go" sign pictured; all 13.5" inserts are repros; should be dated at bottom; existed as 15" inserts for metal frame only.

Sinclair Aircraft: Airplane; should be dated at bottom; few originals known.

Sinclair Dino Gasoline: Dinosaur pictured; larger than originals; if dinosaur is larger than 2", it is a repro; some current repros now being made with smaller dinosaur like original; should be dated at bottom.

Sinclair Ethyl: Red background with Ethyl logo; dated at bottom.

Sinclair H-C Gasoline: "H-C" in center; should be dated at bottom.

Skelly: Design like the one-piece; red, white, and blue; should be dated at the bottom.

Smitholene Aviation Gasoline: With plane; should be dated at bottom; originals are rare.

Sohio Supreme: Repros should be dated at bottom.

Standard: Sign and torch logo; repros should be dated at bottom.

Blue Sunoco: Yellow diamond in center; all 13.5" are repros; should be dated at bottom; existed as 15" inserts for metal frame only or in 12-1/2" glass.

Sunoco Dynafuel: All 13.5" notched lenses are repros; repros should be dated at bottom.

Sunray: With sun; 1991; should be dated at bottom.

Super Test: Repros should be dated at bottom but some undated repros may exist.

Texaco: Large red star with green "T"; should be dated at bottom; very few undated repros exist.

Texaco Diesel Chief: Should be dated at bottom; originals rare.

Texaco Ethyl: Should be dated at bottom; *Note:* early dates like 3-36 or 4-35 would be *early* originals.

Texaco Fire Chief Gasoline: Fire hat and Texaco emblem; should be dated at bottom; originals have hat logo on top.

Texaco Sky Chief: Should be dated at bottom.

Texas Rose Ethyl: Rose and small Ethyl symbol; should be dated at bottom.

Texas Rose Ethyl: Undated; on the reproduction, the diameter of the Ethyl logo is 3.5". And will be approximately .5" above the green outline ring. The original has an Ethyl logo that is approximately 4.5" and almost touches the ring.

T.P. Products: With teepee; all 13.5" inserts are repros; should be dated at bottom; existed as 15" inserts for metal frame only.

Tri-Star Red Hat Gasoline: Red top hat in center; all are repros; should be dated at bottom.

Tulsa Hi-Test: Cattle skull; should be dated at bottom.

Tydol: Should be dated at bottom.

Union Gasoline: Shield; all 13.5" inserts are repros; should be dated at bottom; existed as 15" inserts for metal frame only.

Union 76 Ethyl: Should be dated at bottom; existed as 15" inserts for metal frame only.

Union 76 Gasoline: All 13.5" inserts are repros; should be dated at bottom; existed as 15" inserts for metal frame only.

Wake Up 99: Shows rooster; not dated (circa 1989); originals are not known to exist yet.

White Eagle Gasoline: Large white eagle in center; all are repros; originals one-piece round, 15" metal, or eagle shaped only.

White Rose Gasoline: Boy holding sign; should be dated at bottom.

Note: There is a company producing a series of globes such as those used by Mobil, Shell, Texaco and others. These are 14" tall with a 4" base. They are all plastic (two pieces put together). Such globes never existed as originals, thus it should be noted that all are reproductions. Also, some repro companies now offer half-size glass globes, such as Shell globes, and all of these are repros. Originals are 16-17" tall. Other companies make 1/2 scale globes, miniature globes, etc. Use common sense when determining whether a globe is a reproduction.

Canadian Plastic or Glass Frame Gasoline Globes with 13.5" Glass Inserts

B-A Service Products Nevr-nox: Should be dated at bottom.

Bengal Gasoline: Shows tiger; all are 13.5" inserts are repros; should be dated at bottom; existed as 15" inserts for metal frame only; originals very rare.

High Compression H-C Supertest: Should be dated at bottom; some undated repros may exist.

Maple Leaf: Shows leaf; dated at bottom.

Red Indian: Indian head in headdress; some dated, some not; see Red Indian listing in previous section.

Red Rose Gasoline: Rose pictured; circa 1990; should be dated at bottom.

Silent Chief: Norseman pictured; circa 1990; not dated; originals *extremely* rare.

White Rose: Rose pictured; should be dated at bottom.

Oval Shaped Plastic Body Globes with Glass Inserts

Amoco: Never existed as an original oval globe.

Ford Sales and Service: No originals ovals known; not a gas globe.

Gargoyle: Originals one-piece raised letters.

Pennzoil: No originals known.

13-1/2" Plastic Frame Automotive Globes: (These were the first reproduction globes, from about 1973 or before. The globes are not dated.)

Buick Sales and Service: Valve-in-head motor cars.

Cadillac Authorized Service: Cadillac Shield.

Chevrolet Corvette: Crossed Flags

Chevrolet Sales and Service: Chevy emblem.

Chrysler Sales and Service: Chrysler emblem.

Dodge Brothers Sales and Service: Six-sided star with "DB".

Ford Benzol: Originals extremely rare.

Ford Mustang: With horse.

Ford Sales and Service: Ford script logo.

Lincoln: Black background.

Lincoln: Blue background.

Mercedes-Benz: With emblem.

Nash: With emblem.

Oldsmobile Sales and Service: With emblem.

Plymouth: With flagship.

Pontiac Service: With Indian head.

Studebaker: Studebaker old tire emblem.

Stutz the Car that Made Good in a Day: Stutz Logo

Whizzer Authorized Sales and Service: Whizzer logo.

Note: Some collectors have found original automotive globes. Though very rare, these usually appear in one-piece or metal frame styles. On rarer occasions, glass 13.5" original inserts are found.

Metal Frame Globes with 15" Glass Inserts

(Important! None of the 15" repros are dated, unless noted!)

American: With shield; originals *very* rare.

Ascot: With car; originals extremely rare; if found, probably a repro.

Associated Gas: With oil can trimmed in blue; some dated, some not; originals rare and most known originals trimmed in green.

Associated Aero Type: Chicken wing logo; dated.

Associated White Gold Gas: Green, black, white; originals rare.

Beacon Gasoline: Lighthouse, yellow background; some dated/some not; originals rare.

Central Michigan Bay Petroleum: Leaping lion; all 15" are repros; fantasy globe.

Chevron Gasoline: Winged logo; dated.

Conoco: Fully detailed minuteman; never existed; original has *solid black* minuteman with no detail shown; may or may not be dated.

El Camino: Bell/mission pictured; only *one* original known, extremely rare.

Ethyl Corp.: Repros have yellow rays, not dated; originals have blue rays. Original does not have the word "WITH" at top of globe face.

Ford Sales and Service: Dark blue and white.

Gasoline: Metal perforated inserts; circa 1989; not dated; originals have a sandpaper-like surface over the metal inserts.

General Antiknock Gas: We believe these are not dated; originals are rare.

Gilmore Ethyl: Ethyl logo in center; dated.

Gilmore Blue Green: Should be dated at bottom or top.

Gilmore Fleet: Flag shown; may or may not be dated.
Gilmore Gasoline Monarch of all: Lion pictured; never existed to our knowledge.
Gilmore Red Lion: Some dated, some not.
Grizzly Gasoline: With bear; originals extremely rare.
Hancock "Cock O' the Walk": Shows rooster; some dated, some not; originals rare.
Imperial: Three stars across center; some dated, some not.
Koolmotor High Test Anti Knock Gasolene: Green background; not dated.
Liberty Gasoline: Shows Statue of Liberty; originals rare.
Magnolia Gasoline: Flower in center; all 15" are repros; early 1980s; originals are 16.5" only.
Mileage: Some dated, some not.
Mobile Gasoline: With race car; never existed; all are repros.
Mobilgas: Pegasus; most are dated.
Mohawk Ethyl: "Ethyl" written across bottom; originals *extremely* rare.
Mohawk Gasoline: Indian head shown; some dated, some not.
Phillips Benzo Gas: Originals extremely rare.
Polly Gas: Some dated, some not; yellow border; dated at bottom; originals very rare.
Purol Gasoline: Cast faces with arrow (circa 1981); heavy glass recessed letters; no known originals found yet, but did exist at one time; about a dozen reproduction sets exist; they were made from original mold.
Radio Gasoline: Dated.
Red Crown Gasoline: Crown shown.
Red Parrot Gas: Parrot pictured; not dated; no originals found as of this writing.
Richfield Gas of Power: Dated.
Richlube Motor Oil: Shows car; never existed; all are repros; fantasy piece.
Sareco: Indian pictured; all are repros; fantasy globe.
Seaside Gasoline: Six-sided star with seagulls; some dated, some not.
Shell Gasoline: Repros have red letters with a yellow background with a shell pictured in a red background; originals have yellow letters on red with a shell pictured in a yellow background; originals extremely rare; no mint originals known to exist; check condition.
Signal: Black background with stop light; some dated, some not.
Signal Purrpull: Three made; dated.
Standard Gasoline with Crown: Should be dated.
Sunoco Blue: Yellow diamond in center; most are dated.
Texaco: Leaded glass; reproductions usually appear on a regular metal body; originals always appear on a *very* wide (about 10-12") and unique-looking metal body which should be painted white or green.
Texaco: Red star with green "T" in center; originals rare.
Texaco Ethyl: Repros have red background around white letters; red star with green "T" in center; originals have white background with black letters and are *extremely* rare in 15" metal.
Time Ethyl: With clock, never existed—all are repros.
Union Gasoline: Shield with red and white stripes; originals very rare.
Union 76: Red and black; no shield; never existed.
Union 76 + Ethyl: Red and black; no shield; never existed; some dated, some not.
Union 76 shield: Gold trim; dated or marked.
Veltex: Dated.
White Rose Gasoline: Boy holding sign; most originals have a cone shape notch in bottom of insert.
White Rose No Knock: Boy holding sign; originals rare; most originals have a cone shape notch in bottom of insert.

Canadian 15" Metal Frame Globes

B-A Service Products Nevr-Nox: Dated.
B-A Service Products: Dated.
Peerless Ethyl: Dated.
Esso: With Ethyl logo.
Imperial: With three stars, not dated.
Maple Leaf Co-op Gasoline: Dated.
North Star: Dated.
Red Head Gasoline: Boy pictured; not dated.

One-Piece Gasoline Globes

Note: All repros are the fired-on type unless noted. Many of the baked-on repros have a limited edition number stamped on the inside, i.e., "23 of 100". Originals only had old dates, but most originals were never dated.

Airliner Gasoline: Airplane (all are repros); never existed as an original one-piece globe.
Aviation Gasoline: Airplane pictured; three repros known; circa late 1980s.
Black Hawk Gasoline: Indian (all are repros); never existed as an original globe.
Cities Service: Round; round originals etched only.
Conoco: Triangle; all are repros; never existed as an original one-piece globe.
Conoco Minuteman: Man pictured; all are repros; never existed as an original one-piece globe.
Deep Rock: Deep Rock bow-tie logo; most originals are etched.
Diamond D-X Lubricating Motor Fuel: All are repros; never existed as an original one-piece globe.
Dog-On Good Gas: All are repros; never existed as an original one-piece globe.
Flying A: All are repros; never existed as an original one-piece globe.
Gulf No-Nox: Originals usually etched or raised letters. Repros have blue letters.
Magnolia: With flower; all are repros; never existed as an original one-piece globe with a flower.
Mobilgas: Never existed as an original one-piece globe with horse.
Phillips 66: All are repros; never existed as an original one-piece globe.
Phillips 77 Aviation: All are repros; never existed as an original one-piece globe.
Pure or Purol: Shows arrow; originals very rare.
Richfield Ethyl: All are repros; never existed as an original one-piece globe.
Shell: Clam-shaped; dated "1994" or marked "Repro"; repros very white in color, originals off-white; very few unmarked repros exist.

Shell: Round-shaped; red or yellow, originals are *etched* only; never existed like this.
Shell Gasoline: Never existed with the word "Gasoline" in a one-piece globe; round shape.
Super Shell Ethyl: Clam-shaped; dated "1994" or marked "Repro"; repros very white in color, originals off-white; very few unmarked repros exist.
Super Shell: Clam-shaped; dated "1994" or marked "Repro"; repros very white in color, originals off-white; very few unmarked repros exist.
Sinclair Aircraft: With plane; red background; repros have the rounded wide body style, originals have the narrow body style.
Skelly: Has large "S" logo.
Socony Motor Gasoline: Socony shield logo; reproductions fired-on; originals etched.
Standard Crown: Clear glass or smoky green tinted; all are repros; an amber/brown turned up—this was original but never intended for use on a pump, was probably made just to test the original mold; oddballs like this occasionally surface.
Standard Crowns: *Most* that are repros have regular flange-type glass 6" bases; originals usually have threaded bases with screw-on metal bottoms. However, please note that early lettered original Crowns only have regular flange bases. Originals have 7" bases.
Standard Red Crown: Raised letters; smaller than original; usually not painted, circa 1981. Please note that this globe was reproduced a second time in 1991 from the original mold. Distinct feature is the slight ridge which outlines the four crosses. Raised letters are smaller than original. Unfortunately, originals exist and look the same.
Standard White Crown: With or without hole in top; remember repros have a 6" base; see "Standard Crowns" for details.
Sunoco: All are repros; never existed as one-piece globe.
Texaco: Originals etched or raised letters only.
Texaco Fire Chief: All round ones are repros.
Tioga: All are repros; never existed as an original one-piece globe; originals were on a glass body, glass inserts only.
Tokheim (gas pump): Still no originals known as of this writing.
White Eagle: Originals are etched only.
White Rose: Boy and slate; originals wide body only.

Etched One-Piece Globes
Check body styles in general listings. Most repros are made on the wrong body style.

Buffalo: Buffalo pictured.
Hudson: No originals ever existed.
Red Head Gasoline: No originals ever existed.
Indian/Havoline: With Indian.
Iroquois: With Indian; no originals ever existed.
Mule Gasoline - It Kicks: With mule; fantasy piece.
Red Crown: With Crown.
Sinclair Aircraft: Green background.
Sinclair H-C: With stripes; only one original known as of 1/98.
Sinco: Repro is much smaller than original, has chimney cap; original is regular size one piece, wide body style only with chimney cap; both are etched.
White Plume: Repro made on narrow body; original on rounded wide body only.

Many other one-piece etched repro globes exist. Most were made on the wrong body styles, so know the styles which are listed in this book. In most cases it only takes a few phone calls to sort out these repros. Know the person with whom you are dealing.

One-Piece Automotive Globes
(All repros are fired-on type.)

Buick
Chevrolet Service
Corvette
Ferrari
Ford Truck Service
International Trucks
Rolls-Royce

Note: There may be other "car or service" globes, but originals are extremely rare and usually etched glass.

Repro Gasoline Globe Bodies and Holders

12-1/2" glass bodies are being reproduced, though there are not many around.
13-1/2" plastic bodies in several colors such as red, green, blue, yellow, black, clear, etc. are being reproduced. Also, oval bodies are being reproduced in white, clear and colors. Repro white Capcos are much whiter than originals. They will also bend easily several inches if twisting each half by itself. Originals will bend only about an inch.
13-1/2" glass bodies, narrow as well as the wide "lollipop" styles are being reproduced. Nearly identical to originals—many made from the original mold.
13-1/2" metal frame bodies—99% of all known to exist would be in reproductions.
15" metal frame bodies have been reproduced in steel as well as aluminum. The steel reproductions are nearly identical to originals and difficult to distinguish.
16.5" metal bodies are being reproduced in steel only. Reproductions also nearly identical to originals and difficult to distinguish.

In terms of pricing, reproduction bodies will decrease the value of a gas globe, however in today's market the amount is negligible. Keep in mind that the colored reproduction plastic bodies or the colored reproduction ripple bodies will significantly reduce the price of a globe. So a gas globe with a reproduction colored body, whether plastic or glass ripple would have significantly less value than one with an original body.

Gill ripple bodies are being reproduced; repros have a seam running front to back at top; color not fired onto glass like originals; texture less distinct than originals; flange base only.

Chapter 7. Packing Gasoline Globes for Shipment

It's a loss to the hobby to have even a single globe broken and yet we hear about this happening all too often.

First of all, plan to double box the globe. This is a must! The outer box should be large, at least 20 cubic inches. We can't stress this enough. Most damaged globes are shipped in smaller *flat* boxes. Don't use these—use larger square/cube boxes. Allow at least five to six inches around all sides, top and bottom of the globe for the first box. Styrofoam peanuts are great because they do the job and are light in weight. Once the globe is boxed, place the box in a second larger box. Allow at least four to six inches or more between the smaller box and the larger box. If your first box is 20 cubic inches then the second box should be at least 24 to 26 cubic inches. Again, use Styrofoam peanuts or other packing material. Newspaper works well when other packing material is not available, but it gets heavy. Make sure to line the bottom of each box with at least four inches of soft padding. Place a few loose sheets of cardboard across the top of the inner box for extra protection.

When packing a glass body globe with glass inserts, remove the inserts! Otherwise, you run the risk of damaging the inserts due to vibrations in transport, which can cause them to break. (If you've never removed inserts before, refer to the chapter titled "Types of Globes," which describes the procedure.) Place thin bubble pack, or, if not available, several sheets of newspaper (six to ten) between the two inserts and pack accordingly. Pack the glass body separately or in another area of the box. That way, during shipping it won't bump and break the inserts. By the way, never stack more than six to eight inserts on top of each other when shipping and even when simply storing. The weight alone could break or crack the bottom insert.

When packing plastic body globes with glass inserts, either ship them assembled or take them apart before packing. We've done it both ways with no problem. But when shipping a fully assembled globe, make sure that the three screws on the plastic body are screwed tight. Stuff the inside of the globe with crumpled newspaper only, not packing peanuts.

Metal frame globes with glass inserts should be taken apart. Again, loose inserts and a metal body make a bad combination. If the inserts are very tight or hard to remove, pack them well but make sure the inserts are snug. Place the metal frame globe inserts in a small box placed inside of a bigger box, with the metal body on top. This adds protection to the upper portion of the box in case it is hit from above or something heavy rests on it. Metal bodies are nearly indestructible.

On rare or expensive globes, it is best to use overnight delivery services such as Federal Express. It's not worth taking chances on breakage and although it's more costly, you'd rather not see your globe arrive in "kit" form. Avoid shipping a globe during the extremely cold weather. Globes are more brittle then and break more easily, even plastic globes.

We've heard pros and cons on marking your box as "Fragile" or "Glass". Ours go out unmarked and we rarely have a problem.

Don't forget to insure your globe and hope that it arrives in one piece!

Chapter 8. Restoring Gasoline Globes

Here we will cover the restoration of gasoline pump globes. Not much has changed here since our first book as the same rules still apply. As before, we must answer the frequently asked question: If you have found an etched one-piece globe, is it okay to restore it and, if so, what does that do to its value?

It is an accepted practice among collectors that an old globe—especially an etched, one-piece globe—is similar to an old Model A. It's basically all there, but it my just need a new paint job. Just as the new paint job on a Model A is definitely going to *help* its value, not take away from it, the same goes for a one-piece etched globe.

We are speaking mainly of one-piece globes here because we don't feel there are a lot of globes that can be restored. We'll get into that later. If you have an etched one-piece that has no paint or a bad paint job, it would behoove you to increase its value by putting a good coat of paint on it. We do not feel that this takes away from its value, and that there is little difference in value in a one-piece etched globe with its original paint and one that has been nicely restored—if any difference at all. The value is in the etching, not in the age of the paint.

Now, what other globes can be restored? The other type that can be restored successfully is the metal frame body. This can be done by simply taking the old body and either sandblasting or hand sanding to remove the paint, whichever method is required for the job. Then, apply a coat of good primer and a fast-drying enamel paint. Try to stick with the original color as closely as possible. If you don't know the original color and no one else has any idea, go by the color of the lenses. That is, choose the color that would work best with the lenses.

Some hobbyists have asked about restoring lenses. For the most part, these should not be restored. If there is just a small damaged area that calls for minor repairs, it might be worth the effort. But, generally, once a lens is worn there is little or no value to the piece. You could buy clear glass lenses and make up any beautiful globe you wanted and say that it had been restored—but the piece would then be a reproduction. This theory also applies to fired-on one-piece globes. Again, once a fired-on one-piece has lost its original paint, there is nothing left to show what was on the globe. Sometimes it is possible to identify faint markings on the globe, and try to restore it by adding new paint. But, again, you could take a blank lens or even a one-piece body and claim that it is something it is not. Simply said, once a fired-on globe has lost its paint, it is basically worthless. Some collectors are now restoring just one side of a baked-on one-piece to show what the other side said. This is a good idea, but the globe still has little value.

As for Texaco one-piece cast globes, including the Texaco Ethyl, use caution. These can be restored but the red star was baked on and will *never* look as good as the original paint, so its value will be somewhat less.

What it comes down to is that the only globes that could be restored are the etched one-piece globes, raised letter globes and metal bodies. The raised one-piece globes include the Standard Crowns, even though they don't all have letters, and the Shells, which have recessed letters. These can—and should—be restored.

Etched one-piece globes are fun and fairly easy to restore. Collectors have restored many of these types over the years. To prepare the globe, try to remove as much old paint as possible. Use paint remover and a light metal brush.

Sometimes it takes several coats of paint remover to get all the old paint removed. Then take a steel wool pad or toothbrush and lightly wipe off the globe. Rinse the globe with water and let it dry.

Next, you need to illuminate the globe. Hook up an old wooden base with a light bulb on it to an electrical socket. When you spray on the paint, make sure it goes on evenly. Etched globes *must* be sprayed, as brushing leaves uneven marks on the globe. Work with only a small area at a time, taping off and masking the other areas. Newspaper and masking tape work best. Use a fast drying enamel. If you are painting the globe several different colors, follow the same procedure.

If the word "Gasoline" or another designation is painted around the bottom of the globe, tape off a rectangle around the word and cover with newspaper the areas you don' t want painted. Once each word or design is sprayed to its appropriate color, remove all the tape and newspaper. Using a sharp razor blade, scrape off the excess paint. As you are doing this, periodically keep the globe wet with water. This prevents the loose paint flakes from sticking to the freshly painted surfaces.

Now your globe should look as if it were new-old-stock (NOS).

Raised letter one-piece globes are harder to restore. The old paint is often difficult to remove since it was originally fired on. A professional airbrusher may have more success. If you do attempt it yourself, it may be better to tape off the exact areas that need to be done, unlike for an etched one-piece globe as already described.

As far as broken bases, BB or pellet gun holes and the like are concerned, if someone can fix them with glass, they should be repaired. A professional glass repair person may be the key here. Globes with broken inserts can be made to look much better too. Remarkable advances in glass repair are doing wonders with broken inserts since our first writing.

By the way, ripple glass globes have fired-on paint on the inside of the glass. If you find one that is worn, minimal touch up is okay, but don't go too far. Again, you could take a clear ripple and paint it purple and say, "Guess what I found?!"

As far as Crowns are concerned (non-raised-letter types) we'll leave that up to you. For what it's worth, most collectors won't buy a rarer blue or green Crown that has been repainted. Most collectors wait until an original with good paint comes along. The problem here, though, is originality. Gold, red, and white Crowns are very common. Green, blue, silver-gray or purple Crowns are much rarer and more valuable. A white Crown or faded red one could be made into any other colored globe easily. For instance, you could have a white Crown that is painted green. Much worse is that a later buyer of that globe actually believes he or she has an original green Crown. Repainting a Crown is acceptable in terms of restoring the globe, but keep to its original color. If you sell the globe, honesty dictates that you inform the buyer that it has been repainted.

Chapter 9. Overview of the Petroleum Collectibles Hobby

Thousands of people worldwide are involved in the petroleum collectibles hobby. Beginning in the 1940s people began accumulating paper items, gas pump globes, and early gas pumps on a very limited basis. Only a very small number - less than 10-collectors were active in this era. In the 1960s antique car enthusiasts began to seek out interesting service station items to accent their car collections, and several began to actively build their collections of globes and gas pumps. The 1970s saw the first national-level interest in the hobby, with the introduction of reproduction items to restore gas pumps. The early 1980s saw the beginning of several regularly published periodicals dealing with petroleum collectibles, and by the mid-1980s more books and the first of the now many regional swap meets were being held. The hobby exploded about 1989, with hundreds of new collectors joining in the fun every year. Most areas of the country were within a day's drive of some regional annual event. In 1990 the co-authors of this book first teamed up to produce the first in a series of books designed specifically for today's collector of petroelum memorabilia. This book is number nine in a series that now includes commitments on several others.

Of interest to our readers is the fact that the authors also offer the only complete monthly periodical serving the hobby, Petroleum Collectibles Monthly. When your copy arrives each month, on time, you will find it filled with information on new discoveries, reproduction item alerts, events and auction coverage, a show schedule, a list of displays that are open to the public as well as other places to visit, plus the largest classified ad marketplace serving the hobby. For further information on the hobby, on books available, or to subscribe to Petroleum Collectibles Monthly, contact the authors at:

PETROLEUM COLLECTIBLES MONTHLY
P.O. BOX 556
LaGRANGE, OH 44050-0556
(440) 355-6608
(440) 355-4955 FAX

All of the authors' other books are available through Petroleum Collectibles Monthly, including the self-published "Guide To Gasoline Logos" a complete look at marketing symbols along with a system of dating virtually all petroleum collectibles. This book is intended as a companion to your new copy of "Gas Pump Globes". Contact PCM today for more information.

As a service to readers of this book, we wish to make the following comments about the various types of petroleum collectibles and their relative interest in the hobby today. They are listed in order of popularity.

SIGNS: Signs are considered the most popular petroleum collectible item today. Earliest signs were tin lithographed, by before 1920 porcelain began to see extensive use in the petroleum industry. Plastics, introduced to the industry in 1947 and found in common use after 1960 are much less collectible than the earlier tin and porcelain items. Neon was never extensively used in the gasoline industry, but collectors have recently taken interest in the few neon signs available. Prices range from a mere $25 for a recent plastic item to over $10,000 for the most decorative and artistic procelain signs offered. Many reproductions are available, so be careful and know what you are buying. Texaco signs with unusual graphics are among the most desirable, and to the other end of the spectrum, the Texaco Fire Chief pump signs are perhaps the most commonly found petroleum sign of any kind. Signs were the primary subject of our earlier book, GAS AND OIL SIGNS, released by Motorbooks International in 1996.

GLOBES: Second only to signs are gas pump globes in popularity. From the earliest one piece etched milk glass domes to the plastic bodied globes still used by several companies today, each and every gas pump globe is highly collectible. Globes evolved from the early one piece glass globes, to metal band bodies with glass lenses, to glass bodies with removable lenses, to the plastic bodies formed in halves and sandwiched around lenses. Generally the more graphic the globe, the more desirable. Prices range from around $60 for plain plastic globes to over $15,000 for the most unique. Texaco globes are among the most popular, and indeed the highest price ever recorded for a globe was that received for the cast Texaco Fire Chief Hat globe. Many reproductions are available, so be careful and know what you are buying. Globes were the primary subject of our earlier book, GAS PUMP GLOBES, released by Motorbooks International in 1994.

CANS: Motor oil cans and other product containers first became popular among collectors in the late 1980s. Today the common quart can, used from the early 1930s until the mid-1980s, are the most popular format, but five-quart cans and handy oilers are also very popular as well. As with signs and globes, the more graphic the can, the more desirable. Cans can be found in prices ranging from less than $5 for composite cans with plain graphics to nearly $2000 for the most graphically appealing steel cans. Several reproductions are available, so be careful and know what you are buying. No definitive work on cans has yet been completed.

PUMPS: The first interest surge in petroleum collectibles was in gas pumps, back in the 1960s. Visible pumps were most desirable then, although in recent years visibles and pre-visibles have taken a backseat to ornate clock face and other mechanically intricate and decorative clock-face pumps. Mechanical calculating pumps, particularly with art-deco designs are very popular as well. All pumps manufactured up until the "chrome" era of the 1960s are collectibles. Globes were the primary subject of our earlier book, GAS PUMP COLLECTORS GUIDE , released by Motorbooks International in 1994.

TOYS: Relatively new to the interest in petroleum collectibles is the field of toys. Gasoline related toys have been manufactured since the 1920s, and accurately detailed brand-specific toys have been offered since the 1930s. Some toys can be considered promotionals, while others were simply children's toys available to anyone. Texaco was the first company to regularly use promotional toys as part of their advertising, and they continue to do so today. Prices range from current pricing for new toys, in the $20 range, to well over $1000 for toys made as recently as 15 years ago.

MAPS & PAPER ITEMS: Many collectors, including this writer, started their collections as children collecting road maps. Free road maps were perhaps the most widespread oil company promotion of all times, with millions of maps being given away from 1914 to the present. Most companies abandoned the practice in the 1970s, as did Texaco, however, and all maps are considered collectible. Texaco maps are not as graphically appealling as those from some other brands, and as a result are not very valuable. While maps in general range in price from $1 to about $75, most Texaco maps fall into the $1-$5 range.

INTERNAL CORPORATE ITEMS: Internal corporate items have recently found a place of interest to collectors. Employee pins and badges, award items, trinkets, many of which fall into the "smalls" or "Trinkets" category are addressed below. Corporate publications are of interest as well, and price ranges depend on the type of item involved. Many items are hard to find, since they were never intended for use by the general public.

TRINKETS: Advertising novelties are among the fastest growing in interest of all petroleum collectibles. Salt and pepper shaker sets, oil can banks, radios, and thousands more items area available. Virtually anything that ever had an oil company logo applied is collectible. Categories include items supplied by the oil companies themselves, items sanctioned by the oil companies and personalized by the local distributor or dealer, and items supplied by advertising companies directly to distributors and dealers. Price ranges vary greatly, and prices are rising.

Chapter 10. Major Oil Company Listings: A-M

Please Note: roughly 100 globes shown in this chapter were mounted in bodies other than the original (i.e. a globe originally mounted in a glass body might be seen here mounted in plastic). Check the captions and listings for the accurate body descriptions. Values shown reflect caption descriptions and match the listings of the globes.

AMOCO and PAN-AM

****American Oil Company**

AMOCO GAS, 1915-1932, 15" METAL, $750-1000. *Courtesy of PCM Archives*.

Left: AMOCO, 1940-1961, GILL, $275-450. **Right**: AMERICAN GAS, 1940-1961, GILL, $250-400. *Courtesy of R.V. Witherspoon; Courtesy of PCM Archives.*

Above: AMERICAN GAS, 1926-1961, 15" METAL, $300-475. *Courtesy of R.V. Witherspoon.*

AMERICAN ETHYL, 1933-1934, 15" METAL, $650-850. *Courtesy of Larry Stancil.*

Left: AMOCO, 1932-1961, 15" METAL, $300-500. *Courtesy of PCM Archives.*

ARTEX, 1940-1961, 15" METAL, $275-450. *Courtesy of PCM Archives.*

ARTEX, 1940-1961, GILL, $275-425. *Courtesy of Larry Stancil.*

Left: AMERICAN STRATE, 1922-1938, 15" METAL, $750-1000. *Courtesy of PCM Archives.*

**PAN-AM Southern Corporation

PAN-AM, 1920s, OPE, NO LISTING.
Courtesy of L.F. Perry.

PAN-AM, 1929-1939, 15" METAL, $500-750.
Courtesy of PCM Archives.

PAN-AM ECONOMY, 1939-1947, 15" METAL, $650-850. *Courtesy of PCM Archives.*

PAN-AM GASOLINE/MOTOR OIL, 1925-1928, 15" METAL, $650-900. *Courtesy of PCM Archives.*

BLUE, 1925-1929, 15" METAL, NO LISTING.
Courtesy of L.F. Perry.

PAN-AM QUALITY, 1939-1947, 15" METAL, $650-850. *Courtesy of PCM Archives.*

PAN-AM, 1928-1929, 15" METAL, $600-850.
Courtesy of Larry Stancil.

PAN-AM ORANGE, 1929-1939, 15" METAL, $700-950. *Courtesy of PCM Archives.*

PAN-AM PREMIUM, 1939-1947, 15" METAL, $700-900. *Courtesy of PCM Archives.*

PAN-AM, 1945-1947, BANDED GLASS, $400-600. *Courtesy of PCM Archives.*

PAN-AM TORCH and OVAL, 1947-1956, 12.5" GLASS, $350-550. *Courtesy of Patrick Shimmin.*

AMOCO, 1956-1961, 12.5" GLASS, $250-400. *Courtesy of PCM Archives.*

ASAMERA & AFFILIATES

**Andy-Dandy Oil Co.

ANDY-DANDY OIL CO, 1956-1964, YELLOW RIPPLE, $2500-4000. *Courtesy of Bill Vos.*

**Gasamat

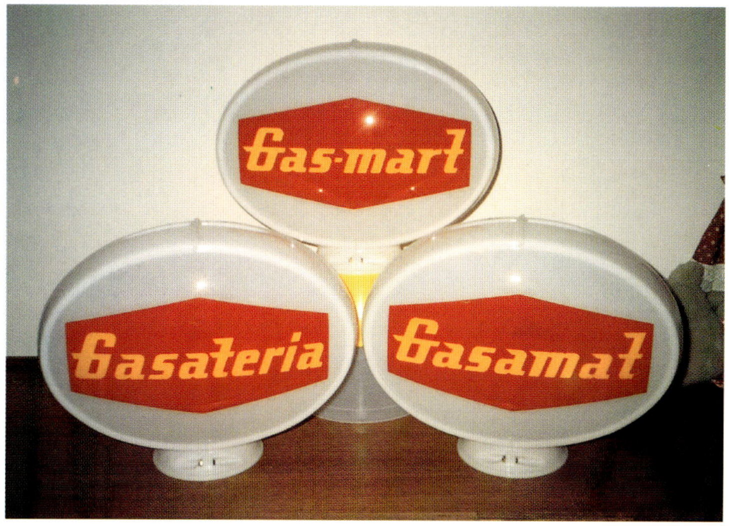

Top: GASMART, 1957-1965, OVAL CAPCO, $200-350.
Left: GASATERIA, 1957, OVAL CAPCO, $200-350.
Right: GASAMAT, 1957-1965, OVAL CAPCO, $200-350.
Courtesy of Lonnie Hop.

ASHLAND & AFFILIATES

**Pepper

AMERICAN, 1956-1961, 12.5" GLASS, $250-400. *Courtesy of PCM Archives.*

WHITE PEPPER, 1924-1928, OPE, $1300-1800. *Courtesy of Thornton Beroth.*

WHITE PEPPER, 1928-1935, 15" METAL, $800-1250. *Courtesy of Jim St. Martin.*

GREEN PEPPER ANTIKNOCK, 1928-1935, 15" METAL, $900-1500. *Courtesy of Jim St. Martin.*

RED PEPPER BETTER ETHYL, 1928-1935, 15" METAL, $900-1400. *Courtesy of Jim St. Martin.*

SUPER PEPPER, 1935-1946, 15" METAL, $800-1050. *Courtesy of Jim St. Martin.*

RED PEPPER ETHYL (EGC), 1935-1946, 15" METAL, $900-1400. *Courtesy of Jim St. Martin.*

TRUCK PEPPER, 1935-1946, 13.5" GLASS, NO LISTING. *Courtesy of PCM Archives.*

****Ashland**

ASHLAND, 1946-1950, 13.5" GLASS, $300-450. *Courtesy of PCM Archives.*

ASHLAND FLYING OCTANES, 1946-1960, 13.5" GLASS, $275-425. *Courtesy of PCM Archives.*

ASHLAND FLYING OCTANES, 1946-1960, CAPCO, $200-300. *Courtesy of PCM Archives.*

ASHLAND ETHYL (EC), 1946-1960, 13.5" GLASS, $300-450. *Courtesy of PCM Archives.*

ASHLAND DIESEL, 1946-1960, CAPCO, $225-325. *Courtesy of PCM Archives.*

ASHLAND GO MIX OUTBOARD FUEL, 1955-1960, CAPCO, $450-650. *Courtesy of PCM Archives.*

ASHLAND FLYING OCTANES BLUE, 1960-1968, CAPCO, $200-300. *Courtesy of Jim Welty.*

ASHLAND A-PLUS, 1960-1968, CAPCO, $150-225. *Courtesy of PCM Archives.*

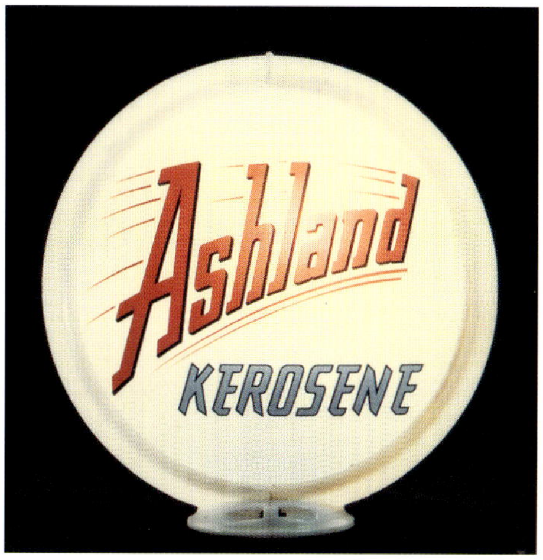

ASHLAND KEROSENE, 1960-1968, CAPCO, $200-300. *Courtesy of Bob Scism.*

****Tri-State**

TRI STATE GASOLINE, 1928-1932, OPB, $1400-1800. *Courtesy of PCM Archives.*

TRI STATE SUPER MOTOR, 1930s, 13.5" GLASS, $300-500. *Courtesy of PCM Archives.*

TRI-STATE ETHYL (EGC), 1930s, 13.5" GLASS, $250-400. *Courtesy of PCM Archives.*

TRI-STATE, 1946-1960, 13.5" GLASS, $225-375. *Courtesy of PCM Archives.*

TRI-STATE ETHYL (EC), 1946-1960, 13.5" GLASS, $250-400. *Courtesy of PCM Archives.*

**National Refining

Left: WHITE ROSE GASOLINE BOY/SLATE, 1920-1930, 15" METAL, $2200-2800.
Right: WHITE ROSE GASOLINE BOY/SLATE, 1926-1930, OP BAKED ON 3PC BODY, $1600-2000. *Courtesy of Bruce Bednar*.

WHITE ROSE GASOLINE BOY/SLATE, 1926-1930, OPB, $2400-3000. *Courtesy of PCM Archives*.

WHITE ROSE GASOLINE BOY/SLATE, 1928-1935, 13.5" GLASS, $1200-1600. *Courtesy of PCM Archives*.

WHITE ROSE ENARCO GAS, 1915-1918, 15" METAL, $650-900. *Courtesy of PCM Archives*.

WHITE ROSE NO KNOCK BOY/SLATE, 1926-1930, OPB, $2500-3200. *Courtesy of PCM Archives*.

WHITE ROSE ETHYL GASOLINE, 1928-1935, 13.5" GLASS, $300-550. *Courtesy of PCM Archives*.

ROYAL GASOLINE-THE NAT.REF.CO., 1928-1935, 14" GILL, $250-450. *Courtesy of PCM Archives.*

WHITE/BOY & SLATE/ROSE, 1936-1940, 13.5" GLASS, $500-850. *Courtesy of PCM Archives.*

ROYAL/GASOLINE-THE NAT.REF.CO., 1936-1940, 13.5" GLASS, $250-375. *Courtesy of PCM Archives.*

SUPER ROYAL, 1936-1940, 13.5" GLASS, $250-375. *Courtesy of PCM Archives.*

NATIONAL/WHITE ROSE/***, 1940-1955, 13.5" GLASS, $250-375. *Courtesy of PCM Archives.*

NATIONAL/WHITE ROSE/ETHYL, 1940-1955, 13.5" GLASS, $250-375. *Courtesy of PCM Archives.*

NATIONAL/ROYAL/***, 1940-1950, 13.5" GLASS, $250-375. *Courtesy of PCM Archives.*

AROUND THE WORLD, 1950-1955, 13.5" GLASS, $3000-4500. *Courtesy of PCM Archives.*

****White Rose**

WHITE/BOY & SLATE/ROSE, 1936-1940, 13.5" GLASS, NO LISTING. *Courtesy of Jim Burris.*

WHITE/ETHYL/ROSE, 1936-1940, 13.5" GLASS, NO LISTING. *Courtesy of Jim Burris.*

ETHYL/WHITE ROSE/ETHYL, 1955-1970, CAPCO, $225-325. *Courtesy of PCM Archives.*

AETNA ON WHITE, 1940s, CAPCO, $175-275. *Courtesy of PCM Archives.*

****White Rose/Mid-West Refining**

WHITE ROSE, 1946-1970, CAPCO, $200-300. *Courtesy of PCM Archives.*

****Aetna Oil**

AETNA MOTOR GAS W/MOUNTAIN, 1930s, 13.5" GLASS, NO LISTING. *Courtesy of PCM Archives.*

AETNA ON GREEN, 1950s, 13.5" GLASS, $225-375. *Courtesy of PCM Archives.*

NEW/WHITE ROSE/ETHYL, 1946-1955, CAPCO, $200-300. *Courtesy of PCM Archives.*

Top Row, Left: (AETNA) MOTO-ZIP, 1940s, 13.5" GLASS, $300-475. *Courtesy of the late Kenny Black.*

Top Row, 2nd From Left: AETNA ETHYL (EGC), 1930s, 13.5" GLASS, $300-425. *Courtesy of PCM Archives.*

Bottom Row, Left: AETNA TOP QUALITY MOTOR OILS, 1930s, 13.5" GLASS, $350-500.

Bottom Row, 2nd From Left: AETNA BENZOL GAS, 1930s, 13.5" GLASS, $450-700.

Bottom Row, 3rd From Left: AETNA XL LUBRICATED ANTI-KNOCK, 1930s, 13.5" GLASS, $400-650.

Bottom Row courtesy of the late Kenny Black.

AETNA ON LIME GREEN, 1950s, CAPCO, $200-300. *Courtesy of PCM Archives.*

FRONTIER DOUBLE REFINED (BOWTIE), 1950-1955, 13.5" GLASS, $250-400. *Courtesy of PCM Archives.*

VALVOLINE GASOLINE, 1930s, 15" METAL, $1100-1600. *Courtesy of PCM Archives.*

AETNA A-PLUS, 1960-1961, CAPCO, $200-275. *Courtesy of PCM Archives.*

FRONTIER ETHYL, 1955-1960, 13.5" GLASS, $225-375. *Courtesy of Larry Stancil.*

VALVOLINE GASOLINE & MOTOR OIL, 1920s, OPE, $2000-2500. *Courtesy of PCM Archives.*

****Frontier Refining**

FRONTIER DOUBLE REFINED (BURST), 1935-1950, 13.5" GLASS, $250-400. *Courtesy of PCM Archives.*

****Valvoline Oil Company**

Left: VALVOLINE AERO, 1920s, 15" METAL, $1200-1700. **Right:** VALVOLINE RED STAR, 1920s, 15" METAL, $1000-1500. *Courtesy of Pete Clark.*

VALVOLINE, 1930s, CAPCO, $350-650. *Courtesy of PCM Archives.*

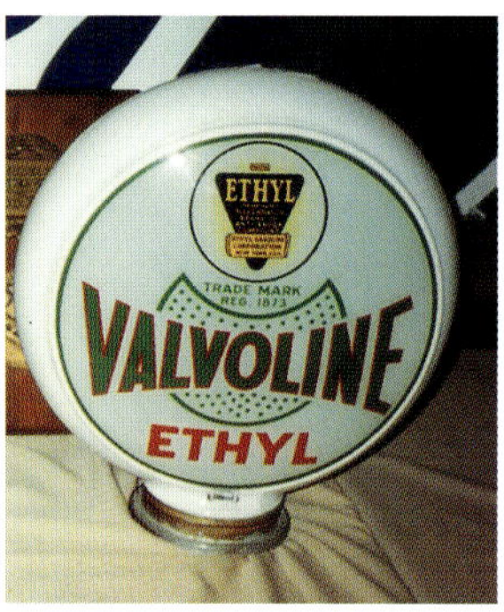

VALVOLINE ETHYL, 1930s, BANDED GLASS, $550-875. *Courtesy of PCM Archives.*

MAGNET, 1930s, 13.5" GLASS, $750-1200. *Courtesy of PCM Archives.*

VALVOLINE GO MIX, 1960-1968, CAPCO, $400-650. *Courtesy of PCM Archives.*

****Freedom Oil Works**

FREEDOM GASOLINE, 1910s, 15" METAL, $850-1200. *Courtesy of PCM Archives.*

FREEDOM MOTOR, 1920s, OPE, NO LISTING. *Courtesy of PCM Archives.*

FREEDOM ETHYL, 1926-1930, OPE, NO LISTING. *Courtesy of Walt Feiger.*

FREEDOM, 1930s, 15" METAL, $650-900. *Courtesy of David Barnhart.*

****Independent Quality Service**

IQS, 1940s, 13.5" GLASS, $300-450. *Courtesy of PCM Archives.*

IQS ETHYL, 1940s, 13.5" GLASS, $325-475. *Courtesy of PCM Archives.*

IQS ETHYL, 1940s, 13.5" GLASS, $325-475. *Courtesy of PCM Archives.*

Dance

DANCE "92" A WHALE OF A BUY, 1940s, 13.5" GLASS, $275-450. *Courtesy of PCM Archives.*

DANCE "100" ETHYL, 1940s, 13.5" GLASS, $275-450. *Courtesy of PCM Archives.*

DANCE PINK GOLDEN POWER PACK, 1960s, CAPCO, $1600-2200. *Courtesy of Jim Welty.*

DANCE BLUE GENUINE ETHYL, 1960s, CAPCO, $1600-2200. *Courtesy of PCM Archives.*

Payless

PAYLESS WHALE OF A BUY, 1952-1960, CAPCO shot on 13.5" GLASS, NO LISTING. *Courtesy of R.V. Witherspoon.*

HI TEST POWER PACK, 1940s, 13.5" GLASS, $200-350. *Courtesy of Bruce Bednar.*

Left: PAYLESS ROAD TESTED REGULAR, 1960-1965, OVAL CAPCO, $250-400. *Courtesy of R.V. Witherspoon.*

PAYLESS NO.2 DIESEL CLEAN FILTERED, 1960-1965, OVAL CAPCO, $400-650. *Courtesy of Glen Miller.*

PAYLESS KEROSENE HEATER OIL, 1960-1965, OVAL CAPCO, $400-650. *Courtesy of PCM Archives.*

Left: PAYLESS EXTRA SPECIAL, 1960-1965, OVAL CAPCO, $250-400. *Courtesy of R.V. Witherspoon.*

****SUPERAMERICA**

Below: PAYLESS GOLDEN THUNDERBIRD SUPER ETHYL, 1960-1965, OVAL CAPCO, $400-650. *Courtesy of R.V. Witherspoon.*

SA (red/blue logo plastic pump topper), 1985-PRES, PLASTIC, $75-125. *Courtesy of PCM Archives.*

**Oskey Bros. Petroleum

NORTH STAR R-TANE (regular), 1955-1965, 13.5" GLASS, $300-500. *Courtesy of Bruce Bednar.*

NORTH STAR DIESEL, 1955-1965, 13.5" GLASS, $300-500. *Courtesy of Bruce Bednar.*

NORTH STAR OVAL, 1965-1980, 2PC PLASTIC OVAL, $200-300. *Courtesy of Lonnie Hop.*

NORTH STAR OVAL, 1965-1980, 1PCPLASTIC OVAL, $200-300. *Courtesy of PCM Archives.*

**Red Head Oil Co.

HI-OCTANE RED HEAD GASOLINE, 1935-1946, 13.5" GLASS, NO LISTING. *Courtesy of Glen Blackmore.*

RED HEAD, 1946-1965, 13.5" GLASS, $300-450. *Courtesy of PCM Archives.*

RED HEAD ETHYL, 1946-1965, 13.5" GLASS, $325-475. *Courtesy of PCM Archives.*

RED HEAD STOVE GASOLINE, 1946-1965, 13.5" GLASS, $350-500. *Courtesy of Bruce Bednar.*

**Southern Oil of NY

ROTARY, 1920s, 15" METAL, $800-1250. *Courtesy of PCM Archives.*

ROTALINE HI COMPRESSTION ANTI-KNOCK GASOLINE, 1920s, 15" METAL, $900-1400. *Courtesy of PCM Archives*.

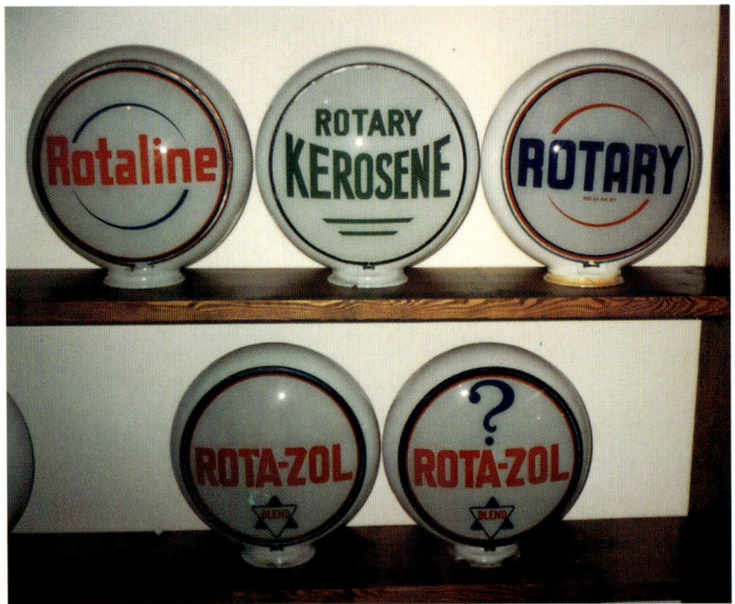

Top Row, Left: ROTALINE, 1946-1960, GILL, $350-550. **Top Row, Center**: ROTARY KEROSENE, 1946-1960, GILL, $300-500. **Top Row, Right**: ROTARY, 1946-1960, GILL, $350-550. **Bottom Row, Left**: ROTA-ZOL BLEND, 1946-1960, GILL, $275-450. **Bottom Row, Right**: ROTA-ZOL BLEND "?", 1946-1960, GILL, $300-500. *Courtesy of Lonnie Hop*.

**Hood Oil Co.

HOOD'S FLO-POWER REGULAR, 1950s, RED CAPCO, $600-1000. *Courtesy of Lonnie Hop*.

**Tresler Oil

TRESLER BENZOL GAS, 1930s, O.P.E., N.L. *Courtesy of PCM Archives*.

COMET DIESEL, 1946-1955, 13.5" GLASS, $375-600. *Courtesy of PCM Archives*.

REGULAR TRESLER COMET, 1955-1964, CAPCO, $250-400. *Courtesy of PCM Archives*.

SILVER COMET, 1955-1964, CAPCO, $850-1250. *Courtesy of Tom Dooley.*

TRESLER COMET REGULAR, 1964-1970, CAPCO, $250-400. *Courtesy of PCM Archives.*

****Cincinnati Oil Works Co.**

WHITE ROSE FILTERED GAS, 1910s, OPE, NO LISTING. *Courtesy of Dan Shaw.*

WHITE ROSE, 1920s-1940s, 15" METAL, $450-700. *Courtesy of PCM Archives.*

ZIP ROSE HIGH COMPRESSION, 1920s-1940s, 15" METAL, $600-900. *Courtesy of PCM Archives.*

****Union Oil Co.**

UNION GASOLINE MOTOR OIL, 1950s, 13.5" GLASS, $400-600. *Courtesy of PCM Archives.*

****Pyroil Chemical Co.**

PYROIL LUBRICATING GASOLINE, 1930s, 14" BANDED GLASS, $500-750. *Courtesy of PCM Archives.*

ATLANTIC & AFFILIATES

****The Atlantic Refining Co.**

Top: ATLANTIC CROSSED ARROWS, 1925-1935, GILL, NO LISTING.
Bottom: ATLANTIC POLARINE GASOLINE, 1915-1920, OPE CHIMNEY, $2200-3500.
Courtesy of R.V. Witherspoon.

ATLANTIC GASOLINE blue lettering, 1918-1933, 16.5" METAL, $600-900.
Courtesy of PCM Archives.

ATLANTIC MOTOR OIL, 1920s, 16.5" METAL, NO LISTING. *Courtesy of PCM Archives.*

ATLANTIC ETHYL (EGC) BLUE BURST, 1927-1936, 15" METAL, $550-800.
Courtesy of PCM Archives.

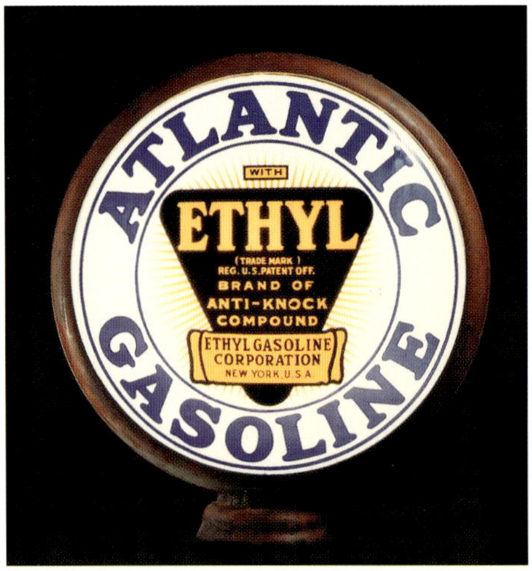

ATLANTIC ETHYL (EGC) YELLOW BURST, 1927-1936, 15" METAL, $550-800.
Courtesy of PCM Archives.

ATLANTIC 68-70, 1925-1928, 15" METAL, $500-750. *Courtesy of PCM Archives.*

ATLANTIC CAPITOL, 1928-1935, 16.5" METAL, $500-750. *Courtesy of PCM Archives.*

ATLANTIC RAYOLITE KEROSENE, 1935-1940, 15" METAL, $700-1200.
Courtesy of PCM Archives.

ATLANTIC WHITE FLASH, 1932-1953, 15" METAL, $450-700. *Courtesy of PCM Archives.*

ATLANTIC HI-ARC, 1936-1953, 15" METAL, $500-750. *Courtesy of PCM Archives.*

ATLANTIC WHITE FLASH, 1932-1953, GILL, $275-450. *Courtesy of R.V. Witherspoon.*

(ATLANTIC) DIESEL FUEL, 1935-1953, 15" METAL, $250-450.
Courtesy of R.V. Witherspoon.

2nd From Left: ATLANTIC HI-ARC, 1936-1953, GILL, $275-450.
3rd From Left: (ATLANTIC) DIESEL FUEL, 1936-1970, GILL, $200-325.
Right: (ATLANTIC) KEROSENE, 1940-1970, GILL, $225-350.
Courtesy of R.V. Witherspoon.

Left: ATLANTIC, 1953-1966, GILL, $200-325.
2nd From Left: ATLANTIC PREMIUM, 1953-1957, GILL, $275-450.
3rd From Left: ATLANTIC IMPERIAL, 1957-1966, GILL, $275-450.
Right: ATLANTIC NEW LOGO, 1966-1970, GILL, $250-375.
Courtesy of R.V. Witherspoon.

CHAMPLIN & AFFILIATES

**Champlin Oil and Refining

CHAMPLIN PRESTO, 1927-1935, OPE, NO LISTING. *Courtesy of PCM Archives.*

CHAMPLIN GASOLINE, 1925-1935, 15" METAL, $550-850. *Courtesy of PCM Archives.*

CHAMPLIN GASOLINE, 1930-1936, 13.5" GLASS, $275-450. *Courtesy of PCM Archives.*

CHAMPLIN PRESTO, 1930-1956, BANDED GLASS, $300-475. *Courtesy of PCM Archives.*

CHAMPLIN, 1956-1962, CAPCO, $175-275. *Courtesy of PCM Archives.*

Below: **Left**: HARBOR PLASTIC LENS, 1960s, CAPCO, $250-400.
Right: CHAMPLIN OBLONG LOGO PLASTIC LENS, 1962-1970, CAPCO, $175-275.
Courtesy of Bruce Bednar.

CHAMPLIN OBLONG LOGO, 1962-1970, CAPCO, $150-250. *Courtesy of PCM Archives.*

**Champlin Oil and Refining/ **Harbor

HARBOR, 1960s, CAPCO, $250-400. *Courtesy of R.V. Witherspoon.*

CHEVRON & AFFILIATES

****Standard Stations, Inc. and Standard Oil/Chevron Dealers**

STANDARD GASOLINE (W/CROWN), 1930-1946, 15" METAL, $900-1400. *Courtesy of PCM Archives.*

STANDARD ETHYL (EGC), 1930-1940, 15" METAL, $750-1250. *Courtesy of PCM Archives.*

STANDARD FLIGHT (W/RED ARROW), 1936-1946, 15" METAL, $650-900. *Courtesy of PCM Archives.*

STANDARD'S SUPREME WITH ETHYL, 1940-1946, 15" METAL, $700-1000. *Courtesy of Tom Rootlieb.*

CHEVRON GASOLINE, 1946-1957, 15" METAL, $800-1400. *Courtesy of PCM Archives.*

CHEVRON SUPREME, 1946-1957, 15" METAL, $800-1400. *Courtesy of Tom Rootlieb.*

****The California Company**

CALSO W/GREEN AND BLUE LINES UNSURPASSED, 1938-1946, 15" METAL, $650-900. *Courtesy of PCM Archives.*

CALSO'S SUPREME WITH ETHYL (EGC), 1938-1946, 15" METAL, $750-1250. *Courtesy of PCM Archives.*

(CALSO) FLIGHT (W/GREEN ARROW), 1938-1946, 15" METAL, $650-900. *Courtesy of PCM Archives.*

**California Oil Company — Northeast Marketing

CALSO GASOLINE, 1946-1949, 13.5" GLASS, $300-500. *Courtesy of Patrick Shimmin.*

CALSO SUPREME, 1946-1949, 13.5" GLASS, $300-500. *Courtesy of Patrick Shimmin.*

CALSO GASOLINE, 1949-1959, CAPCO, $250 350. *Courtesy of PCM Archives.*

CALSO RED BLACK, 1949-1959, CAPCO, $250-350. *Courtesy of R.V. Witherspoon.*

CALSO RED BLACK, 1949-1959, GILL, $300-500. *Courtesy of PCM Archives.*

CHEVRON GASOLINE, 1959-1969, GILL, $350-550. *Courtesy of PCM Archives.*

CHEVRON SUPREME, 1959-1969, CAPCO, $300-450. *Courtesy of PCM Archives.*

CITIES SERVICE & AFFILIATES

**Cities Service Oil Company

CITIES SERVICE OILS ONCE ALWAYS, 1914-1918, OPE ROUND, $2500-3800. *Courtesy of PCM Archives.*

HIGH TEST/KOOLMOTOR/ANTI-KNOCK, 1914-1918, OPE ROUND, $2500-3800. *Courtesy of PCM Archives.*

CITIES SERVICE OILS ONCE ALWAYS, 1918-1936, 11.5" METAL, $850-1300. *Courtesy of Jim Covers.*

CITIES SERVICE OILS ONCE ALWAYS, 1918-1936, 15" METAL, NO LISTING. *Courtesy of PCM Archives.*

CITIES SERVICE OILS AVIATION (overpaint), 1930s, 15" METAL, $700-1100. *Courtesy of PCM Archives.*

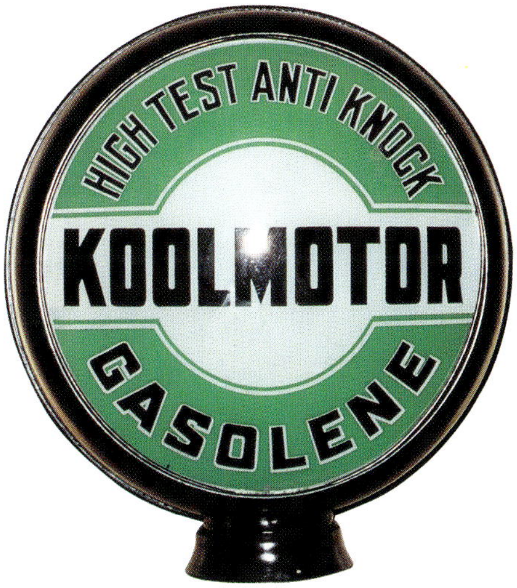

HITEST ANTI KNOCK/KOOLMOTOR/ GASOLENE, 1920-1925, 15" METAL, $650-1000. *Courtesy of R.V. Witherspoon.*

HIGH TEST/KOOLMOTOR/ANTI-KNOCK, 1930, 3 piece clover. $1250-1750 *Courtesy of PCM Archives.*

CITIES SERVICE OILS ONCE ALWAYS, 1926-1934, OPCLOVER, $2300-3200. *Courtesy of PCM Archives.*

CITIES SERVICE OILS ONCE ALWAYS, 1934-1936, 13.5" GLASS, $400-750. *Courtesy of PCM Archives.*

CITIES SERVICE OILS KOOLMOTOR, 1934-1936, 13.5" GLASS, $450-850. *Courtesy of PCM Archives.*

CITIES SERVICE OILS ETHYL, 1934-1936, 13.5" GLASS, $450-850. *Courtesy of PCM Archives.*

CITIES SERVICE OILS ONCE ALWAYS, 1932-1936, GLASS CLOVER, $900-1400. *Courtesy of PCM Archives.*

CITIES SERVICE, 1936-1947, GLASS CLOVER, $850-1350. *Courtesy of PCM Archives.*

CITIES SERVICE KOOLMOTOR, 1936-1947, 14" HULL CLOVER, $1000-1700. *Courtesy of PCM Archives.*

CITIES SERVICE (OILS) [LOGO] KOOLMOTOR, 1934-1936, GLASS CLOVER, $1000-1600. *Courtesy of PCM Archives.*

CITIES SERVICE, 1936-1947, 14" HULL CLOVER, $1000-1600. *Courtesy of PCM Archives.*

CITIES SERVICE ETHYL (EGC), 1936-1947, GLASS CLOVER, $950-1600. *Courtesy of PCM Archives.*

CITIES SERVICE (OILS) [LOGO] ETHYL, 1932-1936, GLASS CLOVER, $950-1500. *Courtesy of PCM Archives.*

CITIES SERVICE KOOLMOTOR, 1936-1947, GLASS CLOVER, $900-1500. *Courtesy of PCM Archives.*

CITIES SERVICE ETHYL (EGC), 1936-1947, 14" HULL CLOVER, $1100-1750. *Courtesy of PCM Archives.*

CITIES SERVICE, 1936-1947, 15" METAL, $600-950. *Courtesy of Lonnie Hop.*

CITIES SERVICE PREMIUM, 1947-1953, RED CAPCO, NO LISTING. *Courtesy of PCM Archives.*

CITIES SERVICE (red), 1953-1960, 13.5" GLASS, $250-325. *Courtesy of PCM Archives.*

CITIES SERVICE KOOLMOTOR, 1936-1947, 15" METAL, $750-1100. *Courtesy of Jim Covers.*

CITIES SERVICE (green), 1953-1960, CAPCO, $200-300. *Courtesy of PCM Archives.*

CITIES SERVICE CLOVER AND TRIANGLE (window), 1932-1960, 15" METAL, $250-400 PER LENS. *Courtesy of Kent Blaine.*

Left: CITIES SERVICE PREMIUM, 1947-1953, 15" METAL, $650-1000.
Right: CITIES SERVICE REGULAR, 1947-1953, 15" METAL, $650-1000.
Courtesy of PCM Archives.

CITIES SERVICE "CARTER HALL CO", 1932-1936, 15" METAL, NO LISTING. *Courtesy of Lonnie Hop.*

LORECO GASOLINE/MOTOR OIL, 15" METAL, $650-900. *Courtesy of PCM Archives.*

LORECO ETHYL, 15" METAL, $400-750. *Courtesy of PCM Archives.*

CITIES SERVICE "CREW-LEVICK" CAST FACE, 1920s, 15" METAL, $1250-1800. *Courtesy of PCM Archives.*

**Warner-Quinlan Company

Left: WARNOCO, 1920-1926, OPE, NO LISTING.
Right: MILEAGE W/OWL, 1927-1930, 15" METAL, $2500-3750.
Courtesy of PCM Archives.

**Shoemaker Coal Co.

BLUE GRASS GASOLINE "SHOEMAKER COAL" CAST, 1920s, 15" METAL, NO LISTING.
Courtesy of John Phippen.

**A.R. Newcombe Oil Co.

KEYSTONE GASOLINE, 1920s, 12" METAL, NO LISTING. *Courtesy of PCM Archives.*

KEYSTONE GASOLINE, $650-900.
Courtesy of PCM Archives.

****Jenney Manufacturing Co.**

JENNEY AUTO OILS/GASOLINE, 1920s, 15" METAL, $3500-5000. *Courtesy of PCM Archives.*

JENNEY ETHYL GASOLINE (EGC), 1920s, 15" METAL, $900-1400. *Courtesy of PCM Archives.*

CUT PRICE GAS, 1930-1940, 15" METAL, $300-500. *Courtesy of PCM Archives.*

JENNY AERO SOLVENIZED, 1935-1940, GILL, $3500-4500. *Courtesy of PCM Archives.*

JENNY AERO, 1940-1960, GILL, $300-500. *Courtesy of PCM Archives.*

JENNY SUPER AERO, 1940-1960, GILL, $300-500. *Courtesy of PCM Archives.*

****Winona Oil Company**

POMIAC GASOLINE, 1920s, OPE, NO LISTING. *Courtesy of PCM Archives.*

POMIAC, 1920s, 15" METAL, $400-600. *Courtesy of PCM Archives.*

****Milton Oil Co.**

DIXCEL GASOLINE, 1930s, 15" METAL, $650-950. *Courtesy of PCM Archives.*

A TRIAL TELLS/DIXCEL/GASOLINE, 1930s, 13.5" GLASS, $350-550. *Courtesy of PCM Archives.*

PIONEER OIL CO/DIXCEL/GASOLINE, 1940s, 13.5" GLASS, $350-550. *Courtesy of PCM Archives.*

PIONEER OIL CO/MOTOR/GASOLINE, 1940s, 13.5" GLASS, $350-550. *Courtesy of Chuck Dudley.*

PIONEER OIL CO MOTOR GASOLINE, 1940s, 13.5" GLASS, $350-550. *Courtesy of PCM Archives.*

DIXCEL PREMIUM ETHYL (EC), 1940s, 13.5" GLASS, $350-550. *Courtesy of PCM Archives.*

****Pioneer**

PIONEER, 1950s, CAPCO, $175-250. *Courtesy of PCM Archives.*

PIONEER REGULAR, 1955-1965, CAPCO, $175-250. *Courtesy of PCM Archives.*

****Milton-Savex**

SAVE WITH SAVEX, 1946-1970, 13.5" GLASS, $250-400. *Courtesy of PCM Archives.*

SAVE ETHYL, 1946-1970, 13.5" GLASS, $250-400. *Courtesy of PCM Archives.*

****Valley Oil Co.**

VALOCO, 1930s, BANDED GLASS, $450-750.
Courtesy of PCM Archives.

CLARK & AFFILIATES

****Clark Oil and Refining Corp.**

CLARK HIGH GRADE GASOLINE, 1930s, 15" METAL - FLAT, NO LISTING.
Courtesy of Larry Starkweather.

CLARK SUPER 100, 1946-1956, 13.5" GLASS, $350-500. *Courtesy of George Johnson.*

CLARK SUPER 100, 1936-1956, ORANGE CAPCO, $300-450. *Courtesy of Jim Welty.*

CLARK, 1956-1982, 13.5" GLASS, $225-325.
Courtesy of PCM Archives.

****Owens Oil Co.**

OCO PEP REGULAR GAS, $250-400.
Courtesy of PCM Archives.

OCO PEP ETHYL (EC), 1950s, CAPCO, $250-400. *Courtesy of Lonnie Hop.*

COLONIAL & AFFILIATES

****Colonial Oil Industries**

COLONIAL OIL INDUSTRIES, 1946-1956, 13.5" GLASS, NO LISTING.
Courtesy of Bruce Bednar.

COLONIAL KEROSENE, 1956-1970, CAPCO, $350-550. *Courtesy of Larry Stancil.*

Left: COLONIAL ROYAL EAGLE, 1956-1970, CAPCO, $350-500.
Center: COLONIAL EAGLE, 1956-1970, CAPCO, $350-500.
Right: COLONIAL GOLDEN EAGLE, 1956-1962, 13.5" GLASS, $450-700.
Courtesy of Jim Welty.

COLONIAL DIESEL, 1956-1970, CAPCO, $350-550. *Courtesy of PCM Archives.*

CONOCO & AFFILIATES
****Continental Oil Company**

CONOCO MINUTEMAN GASOLINE, 1913-1929, 15" METAL, $2500-3300. *Courtesy of John Logsdon.*

CONOCO TRIANGLE W/GREEN OUTLINE, 1929-1950, 13.5" GLASS, $500-750. *Courtesy of PCM Archives.*

CONOCO MINUTEMAN ETHYL, 1927-1929, 15" METAL, $1500-2200. *Courtesy of PCM Archives.*

CONOCO TRIANGLE W/GREEN OUTLINE, 1933-1950, GREEN CAPCO, $250-400. *Courtesy of PCM Archives.*

CONOCO ETHYL (EGC), 1929-1941, 13.5" GLASS, $500-750. *Courtesy of PCM Archives.*

CONOCO ETHYL (EGC), 1933-1941, RED CAPCO, $250-400. *Courtesy of Bob Drake.*

**Marland Oil Company

MARLAND GASOLINE, 1920s, OPE, $1800-2600. *Courtesy of PCM Archives.*

MARLAND GASOLINE (small Marland Oils), 1920s, OPE, $1800-2600. *Courtesy of PCM Archives.*

MARLAND GASOLINE, 1927-1929, GILL, $500-850. *Courtesy of PCM Archives.*

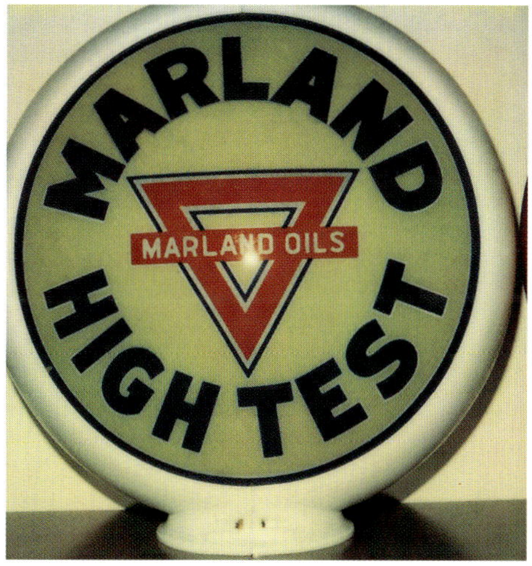

MARLAND HI-TEST, 1927-1929, GILL, $550-900. *Courtesy of PCM Archives.*

**Western

WESTERN, 1960-1969, SHAPED PLASTIC, $200-300. *Courtesy of PCM Archives.*

**Zenith Gas Pool

MILEAGE GASOLINE, 1930s, 15" METAL, $1800-2300. *Courtesy of PCM Archives.*

MILEAGE (w/ wheel), 1940-1949, GILL, $600-900. *Courtesy of Boots Turek.*

MILEAGE, 1949-1960, GILL, $200-350. *Courtesy of PCM Archives.*

SUPER MILEAGE, 1949-1960, GILL, $225-375. *Courtesy of PCM Archives.*

MILEAGE, 1960-1969, SHAPED PLASTIC, $200-300. *Courtesy of PCM Archives.*

**Direct Service Oil Co.

SUPER MAJOR EHTYL, 1940s, 13.5" GLASS, $250-450. *Courtesy of Mark Rime.*

**Malco Refining

MALCO W/NM LOGO, 1940s, 13.5" GLASS, $350-550. *Courtesy of PCM Archives.*

**Valley Oil Company

VALLEY, 1950s, CAPCO, $225-375. *Courtesy of PCM Archives.*

**Kriegler

KRIEGER "KRIEG-O-LENE", 1930s, 16.5" METAL, NO LISTING. *Courtesy of PCM Archives.*

CO-OPERATIVE ASSOCIATIONS

**Farmland Industries

UNION CERTIFIED, 1929-1931, GILL, $300-500. *Courtesy of PCM Archives.*

USE/CO-OP/PRODUCTS, 1930s, GILL, $300-500. *Courtesy of PCM Archives.*

USE/CO-OP ETHYL (EGC), 1930s, GILL, $300-550. *Courtesy of PCM Archives.*

CO-OP WHITE ON RED, 1940s, 13.5" GLASS, $225-350. *Courtesy of PCM Archives.*

CO-OP RED ON WHITE, 1940s, GILL, $225-400. *Courtesy of PCM Archives*

CO-OP GREEN ON WHITE, 1940s, GILL, $225-375. *Courtesy of PCM Archives.*

CO-OP RED ON BLUE, 1940s, GILL, $275-450. *Courtesy of PCM Archives.*

CO-OP DOUBLE CIRCLE, 1962-1985, PLASTIC, $200-325. *Courtesy of PCM Archives.*

**Falcon

FALCON GASOLINE, 1946-1955, GILL, NO LISTING. *Courtesy of PCM Archives.*

FALCON, 1955-1970, CAPCO, $800-1250. *Courtesy of PCM Archives.*

FALCON ETHYL, 1955-1970, CAPCO, $800-1250. *Courtesy of PCM Archives.*

**Union Oil Company

UNION OIL CO FARMCO, 1930s, GILL, $300-450. *Courtesy of PCM Archives.*

**Farmers Union Central Exchange

FARMERS UNION COOPERATIVE STRENGTH, 1932-1934 , GILL, NO LISTING. *Courtesy of PCM Archives.*

FARMERS UNION GASOLINE, 1934-1938, 13.5" GLASS, $900-1500. *Courtesy of PCM Archives.*

FARMERS UNION GASOLINE W/CO-OP seal, 1938-1952, 13.5" GLASS, $450-750. *Courtesy of PCM Archives.*

FARMERS UNION ETHYL, 1942-1952, 13.5" GLASS, $450-750. *Courtesy of PCM Archives.*

FARMERS UNION FULL POWER, 1938-1952, GILL, $500-750. *Courtesy of PCM Archives.*

Left: (CENEX) CO-OP SHIELD, 1952-1972, PLASTIC, $250-450.
Center: (CENEX) SUPER CO-OP SHIELD, 1952-1972, PLASTIC, $250-450.
Right: (CENEX) DIESEL CO-OP SHIELD, 1952-1972, PLASTIC, $300-500.
Courtesy of Lonnie Hop

**Midland Cooperatives

USE CO-OP GASOLINE (red/yellow), 1926-1930, 15" METAL, $1800-2500. *Courtesy of PCM Archives.*

USE CO-OP GASOLINE, 1930s, 13.5" GLASS, $750-1250. *Courtesy of Charlie Wright*

HI TEST/CO-OP/ANTI NOK, 1930s, 13.5" GLASS, $800-1300. *Courtesy of Charlie Wright*

CO-OP COMPETITIVE GASOLINE, 1930s, 13.5" GLASS, $750-1250. *Courtesy of PCM Archives.*

MIDLAND/CO-OP/WHOLESALE, 1940-1955, GILL, $800-1250. *Courtesy of PCM Archives.*

CO-OP SUPER OCTANE GASOLINE, 1940-1955, GILL, $275-450. *Courtesy of PCM Archives.*

MIDLAND "M" NO BOX, 1960-1987, CAPCO, $125-225. *Courtesy of PCM Archives.*

****Farmers Educational and Co-Operative State Union**

FARMERS UNION W/PLOW SCENE, 1920s, 15" METAL, $2000-2800. *Courtesy of PCM Archives.*

FARMERS UNION CO-OPERATIVE HIGH OCTANE, 1930s, 13.5" GLASS, $250-400. *Courtesy of PCM Archives.*

STATEX, 1940-1958, 13.5" GLASS, $225-375. *Courtesy of PCM Archives.*

FELCO, 1958-1970, CAPCO, $150-250. *Courtesy of PCM Archives.*

**Southern States Co-Operatives, Inc.

SOUTHERN STATES COOPERATIVES, 1934-1940, 15" METAL, $900-1500. *Courtesy of PCM Archives.*

**Tennessee Co-Operatives

CO-OP REGULAR, 1946-1970, 13.5" GLASS, NO LISTING. *Courtesy of PCM Archives.*

**Growmark Industries

ANTI-KNOCK, 1930s, 15" METAL, $350-500. *Courtesy of PCM Archives.*

ALADDIN GASOLINE, 1940s, BANDED GLASS, $500-850. *Courtesy of PCM Archives.*

ILLINOIS FARM SUPPLY COMPANY, 1950-1958, CAPCO, $200-300. *Courtesy of PCM Archives.*

ALADDIN GASOLINE ILL. FARM SUPPLY CO., 1930s, 15" METAL, $750-1250. *Courtesy of PCM Archives.*

**Missouri Farmers Association

MFA GASOLINE MOTOR OILS GREASES, 1946-1960, 13.5" GLASS, $300-475. *Courtesy of PCM Archives.*

MFA, 1960-1970, CAPCO, $225-325. *Courtesy of PCM Archives.*

**Farmers Gas and Oil Company of Michigan

FARMERS/CO-OP/PETROLEUM, 1930s, 15" METAL, $600-850. *Courtesy of PCM Archives.*

FARMERS/HI OCTANE/CO-OP, 1940s, GILL, $250-450. *Courtesy of PCM Archives.*

**Central Co-Operative Wholesale

CO-OP "AFFIL W/NATL CO-OP INC." (white on red), 1930s, 15" METAL, $500-800. *Courtesy of PCM Archives.*

CO-OP "AFFIL W/NATL CO-OP INC." (white on red), 1930s, GILL, $300-450. *Courtesy of PCM Archives.*

CO-OP "AFFIL W/NATL CO-OP INC." (white on blue), 1930s, 15" METAL, $550-850. *Courtesy of PCM Archives.*

CO-OP "AFFIL W/NATL CO-OP INC." (white on blue), 1930s, GILL, $350-500. *Courtesy of PCM Archives.*

**Co-Operative GLF Farm Supplies, Inc.

GLF QUALITY w/painted steel faces, 1930s, 15" METAL, $800-1250. *Courtesy of PCM Archives.*

GLF QUALITY (white lettering on red), 1946-1964, 13.5" GLASS, $300-500. *Courtesy of PCM Archives.*

**Landmark Co-Operatives

QUALITY/FARM BUREAU/PRODUCTS (ON SHIELD), 1940s, 13.5" GLASS, $250-450. *Courtesy of PCM Archives.*

LANDMARK FYR-ZON GASOLINE, 1960s, 13.5" GLASS, $300-500. *Courtesy of PCM Archives.*

FYR-ZON GASOLINE, 1950s, CAPCO, $200-300. *Courtesy of PCM Archives.*

****Farmers Union Co-Operative Organization of Iowa**

FARMOIL W/PLOW SCENE, 1920s, 15" METAL, $1750-2300. *Courtesy of PCM Archives.*

****Iowa Farm Service Company**

FARM BUREAU GASOLINE, 1930s, GILL, NO LISTING. *Courtesy of PCM Archives.*

Top Left: FARM BUREAU SPECIAL, 1940s, 14" GILL, $250-450.
Top Right: FARM BUREAU HI-COMPRESSION, 1940s, 14" GILL, $350-550.
Bottom Four: FARM BUREAU OILOGAS, 1940s, 14" GILL, $250-550.
Courtesy of PCM Archives.

****UFA Central Co-Operatives Association, Ltd.**

MAPLE LEAF GASOLINE, 1930s, 15" METAL, $1800-2500. *Courtesy* of Don Loughren

MAPLE LEAF PREMIUM, 1930s, 15" METAL, $1800-2500. *Courtesy* of Don Loughren

****Kansas Farmers Union**

K.F.U. 70, 1940s, GILL, NO LISTING. *Courtesy of PCM Archives.*

CROWN & AFFILIATES
****Crown Central Petroleum**

LEFT: CROWN GAS-OIL, 1933-1956, GILL, $300-475. **RIGHT**: CROWNZOL, 1933-1956, GILL, $300-475. *COURTESY OF PCM ARCHIVES.*

CROWNZOL WITH "HI-OCTANE USA" DECAL, 1950s, GILL, $325-500. *Courtesy of Jim Welty.*

CROWN SILVER, 1956-1973, GILL, $300-450. *Courtesy of PCM Archives.*

CROWN GOLD, 1956-1973, GILL, $325-475. *Courtesy of PCM Archives.*

CROWN DIESEL, 1956-1973, GILL, $275-425. *Courtesy of PCM Archives.*

****Thrift**

THRIFT PREMIUM, 1950s, GILL, $300-425. *Courtesy of PCM Archives.*

CRYSTAL & AFFILIATES
****Stone Oil Co.**

STONES REGULAR, 1950s, CAPCO, $250-350. *Courtesy of Bruce Bednar.*

STONES ETHYL, 1950s, CAPCO, $250-350. *Courtesy of Bruce Bednar.*

Platolene

500 R. J. OIL AND REFINING CO., 1940s, 13.5" GLASS, $550-850. *Courtesy* of Rick Bonenberger.

PLATOLENE 500, 1960s, OVAL CAPCO, $350-600. *Courtesy of Lonnie Hop.*

DERBY & AFFILIATES

****Derby Oil Co.**

DERBY SUPER RATIO MOTOR FUEL, 1926-1935, 15" METAL, $500-850. *Courtesy of PCM Archives.*

DERBY GASOLINE, 1935-1946, GILL, $275-450. *Courtesy of John Logsdon.*

DERBY GASOLINE, 1935-1946, GILL, $275-450. *Courtesy of PCM Archives.*

DERBY'S FLEXGAS ETHYL large (EC), 946-1950, GILL, $275-450. *Courtesy of PCM Archives.*

DERBY'S FLEXGAS ETHYL large (EC), 1946-1950, 13.5" GLASS, $275-450. *Courtesy of Edwin C. Schwartz.*

DERBY, 1950-1970, GILL, $275-450. *ourtesy of Jason Henderson.*

DERBY, 1965-1970, GILL PLASTIC LENS, $250-400. *Courtesy of Jim Masson.*

DERBY DIESEL, 1950-1970, GILL, $250-450. *Courtesy of PCM Archives.*

****Sovereign Service**

SOVEREIGN "SS", 1937-1954, GILL, $300-450. *Courtesy of PCM Archives.*

SOVEREIGN ETHYL, 1937-1954, GILL, $300-450. *Courtesy of PCM Archives.*

SOVEREIGN QUALITY, 1937-1954, GILL, $300-450. *Courtesy of PCM Archives.*

KANOTEX SOVEREIGN QUALITY, 1937-1952, GILL, $400-600. *Courtesy of PCM Archives.*

DIXIE DISTRIBUTORS

****Dixie Distributors, Inc.**

DIXIE OILS GASOLINE, 1926-1928, OPE, $1500-2200. *Courtesy of PCM Archives.*

DIXIE POWER TO PASS, 1928-1931, BLUE OVAL, $2200-3000. *Courtesy of Lonnie Hop.*

DIXIE GASOLINE BENZOL, 1928-1931, GREEN OVAL, $2400-3200. *Courtesy of Lonnie Hop.*

DIXIE KNOCK OUT, 1928-1931, WHITE OVAL, $2300-3100. *Courtesy of Lonnie Hop.*

DIXIE LIBERTY, 1928-1931, YELLOW OVAL, $2400-3200. *Courtesy of Lonnie Hop*.

DIXIE NEVERNOX, 1928-1931, RED OVAL, $2400-3200. *Courtesy of Lonnie Hop*.

DIXIE GASOLINE, 1920s-1930s, 15" METAL, $750-1200. *Courtesy of Lonnie Hop*.

DIXIE GASOLINE POWER TO PASS, 1931-1940, 15" METAL, $800-1400. *Courtesy of PCM Archives*.

DIXIE GASOLINE POWER TO PASS, 1931-1960, YELLOW RIPPLE, $1000-1500. *Courtesy of PCM Archives*.

DIXIE GASOLINE POWER TO PASS, 1941-1960, CAPCO, $350-550. *Courtesy of PCM Archives*.

DIXIE ETHYL (EC), 1941-1949, YELLOW CAPCO, $300-500. *Courtesy of R.V. Witherspoon*.

DIXIE BLUE, 1930-1932, 13.5" GLASS, $250-400. *Courtesy of PCM Archives*.

****Burford Oil Company**

CRYSTAL, 1930s, OPB, NO LISTING. *Courtcsy of Nocl Lcvy.*

CRYSTAL UPSIDE DOWN, 1930s, OPB, NO LISTING. *Courtesy of Noel Levy.*

SILENCER ETHYL (EGC), 1930s, 15" METAL, $600-850. *Courtesy of PCM Archives.*

DIXIE SILENCER STOPS KNOCKS, 1940s, YELLOW CAPCO, $300-550. *Courtesy of Noel Levy.*

SILENCER DIXIE ETHYL (EGC), 1940s, RED CAPCO, $275-500. *Courtesy of Noel Levy.*

DX & AFFILIATES
Mid-Continent Petroleum

NEVR NOX GASOLENE DIAMOND PRODUCT, 1920s, OPE, $1700-2600. *Courtesy of PCM Archives.*

NEVR NOX "PEPPY POWER", 1920s, OPE, $2000-2800. *Courtesy of PCM Archives.*

NEVR NOX GASOLENE NON-POISONOUS, 1920s, OPE, $1500-2000. *Courtesy of PCM Archives.*

DIAMOND NITRO GASOLINE, 1920s, OPE, NO LISTING. *Courtesy of PCM Archives.*

DIAMOND, 1920s, OPE, $1000-1600. *Courtesy of Bruce Bednar.*

DIAMOND, 1925-1933, OPE, $750-1250.
Courtesy of PCM Archives.

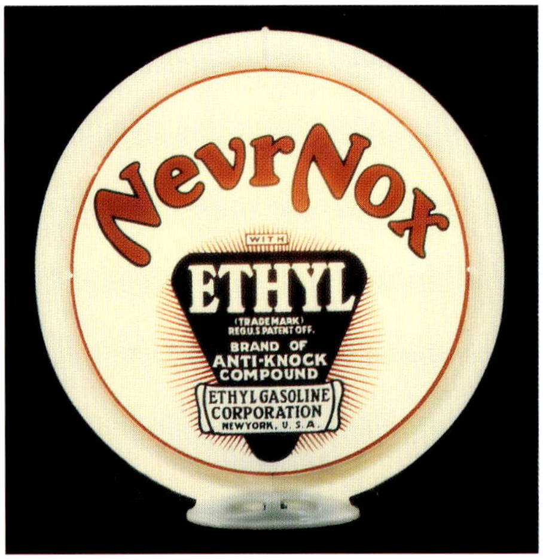

NEVRNOX ETHYL, 1933-1936, 13.5" GLASS, $250-475. *Courtesy of PCM Archives.*

POWER.white, 1936-1940, 13.5" GLASS, $250-350. *Courtesy of PCM Archives.*

DIAMOND, 1930-1933, 13.5" GLASS, $200-400. *Courtesy of PCM Archives.*

DX LUBRICATING MOTOR FUEL.WHITE, 1936-1946, BANDED, $225-350.
Courtesy of Marvin Allie.

POWER "G".white, 1940-1946, BANDED, $275-450. *Courtesy of PCM Archives.*

DIAMOND DX LUBRICATING MOTOR FUEL, 1933-1936, 13.5" GLASS, $250-450.
Courtesy of PCM Archives.

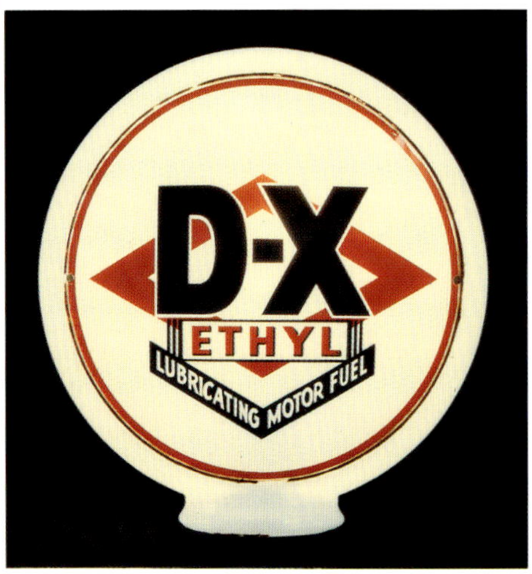

DX ETHYL LUBRICATING MOTOR FUEL, 1936-1946, BANDED, $300-475.
Courtesy of PCM Archives.

DX LUBRICATING GASOLINE, 1946-1957, CAPCO, $200-325. *Courtesy of PCM Archives.*

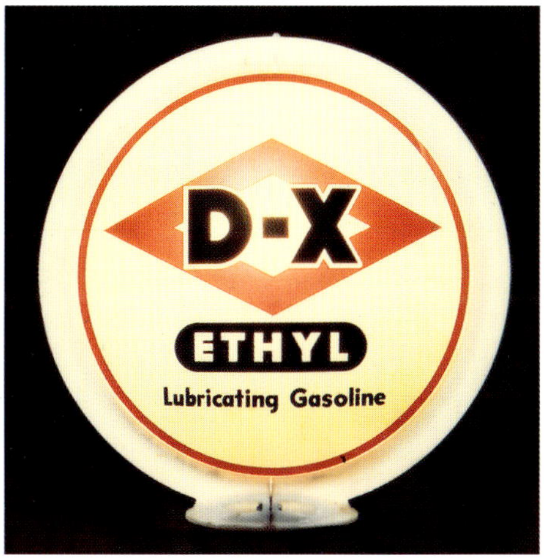

DX ETHYL, 1946-1955, CAPCO, $225-350.
Courtesy of PCM Archives.

DX DIESEL, 1946-1957, CAPCO, $200-300.
Courtesy of PCM Archives.

DX BORON, 1955-1957, CAPCO, $150-275.
Courtesy of PCM Archives.

DX LUBRICATING GASOLINE - 1957 LOGO, 1957-1968, CAPCO, $125-225.
Courtesy of PCM Archives.

DX BORON-1957 LOGO, 1957-1968, CAPCO, $125-225. *Courtesy of PCM Archives.*

DX DIESEL - 1957 LOGO, 1957-1968, CAPCO, $200-325. *Courtesy of PCM Archives.*

DX MARINE - 1957 LOGO, 1957-1968, CAPCO, $450-850. *Courtesy of Dick Thomas.*

****Sunray Oil**

SUNRAY, 1930-1940, ORANGE RIPPLE, NO LISTING. *Courtesy of PCM Archives.*

SUNRAY GASOLINE, 1950-1955, ORANGE CAPCO, $750-1100.
Courtesy of PCM Archives.

**Premier

PREMIER MOTOR, 1935-1955, 13.5" GLASS, $225-375. *Courtesy of PCM Archives.*

PREMIER DIESEL FUEL, 1950-1965, CAPCO, $200-275. *Courtesy of PCM Archives.*

**Barnsdall Refining Co.

BARNSDALL DOMED GLOBE, early 1920s, OPE, $2500-3500. *Courtesy of PCM Archives.*

BARNSDALL MOTOR FUEL, 1930-1941, 13.5" GLASS, $300-475. *Courtesy of PCM Archives.*

Left: SUPER GAS/BE/B/SQUARE/PREMIUM, 1940-1955, 13.5" GLASS, $300-450.
2nd From Left: BARNSDALE/BE/B/SQUARE/GASOLINE, 1930-1941, 13.5" GLASS, $300-475. *Courtesy of PCM Archives.*

**Bareco—Barnsdall

Center: BARNSDALL/SUPER-GAS/ETHYL (EGC), 1933-1940, 13.5" GLASS, $300-475.
Right: BARECO/BE/B/SQUARE/GASOLINE, 1940-1955, 13.5" GLASS, $300-450.
Courtesy of PCM Archives.

BARECO/SUPER-GAS/ETHYL (EC), 1940-1955, 13.5" GLASS, $300-450.
Courtesy of PCM Archives.

**Monarch Manufacturing Co.

MONAMOTOR GASOLINE, 1920s, OPE, $2000-2600. *Courtesy of PCM Archives.*

****Sweney Gasoline and Oil Co.**

SWEENEY MOTOR GASOLINE, 1930s, 15" METAL, NO LISTING. *Courtesy PCM Archives.*

ERICKSON & AFFILIATES
****Erickson Oil Co.**

HOLIDAY (logo shaped), 1962-1965, PLASTIC 2pc, NO LISTING. *Courtesy of PCM Archives.*

ERICKSON'S 101 SUPER REGULAR, 1940-1960, GLASS OVAL, $600-950. *Courtesy of PCM Archives.*

ERICKSON'S 110 HI GRADE ETHYL, 1940-1960, GLASS OVAL, $600-950. *Courtesy of PCM Archives.*

ERICKSON'S RANGE OIL HIGHEST QUALITY, 1940-1960, GLASS OVAL, $650-1000. *Courtesy of Ed Sinclair.*

****Erickson Oil Products**

ERICKSON OIL PRODUCTS, 1952-1970, OVAL CAPCO, $250-350. *Courtesy of PCM Archives.*

EXXON & AFFILIATES
Standard Oil Company of New Jersey

STANDARD MOTOR GASOLINE (CONVEX SIDES), 1912-1918, OPE CHIMENY, $3000-4000. *Courtesy of PCM Archives.*

STANDARD MOTOR GASOLINE (RECESSED SIDES), 1912-1918, OPE CHIMENY, $3000-4000. *Courtesy of PCM Archives.*

STANDARD MOTOR GASOLINE, 1918-1923, 15" METAL, $1200-1500. *Courtesy of PCM Archives.*

"STANDARD" ETHYL (EGC) GASOLINE, 1924-1926, 15" METAL, NO LISTING. *Courtesy of PCM Archives.*

STANDARD BAR AND CIRCLE, 1924-1933, 15" METAL, $400-600. *Courtesy of PCM Archives.*

ACTO, 1930-1939, 16.5" METAL, $300-450. *Courtesy of PCM Archives.*

"STANDARD" WHITE, 1924-1930, 16.5" METAL, $400-650. *Courtesy of PCM Archives.*

STANDARD BAR AND CIRCLE REVERSE, 1924-1933, 16.5" METAL, $400-650. *Courtesy of PCM Archives.*

ESSO, 1933-1972, 15" METAL, $300-450. *Courtesy of R.V. Witherspoon.*

STANDARD BAR AND CIRCLE, 1924-1933, 16.5" METAL, $400-600. *Courtesy of PCM Archives.*

ESSO SCRIPT, 1926-1933, 15" METAL, $700-950. *Courtesy of PCM Archives.*

NEW ESSO, 1938 DECAL, 15" METAL, $350-550. *Courtesy of PCM Archives.*

Left: ESSOLENE, 1933-1939, 16.5" METAL, $350-500.
Right: ESSO PLUS, 1962-1972, 13.5" GLASS, $300-500.
Courtesy of Bruce Bednar.

ESSO KEROSENE, 1940-1972, 16.5" METAL, $400-650. *Courtesy of Bruce Bednar.*

ESSO EXTRA, 1939-1952, 15" METAL, $350-500. *Courtesy of Calloway Oil Co.*

ESSO AVIATION, 1933-1972, 13.5" GLASS, $750-1250. *Courtesy of PCM Archives.*

ESSO MARINE, 1940-1972, 15" METAL, $800-1250. *Courtesy of PCM Archives.*

ESSO (new wide band logo), 1965-1972, GILL, $400-650. *Courtesy of PCM Archives.*

Left: ESSO EXTRA UNDERLINED, 1952-1972, 15" METAL, $350-500.
Right: ESSO EXTRA, 1950-1952, 13.5" GLASS, $350-500.
Courtesy of Bruce Bednar.

**Exxon

EXXON, 1972-1980, CAPCO, $250-350.
Courtesy of PCM Archives.

EXXON EXTRA, 1972-1980, CAPCO, $250-350.
Courtesy of PCM Archives.

EXXON DIESEL, 1972-1980, CAPCO, $250-350.
Courtesy of PCM Archives.

**Humble Oil and Refining

HUMBLE OIL AND REF/HUMBLE OILS, 1919-1932, 15" METAL, $900-1400.
Courtesy of PCM Archives.

FLASHLIKE ETHYL, 1929-1932, 15" METAL, $500-800. *Courtesy of Bob Rymer.*

HUMBLE BAR AND CIRCLE, 1932-1935, 15" METAL, $600-900. *Courtesy of Bob Rymer.*

HUMBLE, 1935-1945, 15" METAL, $500-850.
Courtesy of Bob Rymer.

THRIFTANE, 1939-1947, 15" METAL, $500-750.
Courtesy of J. Sam McIntyre.

**Colonial Beacon Oil Company

COLONIAL GAS, 1920-1930, OCTAGON METAL, $650-900. *Courtesy of Dwight Sawyer.*

COLONIAL ETHYL, 1926-1930, OCTAGON METAL, $650-900. *Courtesy of Dwight Sawyer.*

COLONIAL WINTERGAS, 1926-1930, OCTAGON METAL, $700-950. *Courtesy of Dwight Sawyer.*

COLONIAL BAR AND CIRCLE, 1930-1933, 15" METAL LH GREEN, $450-650. *Courtesy of Dwight Sawyer.*

**Powerine Oil Co.

BEARCAT MOTOR FUEL, 1920s, 15" METAL, NO LISTING. *Courtesy of A.J. Vogel.*

**Northwest Refining Co.

GRIZZLY DUBBS CRACKED, 1930s-1946, CAPCO, $2000-2500. *Courtesy of PCM Archives.*

**Yale Oil Co.

LITENING "B", 1920s, 15" METAL, $550-750. *Courtesy of PCM Archives.*

LITENING ETHYL, 1938-1946, CLEAR RIPPLE, $900-1300. *Courtesy of PCM Archives.*

WILDFIRE, 1938-1946, CLEAR RIPPLE, $1500-1900. *Courtesy of PCM Archives.*

**Oval E

Left: OVAL-E, 1946-1950, RED CAPCO, $250-400.
Right: OVAL-E EXTRA, 1946-1950, BLUE CAPCO, $300-450.
Courtesy of Bob Rymer.

**Carter Oil Co.

CARTER, 1946-1962, BLUE CAPCO, $300-400.
Courtesy of Bob Rymer.

CARTER EXTRA, 1946-1962, BLUE CAPCO, $350-450. *Courtesy of Bob Rymer.*

**Pate Oil Co.

(PATE) AIR GLIDE (shield), 1946-1956, 13.5" GLASS, $600-800. *Courtesy of PCM Archives.*

PATE DIESEL FUEL, 1956-1961, 13.5" GLASS, $300-475. *Courtesy of Scott Anderson.*

PATE RANGE OIL, 1956-1961, 13.5" GLASS, $300-500. *Courtesy of Scott Anderson.*

**Oklahoma Oil Company

OKLAHOMA WHITE, 1930s, OPP, $400-600.
Courtesy of PCM Archives.

**Quality Gasoline Company

BONDED 68, 1940-1946, 13.5" GLASS, $750-1000. *Courtesy of PCM Archives.*

BONDED 70, 1940-1946, 13.5" GLASS, $750-1000. *Courtesy of PCM Archives.*

BONDED 78, 1940-1946, 13.5" GLASS, $750-1000. *Courtesy of PCM Archives.*

BONDED ETHYL, 1946-1955, 13.5" GLASS, $750-1000. *Courtesy of Pat Quail.*

BONDED REGULAR, 1955-1958, PLASTIC, $300-450. *Courtesy of Jim Welty.*

**Johnson Oil Refining Co.

JOHNSON TIME TELLS GASOLENE, 1930s, 15" METAL, $2200-2800. *Courtesy of PCM Archives.*

JOHNSON TIME TELLS GASOLENE, 1930s, 13.5" GLASS, $1500-2000. *Courtesy of PCM Archives.*

JOHNSON TIME TELLS ETHYL GASOLENE, 1930s, 15" METAL, $1500-2000. *Courtesy of PCM Archives.*

JOHNSON WINGED 70 TIME TELLS GASOLENE, 1934-1940, 13.5" GLASS, $800-1250. *Courtesy of PCM Archives.*

Left: JOHNSON BLACK/WHITE/ ORANGE, 1940s, CAPCO, $200-300.
Right: JOHNSON WINGED 70 GASOLENE, 1940s, 13.5" GLASS, $750-1200. *Courtesy of PCM Archives.*

**Hoosier Petroleum Corp.

HOOSIER PETE 100, 1930s-1958, 13.5" GLASS, $500-750. *Courtesy of PCM Archives.*

HOOSIER PETE FUEL OIL, 1930s-1958, 13.5" GLASS, $600-850. *Courtesy of PCM Archives.*

**Ryan's

RYAN'S JET KEROSENE, 1950s, 13.5" GLASS, $400-650. *Courtesy of PCM Archives.*

**Southern Oil Co.

DEEM, 1955-1962, CAPCO, $200-325. *Courtesy of PCM Archives.*

**Wood Oil Company

WOOD HI TEST, 1930s-1940s, 13.5" GLASS, No Listing. *Courtesy of PCM Archives.*

**Imperial Oil Limited

PREMIER, 1911-1918, OPE, NO LISTING. *Courtesy of PCM Archives.*

Left: IMPERIAL PREMIER, 1918-1925, OPE RW, $600-1000.
Right: IMPERIAL QUEEN, 1918-1925, OPE RW, $800-1400. *Courtesy of PCM Archives.*

ESSO EXTRA ETHYL, 1938-1950, 15" METAL HP RED, $750-1000. *Courtesy of PCM Archives.*

Left: IMPERIAL ETHYL, 1930-1938, 15" METAL HP RED, $650-900.
Right: PREMIER GASOLINE, 1925-1930, OPB W, $900-1150. *Courtesy of PCM Archives.*

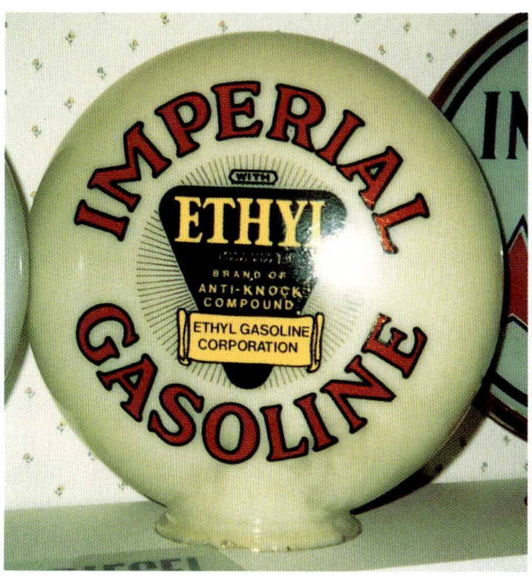

IMPERIAL ETHYL, 1926-1930, OPB W, $600-900. *Courtesy of PCM Archives.*

ESSO ETHYL, 1938-1950, 15" METAL HP RED, $500-750. *Courtesy of PCM Archives.*

ACTO (red letters with blue border), 1930-1940, 15" METAL HP RED, $300-500. *Courtesy of PCM Archives.*

FINA & AFFILIATES

**Panhandle Oil Co.

PANHANDLE PRODUCING REFINING, 1923-1930s, 15" METAL, $900-1500. *Courtesy of Cliff Lauten.*

PANHANDLE SPECIAL, 1930s, 15" METAL, $800-1300. *Courtesy of PCM Archives.*

PANHANDLE NOXLESS, 1930s, 15" METAL, $800-1300. *Courtesy of PCM Archives.*

PANHANDLE, 1940s, 15" METAL, $800-1400. *Courtesy of PCM Archives.*

**American Liberty

AMERICAN LIBERTY REGULAR, 1940s, GILL, $1250-1700. *Courtesy of PCM Archives.*

AMLICO REGULAR, 1950s, CAPCO, $200-300. *Courtesy of PCM Archives.*

AMLICO PREMIUM, 1950s, CAPCO, $200-300. *Courtesy of PCM Archives.*

**Petro-Atlas

THE ELDORADO/ ELRECO/ REFINING COMPANY, 1930-1946, GILL, $300-550. *Courtesy of Steve Tobin.*

SPECIAL/THE EL DORADO/ ELRECO/REFINING CO./GASOLINE, 1930-1946, GILL, $325-575. *Courtesy of PCM Archives.*

ELRECO REFORMED GASOLINE, 1930-1946, GILL, $275-525. *Courtesy of PCM Archives.*

ELRECO REFORMED ETHYL, 1930-1946, GILL, $275-525. *Courtesy of PCM Archives.*

ELRECO PREMIUM, 1960-1946, GILL, $250-500. *Courtesy of PCM Archives.*

BUY MILES NOT GALLONS/ELRECO/REGULAR, 1946-1955, GILL, $300-500. *Courtesy of PCM Archives.*

BUY MILES NOT GALLONS/ELRECO/PREMIUM, 1946-1955, GILL, $300-500. *Courtesy of PCM Archives.*

**Colonial Oil Co.

COLONIAL PREMIUM, 1950s, CAPCO, $175-275. *Courtesy of PCM Archives.*

**Cosden Petroleum

COSDEN GASOLINE OILS, 1925-1930, OPE, $1600-2200. *Courtesy of PCM Archives.*

COSDEN LIQUID GAS OILS, 1930-1938, 13.5" GLASS, $400-650. *Courtesy of PCM Archives.*

COSDEN HIGHER OCTANE, 1938-1950, 13.5" GLASS, $400-625. *Courtesy of PCM Archives.*

COSDEN-GREEN/RED, 1950-1960, CAPCO. $175-225. *Courtesy of PCM Archives*.

COSDEN (red & white), 1960-1963, CAPCO, $150-250. *Courtesy of PCM Archives*.

**Onyx

ONYX REGULAR GASOLINE, 1920s, 15" METAL, $750-1100. *Courtesy of PCM Archives*.

**Wides Oil Company

SAVE WITH/WIDES/GAS FOR LESS, 1955-1970, 13.5" GLASS, $250-450. *Courtesy of PCM Archives*.

FLYING A/GETTY
**Tidewater Oil Company

TYDOL, 1910s, OPC, $1100-1800. *Courtesy of PCM Archives.*

TYDOL T/W/O CAST FACES, 1920-1930, 15" METAL, H/L BLACK, $900-1400. *Courtesy of PCM Archives.*

TYDOL ETHYL CAST FACES (EGC), 1927-1930, 15" METAL, H/L BLACK, $950-1450. *Courtesy of Kyle Moore.*

TYDOL, 1925-1941, 15" METAL, H/L SILVER, $350-550. *Courtesy of PCM Archives.*

TYDOL W/FLYING A DECAL, 1941-1942, 15" METAL, H/L BLACK, $500-700. *Courtesy of PCM Archives.*

TYDOL ARCHED OVER ETHYL (EC), 1941-1942, "13.5"GLASS W", $350-600. *Courtesy of R.V. Witherspoon.*

TYDOL ARCHED OVER ETHYL (EGC), 1935-1941, 15" METAL, H/L BLACK, $600-850. *Courtesy of PCM Archives.*

TYDOL A W/ORANGE BAND, 1941-1942, GILL, $400-650. *Courtesy of PCM Archives.*

TYDOL "A" red and black, 1942-1946, GILL, $400-650. *Courtesy of PCM Archives.*

TIDEX, 1935-1941, 15" METAL, H/L BLACK, $300-550. *Courtesy of PCM Archives.*

TYDOL A W/ORANGE BAND, 1941-1942, 13.5" GLASS, W, $375-600. *Courtesy of PCM Archives.*

TYDOL ETHYL W/AIRPLANE, 1942-1946, GILL, $1000-1500. *Courtesy of PCM Archives.*

TYDOL ETHYL (BLACK/RED), 1946-1947, CAPCO, $175-250. *Courtesy of PCM Archives.*

TYDOL FLYING A ETHYL red and black, 1947-1949, GILL, $400-600. *Courtesy of PCM Archives.*

TYDOL A DIESEL FUEL, 1949-1953, 13.5" GLASS, $450-700. *Courtesy of PCM Archives.*

TYDOL ETHYL, 1935-1941, OP SHOEBOX, NO LISTING. *Courtesy of Lonnie Hop.*

TYDOL, 1935-1941, SHOEBOX, $450-750. *Courtesy of PCM Archives.*

TIDEX, 1935-1949, SHOEBOX, $450-750. *Courtesy of PCM Archives.*

TYDOL "A", 1941-1942, SHOEBOX, $450-750. *Courtesy of PCM Archives.*

TYDOL ETHYL, 1942-1947, SHOEBOX, $500-800. *Courtesy of PCM Archives.*

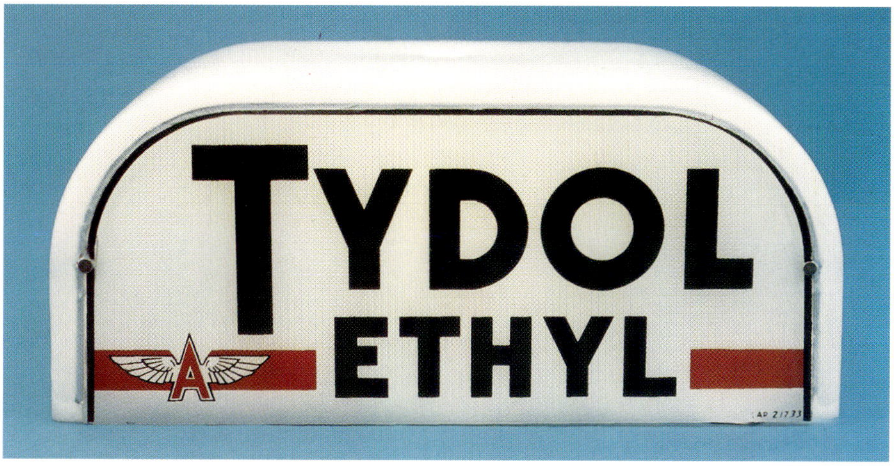

TYDOL "A" ETHYL, 1946-1947, SHOEBOX, $500-800. *Courtesy of PCM Archives.*

TYDOL FLYING A, 1949-1956, SHOEBOX, $500-800. *Courtesy of PCM Archives.*

Left: FLYING A GASOLINE (R/W LARGE TYPE), 1946-1956, 13.5" GLASS, $300-500.
Right: ASSOCIATED GASOLINE W/ETHYL (EGC), 1926-1932, 15" METAL, HP RED, $800-1250.
Courtesy of David Mancuso.

FLYING A GASOLINE, 1956-1960, SHOEBOX, $400-650. *Courtesy of Lonnie Hop.*

****Associated Oil Company**

ASSOCIATED GASOLINE, 1915-1932, 15" METAL, HP RED, $2500-3500. *Courtesy of PCM Archives.*

ASSOCIATED FLYING A "AERO TYPE", 1933-1940, 15" METAL, HP RED, $1200-1800.
Courtesy of PCM Archives.

ASSOCIATED GASOLINE W/ETHYL (EGC), 1926-1932, 15" METAL, HP RED, $800-1250. *Courtesy of PCM Archives.*

ASSOCIATED AVIATION ETHYL, 1932-1940, 15" METAL, HP RED, $1600-2200. *Courtesy of PCM Archives.*

ASSOCIATED WHITE GOLD GASOLINE, 1920s, 15" METAL, HP GREEN, $1100-1500. *Courtesy of PCM Archives.*

FLYING A GASOLINE (R/W LARGE TYPE), 1940-1946, 15" METAL, HP RED, $800-1200. *Courtesy of PCM Archives.*

FLYING A GASOLINE (R/W LARGE TYPE), 1946-1956, GILL, $350-550. *Courtesy of PCM Archives.*

FLYING A ETHYL (BLK) GASOLINE (WHITE), 1940-1946, 15" METAL, HP RED, $900-1400. *Courtesy of PCM Archives.*

FLYING A ETHYL (BLK) GASOLINE (WHITE), 1946-1956, 13.5" GLASS, $350-600. *Courtesy of PCM Archives.*

FLYING A GASOLINE (R/W SMALL TYPE), 1946-1956, CAPCO, $275-400. *Courtesy of PCM Archives.*

FLYING A ETHYL GASOLINE (W/R SM TYPE), 1946-1956, CAPCO, $300-450. *Courtesy of PCM Archives.*

**Tidewater Associated Oil Company —Flying A Marketing

FLYING A SUPER EXTRA, 1956-1970, GILL, $300-500. *Courtesy of PCM Archives.*

**Seaside Oil Company

GRADE NON-CARBONIZING, 1920s, 15" METAL, $1600-2100. *Courtesy of PCM Archives.*

SEASIDE SILVER GULL, 1930-1950, 15" METAL, $2000-2500. *Courtesy of PCM Archives.*

SEASIDE ETHYL, 1935-1950, 15" METAL, $2000-2500. *Courtesy of Dwaine Buck.*

SEASIDE ETHYL (w/decal), 1935-1950, 15" METAL, $2000-2600. *Courtesy of Dwaine Buck.*

**Mohawk Oil Co.

MOHAWK GASOLINE, 1930s, 15" METAL, HP RED, $3500-5000. *Courtesy of Autopia.*

TIDAL, 1920s, 15" METAL, $300-500. *Courtesy of PCM Archives.*

NORTHWESTERN NOCO GASOLINE, 1920s, 15" METAL, BLACK, $700-1100. *Courtesy of Charles Wright.*

GULF & AFFILIATES
****Gulf Oil Co.**

GULF NO-NOX, 1925-1930, OPE, W, $1000-1650. *Courtesy of PCM Archives.*

GULF, 1930-1935, OPC, $500-800. *Courtesy of PCM Archives.*

Left: THAT GOOD GULF GASOLINE, 1925-1930, OPE, N, $750-1100.
Right: THAT GOOD GULF GASOLINE, 1925-1930, OPE, W, $600-950. *Courtesy of Bruce Bednar.*

GULF, 1930-1935, OPB, N, $450-750. *Courtesy of PCM Archives.*

GULF NO NOX, 1925-1930, OPC, W/N, $1000-1650. *Courtesy of PCM Archives.*

GULF KEROSENE, 1920-1930, OPE, W, $1500-2500. *Courtesy of PCM Archives.*

GULF, 1935-1952, 13.5" GLASS, N, $250-450. *Courtesy of PCM Archives.*

GULF, 1935-1952, 12.5" GLASS, W, $300-500. *Courtesy of PCM Archives.*

MARINE WHITE-SPECIAL FOR MARINE USE, 1940-1947, 13.5" GLASS, N, $900-1400. *Courtesy of PCM Archives.*

GULF DIESEL, 1946-1952, 12.5" GLASS, W, $400-650. *Courtesy of PCM Archives.*

GULF DIESELECT, 1946-1952, 12.5" GLASS, W, $450-700. *Courtesy of PCM Archives.*

GULF MARINE WHITE GASOLINE, 1946-1952, 12.5" GLASS, W, $550-900. *Courtesy of PCM Archives.*

GULF, 1930s, 18" NEON, NO LISTING. *Courtesy of PCM Archives.*

****Wilshire Oil Co.**

POLLY GAS, 1920s-1940, 15" METAL, $5000-7000. *Courtesy of PCM Archives.*

BUY ECONOMY GASOLINE, 1920s-1930, 15" METAL, $500-850. *Courtesy of PCM Archives.*

ECONOMY ANTI-KNOCK GASOLINE, 1930-1940, 15" METAL, $1500-1800. *Courtesy of PCM Archives.*

RADIO GASOLINE, 1930s-1940, 15" METAL, $1200-1800. *Courtesy of PCM Archives.*

**Sunset Oil Company

SUNSET GASOLINE (w/sunset scene), 1935-1940s, 15" METAL, NO LISTING. *Courtesy of PCM Archives.*

MARINE (no sunset), 1930-1935, 15" METAL, $1900-2800. *Courtesy of PCM Archives.*

ASCOT W/RACE CAR, 1930-1937, 15" METAL, $3500-5000. *Courtesy of PCM Archives.*

SUNSET DE LUXE, 1937-1940s, 15" METAL, NO LISTING. *Courtesy of PCM Archives.*

SUNSET ETHYL, 1937-1940s, 15" METAL, NO LISTING. *Courtesy of PCM Archives.*

**L.L. Coryell & Sons

CORYELL 70 SUPER, 1930s, 15" METAL, $700-1000. *Courtesy of PCM Archives.*

CORYELL 70 SUPER, 1940s, 13.5" GLASS, $300-500. *Courtesy* of Harold Davis.

CORYELL ETHYL SENIOR (EC), 1940s, 13.5" GLASS, $300-500. *Courtesy of PCM Archives.*

SUNSET THRIFTY (w/Scotsman), 1940s, 15" METAL, NO LISTING. *Courtesy of PCM Archives.*

**Superior Oil Co.

SUPERIOR ETHYL (EC) GASOLINE, 1946-1968, CAPCO, $200-300. *Courtesy of PCM Archives.*

**Smith Oil Co.

SMITH PARAMOUNT GASOLINE, 1930s, 14" GLASS, $350-550. *Courtesy of PCM Archives.*

**British American Oil Co., Ltd.

BRITISH AMERICAN SUPER POWER, 1928-1932, OPB, $750-1250. *Courtesy of Edwin C. Schwartz.*

BA MOTOR, 1940-1954, 13.5" GLASS, W, $300-500. *Courtesy of Don Loughren.*

BA NEVR-NOX, 1940-1954, 13.5" GLASS, W, $350-550. *Courtesy of PCM Archives.*

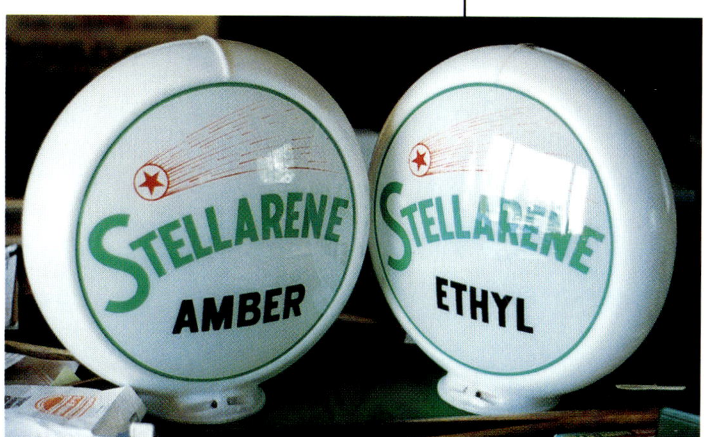

Left: STELLARENE AMBER, 1950s, CAPCO, $300-500. **Right:** STELLARENE ETHYL, 1950s, CAPCO, $300-500. *Courtesy of Don Loughren.*

**Hi-Way Refineries, Ltd.

HI WAY GASOLINE, 1930s, 15" METAL, NO LISTING. *Courtesy of Don Loughren.*

**Puritan Oil Company, Ltd.

STELLARENE NON KNOCK, 1940s, 13.5" GLASS, $400-650. *Courtesy of Don Loughren.*

**Gas and Oil Products, Ltd.

GOP MIRACLE 99, 1950s, 15" METAL, $600-850. *Courtesy of Don Loughren.*

MERCURY 99 ETHYL, 1950s, 15" METAL, $900-1300. *Courtesy of Don Loughren.*

HESS & AFFILIATES
**Olixir

OLIXIR, 1930s, 15" METAL, BLACK, $400-600. *Courtesy of Bruce Bednar.*

**Meadville Corp.

MERIT, 1950s, GILL, $300-500. *Courtesy of Jim Welty.*

MERIT ETHYL (EGC), 1950s, GILL, $350-550. *Courtesy of PCM Archives.*

HUDSON & AFFILIATES
**Silver King

SILVER KING, 1930s, 15" METAL, $400-650. *Courtesy of PCM Archives.*

**Hudson Oil Co.

HUDSON GASOLINE, 1933-1946, GILL, $900-1300. *Courtesy of PCM Archives.*

HUDSON TRANSPORT GASOLINE, 1933-1946, GILL, $1000-1500. *Courtesy of PCM Archives.*

AAA-1 ETHYL, 1940-1946, GILL, $400-650.
Courtesy of PCM Archives.

HUDSON REGULAR, 1955-1962, ORANGE CAPCO, $700-1200.
Courtesy of PCM Archives.

HUDSON ETHYL GASOLINE (EC), 1955-1962, RED CAPCO, $800-1100.
Courtesy of PCM Archives.

HUDSON ETHYL (EC), 1955-1962, RED CAPCO, $800-1100.
Courtesy of PCM Archives.

HUDSON KEROSENE, 1950s, CAPCO, $400-600. *Courtesy of Lonnie Hop.*

****Hornet Oil Company**

HORNET GASOLINE, 1950s, CAPCO, $250-350. *Courtesy of PCM Archives.*

HUSKY & AFFILIATES
****Western Oil and Fuel**

HUSKY "QUALITY PETROLEUM PRODUCTS", 1935-1940, 13.5" GLASS, $1700-2300.
Courtesy of Steve Parker.

HUSKY w/northern lights, 1940-1960, 13.5" GLASS, $1800-2400.
Courtesy of PCM Archives.

HUSKY w/northern lights.

Left: HUSKY HI-POWER, 1954-1960, CAPCO, $1200-1600.
Courtesy of PCM Archives.
Right: HUSKY w/northern lights, 1940-1960, GILL, $1700-2500.
Courtesy of Collectors' Auction Service.

**H. Earl Clack, Inc.

HI-POWER GASOLINE-HECCOLENE OIL, 1930-1954, 15" METAL, LP, $650-1000. *Courtesy of PCM Archives.*

**Husky-Hi Power, Inc.

HUSKY HI-POWER, 1954-1960, 15" METAL, $1800-2400. *Courtesy of PCM Archives.*

**Frontier Oil

FRONTIER RARIN' TO GO, 1935-1946, GILL, $1400-2000. *Courtesy of PCM Archives.*

FRONTIER RARIN' TO GO, 1935-1946, GILL, $1400-2000. *Courtesy of Todd Sime.*

FRONTIER GAS, 1946-1956, 13.5" GLASS, $900-1500. *Courtesy of PCM Archives.*

FRONTIER DIESEL, 1946-1956, 13.5" GLASS, $1000-1600. *Courtesy of Lou Fredrick.*

FRONTIER RARIN' TO GO, 1956-1970, CAPCO, $600-900. *Courtesy of PCM Archives.*

FRONTIER FLEET FUEL DIESEL, 1956-1970, CAPCO, $300-500. *Courtesy of Bruce Bednar.*

**Western States Petroleum

BEELINE GASOLINE, 1946-1965, CAPCO, $1200-1750. *Courtesy of R.V. Witherspoon.*

INDEPENDENT OIL MEN OF AMERICA (IOMA)

Generic Red Hat Globes

RED / HAT GASOLINE, 1925-1932, 15" METAL, HP RED, $1800-2300.
Courtesy of PCM Archives.

High Hat/Jones Brothers Oil Co./Weeks Oil Co.

Top: HIGH HAT GASOLINE W/HAT AND LOGO, 1920s, 15" METAL, NO LISTING.
Left: JONES BROS. RED HAT, 1920s, 15" METAL, HP RED, $2000-2500.
Courtesy of J. Span.
Right: WEEKS OIL CO., INC. RED HAT GASOLINE, 1925-1932, 15" METAL, HP RED, $2000-2500. *Courtesy of J. Span.*

Blake Oil Company

BLAKE OIL CO. GASOLINE, 1920s, OPE, RW, $2000-2600. *Courtesy of PCM Archives.*

Consumers Oil Co.

CONSUMERS OIL CO. WINCHESTER, TENN, 1920s, OPE, N, $2000-2600.
Courtesy of PCM Archives.

CSC RED HAT GASOLINE, 1925-1932, OPE, RW, $3500-5000.
Courtesy of PCM Archives.

Gerkins Oil Company

GERKINS QUALITY GASOLINE, 1925-1935, 15" METAL, HP RED, $1000-1500. *Courtesy of Ron Harrah.*

Gibson Brothers Oil Co.

GIBSON BROS OIL CO, 1925-1935, 15" METAL, HP RED, $1000-1500. *Courtesy of PCM Archives.*

H.S. Goodell & Sons

H.S. GOODELL RED HAT GASOLINE, 1925-1935, 15" METAL, HP RED, $1900-2500. *Courtesy of Ron Harrah.*

Home Oil Company

HOME OIL CO. RED HAT GASOLINE, 1925-1932, 15" METAL, HP RED, $2000-2500. *Courtesy of PCM Archives.*

Independent Oil Company

INDEPENDENT OIL CO. SEYMOUR, MO, 1925-1940, 15" METAL, HP RED, $1000-1500. *Courtesy of PCM Archives.*

Matheson Oil Company

MATHESON GASOLINE, 1920s, 15" METAL, HP RED, $1200-1700. *Courtesy of R.V. Witherspoon.*

MATHESON RED HAT GASOLINE, 925-1932, OPE, RW, $3500-5000. *Courtesy of PCM Archives.*

Musco Oil Company

MUSCO RED HAT GASOLINE, 1925-1932, 15"METAL, HP RED, $2000-2500. *Courtesy of PCM Archives.*

Northland Gasoline

NORTHLAND GOLDEN FLASH, 1930s, 13.5" GLASS, W, $350-550. *Courtesy of Dennis Beckstrom.*

O'Neils Paint & Hardware Co.

O'NEILS EVERY DROP TESTED, 1920s, 15" METAL, HP RED, $1500-2000. *Courtesy of Lonnie Hop.*

Pendergast Oil Company

PENDERS RED HAT GASOLINE, 1925-1932, 15" METAL, HP RED, $2000-2500. *Courtesy of R.V. Witherspoon.*

Petersons Oil Company

PETERSONS OIL CO. GASOLINE, 1920s, OPE, RW, $2000-2600. *Courtesy of PCM Archives.*

PETERSONS OIL CO. GASOLINE, 1925-1935, 15" METAL, HP RED, $1100-1600. *Courtesy of PCM Archives.*

Quincy

QUINCY RED HAT GASOLINE, 1920s, 15" METAL, HP RED, $2000-2500. *Courtesy of Tom Higgins.*

Left: QUINCY POW-R-PLUS (w/eagle), 1930s, 15" METAL, HP RED, $1200-1750. **Right:** QUINCY POW-R-PLUS ETHYL, 1930s, 15" METAL, HP RED, $600-850. *Courtesy of Tom Higgins.*

QUINCY POWERFUL GASOLINE, 1930s, 15" METAL, HP RED, $600-850. *Courtesy of Tom Higgins.*

Royal and Royal 400

ROYAL GASOLINE, 1920s, OPE, RW, $2100-2700. *Courtesy of PCM Archives.*

ROYAL 400 RED HAT GASOLINE, 1930s, 13.5" GLASS, W, $1600-2200. *Courtesy of Glenn Nelson.*

Swanson Oil Corporation

SWANSON OIL CORPORATION GASOLINE (METAL), 1920S, 15" METAL, HP RED, $1200-1700. *Courtesy of PCM Archives.*

Westland Oil Co.

WESTLAND GASOLINE (w/red hat), 1920s, OPE, RW, $3500-5000. *Courtesy of PCM Archives.*

Top: BUFFALO GASOLINE, 1930s-1950s, 13.5" GLASS, WN, $1600-2200.
Bottom: WESTLAND GASOLINE (w/buffalo), 1930s, 13.5" GLASS, WN, $1800-2400. *Courtesy of PCM Archives.*

ETHYL GASOLINE (w/buffalo), 1930s-1950s, 13.5" GLASS, WN, $1400-2000. *Courtesy of PCM Archives.*

BUFFALO GASOLINE, 1955-1970, 13.5" GLASS, WN, $1700-2300. *Courtesy of Kim Pyatt.*

BUFFALO PREMIUM GASOLINE, 1955-1970, 13.5" GLASS, WN, $350-550. *Courtesy of PCM Archives.*

WESTLAND DIESEL, 1955-1970, 13.5" GLASS, WN, $350-550. *Courtesy of Bruce Bednar.*

KENDALL & AFFILIATES

****Kendall Oil**

KENDALL SPHERE SHAPE, 1922-1925, OPE, $750-1100. *Courtesy of PCM Archives.*

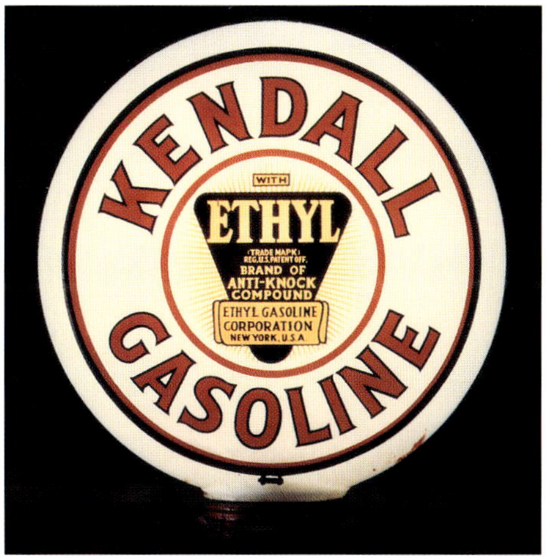

KENDALL ETHYL, 1935-1936, GILL, $225-425. *Courtesy of PCM Archives.*

KENDALL DELUXE (EARLY THIN TYPE), 1936-1946, 14" GILL, $300-500. *Courtesy of PCM Archives.*

KENDALL POLLY POWER, 1946-1970, CAPCO, $225-325. *Courtesy of PCM Archives.*

KENDALL GASOLINE W/ AIRPLANE, 1926-1935, 15" METAL, HP, RED, $4500-6000. *Courtesy of PCM Archives.*

****Waverly Oil Works**

WAVERLY GASOLINE "SPHERE", 1910s, OPE, NO LISTING. *Courtesy of PCM Archives.*

WOW ANTI-KNOCK GASOLINE, 1920s, OPE, N, NO LISTING. *Courtesy of Collectors' Auction Service.*

WAVERLY BLUE MOON, 1920s, 2PC GLASS, N, $850-1250. *Courtesy of PCM Archives.*

WAVERLY GAS, 1920s, 15" METAL, $600-900. *Courtesy of PCM Archives.*

GOLDEN QUALITY / ECONOMY, 1930s, 13.5" GLASS, N, $250-350. *Courtesy of David & Doug Geiger.*

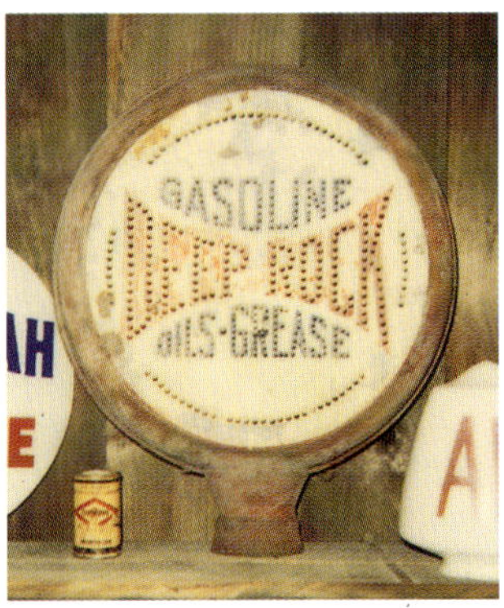

GASOLINE/DEEP ROCK/OILS-GREASES, 1913-1920, 15" METAL, $800-1200. *Courtesy of Kurt Block.*

KERR McGEE & AFFILIATES

****Deep Rock Oil Co.**

GASOLINE/DEEP ROCK/MOTOR OIL, 1920-1928, OPE, N, $1200-1500. *Courtesy of PCM Archives.*

GASOLINE/DEEP ROCK/OILS-GREASES, 1920-1928, OPE, N, $1350-1700. *Courtesy of PCM Archives.*

Left: DEEP ROCK W/RINGS, 1932-1934, 13.5" GLASS, $300-500.
Right: DEEP ROCK GREEN GASOLINE, 1936-1950, 13.5" GLASS, $350-550. *Courtesy of PCM Archives.*

GASOLINE/DEEP ROCK/MOTOR OIL, 1913-1920, 15" METAL, $750-1000. *Courtesy of PCM Archives.*

GASOLINE/DEEP ROCK/MOTOR OIL, 1928-1932, 13.5" GLASS, $300-550. *Courtesy of PCM Archives.*

KANT NOCK WITH ETHYL (EGC), 1928-1932, 13.5" GLASS, $350-550. *Courtesy of PCM Archives.*

Left: SUPER KANT NOCK GASOLINE, 1932-1936, 13.5" GLASS, $350-600.
Right: DEEP ROCK SUPER GASOLINE, 1936-1950, 13.5" GLASS, $300-450.
Courtesy of George Johnston.

DEEP ROCK ETHYL GASOLINE, 1936-1942, 13.5" GLASS, $300-450.
Courtesy of PCM Archives.

DEEP ROCK ETHYL GASOLINE, 1942-1950, 13.5" GLASS, $300-450.
Courtesy of PCM Archives.

DEEP ROCK, 1950-1972, CAPCO, $175-$250.
Courtesy of PCM Archives.

DEEP ROCK WITH ARROWS, 1972-1980, CAPCO, $250-350. *Courtesy of PCM Archives.*

**Kermac

KERR MCGEE, 1965-1970, CAPCO, $200-325.
Courtesy of PCM Archives.

**Triangle Refineries

TRIANGLE GASOLINE W/CLOVER, 1940s, 13.5" GLASS, $500-750.
Courtesy of PCM Archives.

CLOVERLEAF GASOLINE, 1960s, "13.5" GREEN CAPCO", $350-500.
Courtesy of Lonnie Hop.

**Southport Oil Corporation

SOUTHPORT REGULAR GASOLINE HIOTANE, 1940s, 13.5" GLASS, $400-700. *Courtesy of George Kraus.*

SOUTHPORT SUPER GASOLINE HIOTANE, 1940s, 13.5" GLASS, $400-700. *Courtesy of Michael Haggard.*

**Mileage Mart

SAVE GO MILEAGE MART, 1955-1962, 13.5" CAPCO, $950-1400. *Courtesy of Rick Bonenberger.*

**Trackside Service Stations

TRACKSIDE/ECONOMY/MOTOR, 1930s, 15" METAL, $600-850. *Courtesy of PCM Archives.*

TRACKSIDE REGULAR, 1955-1965, 13.5" GLASS, $300-500. *Courtesy of PCM Archives.*

**Peoples Oil Co.

PEOPLES GENUINE POWER PLUS, 1950s, 13.5" GLASS, W, $400-650. *Courtesy of Jack Thomas.*

POWER OIL CO. GOLDEN FLASH GASOLINE, 1950s, 13.5" GLASS, $400-650. *Courtesy of PCM Archives.*

**Wallace and Walco

WALCO, 1950s, CAPCO, $150-225. *Courtesy of PCM Archives.*

**Mutual Oil Co.

Left: MUTUAL, 1931-1957, 15" METAL, $1250-1750. **Right**: MUTUAL ETHYL, 1931-1942, 15" METAL, $600-850. *Courtesy of Bruce Bednar & Courtesy of PCM Archives.*

MUTUAL, 1957-1970, 13.5" GLASS, $750-1000. *Courtesy of Chris Barnes.*

**Golden Rule

GOLDEN RULE, 1940s, 13.5" GLASS, W, $300-500. *Courtesy of Chris Barnes.*

Right: VAUGHT'S SUPER REGULAR, 1940s, 13.5" GLASS, $400-600. *Courtesy of PCM Archives.*

GOLDEN RULE ETHYL, 1940s, 13.5" GLASS, W, $300-450. *Courtesy of Chris Barnes.*

**Sparky

SPARKY DYNAMO ETHYL, 1950s, CAPCO, $300-450. *Courtesy of Chris Barnes.*

**Vaught

VAUGHT'S BLUE BLAZE KEROSENE, 1940s, 13.5" GLASS, $400-600. *Courtesy of PCM Archives.*

LION OIL COMPANY

LION, 1924-1930, OPE, N, $600-1100. *Courtesy of Kent Blaine.*

LION ONE PIECE W/LION, 1926-1930, OPE, N, $3500-5000. *Courtesy of PCM Archives.*

LION PETROLEUM PRODUCTS, 1930-1946, 12" METAL, $1800-2500. *Courtesy of PCM Archives.*

LION ETHYL (EGC), 1930-1942, 13.5" GLASS, W, $500-850. *Courtesy of Kent Blaine.*

LION, 1946-1960, 13.5" GLASS, W, $750-1000. *Courtesy of R.V. Witherspoon.*

LION KNIX KNOX, 1946-1960, 13.5" GLASS, W, $1200-1600. *Courtesy of Bob Davis.*

MAPCO & AFFILIATES
****Red Ace Petroleum**

RED ACE, 1946-1965, 13.5" GLASS, $300-550. *Courtesy of PCM Archives.*

****Capitol Oil Company**

RED BIRD, 1950s, 13.5" GLASS, NO LISTING. *Courtesy of PCM Archives.*

MARATHON & AFFILIATES
****Lincoln Oil Refining Co.**

GASOLINE LINCO OILS, 1924-1934, 15" METAL, HP BLUE, $900-1300. *Courtesy of PCM Archives.*

LINCO, 1934-1939, 13.5" GLASS, W, $250-450. *Courtesy of PCM Archives.*

****Red Fox Oil Company**

RED FOX GASOLINE, 1920-1925, 15" METAL, LP CREAM, NO LISTING. *Courtesy of PCM Archives.*

**Transcontinental Oil Co.

MARATHON, 1920-1938, 15" METAL, LP GREEN, $1200-1500. *Courtesy of PCM Archives.*

MULTIPOWER, 1928-1935, 15" METAL, LP BLACK, NO LISTING. *Courtesy of PCM Archives.*

**The Ohio Oil Company

MARATHON BEST IN THE LONG RUN (deco logo), 1935-1938, 13.5" GLASS, N, $900-1350. *Courtesy of PCM Archives.*

(MARATHON) ETHYL w/runner logo, 1938-1940, 13.5" GLASS, N, $550-800. *Courtesy of PCM Archives.*

MARATHON w/runner logo, 1940-1954, CAPCO, $400-650. *Courtesy of PCM Archives.*

MARATHON MILE MAKER, 1946-1962, 13.5" GLASS, N, $500-800. *Courtesy of PCM Archives.*

MARATHON SUPER M, 1954-1962, 13.5" GLASS, N, $450-750. *Courtesy of PCM Archives.*

MARATHON DIESEL #1, 1946-1962, 13.5" GLASS, N, $300-500. *Courtesy of PCM Archives.*

MARATHON DIESEL #2, 1946-1962, 13.5" GLASS, N, $300-500. *Courtesy of PCM Archives.*

MARATHON REGULAR MILE MAKER, 1962-1970, CAPCO, $150-250. *Courtesy of PCM Archives.*

MARATHON PREMIUM SUPER-M, 1962-1970, CAPCO, $150-250. *Courtesy of PCM Archives.*

MARATHON DIESEL FUEL, 1962-1970, CAPCO, $175-275. *Courtesy of PCM Archives.*

MARAFUEL, 1962-1970, CAPCO, $200-300. *Courtesy of PCM Archives.*

(MARATHON) MILE MAKER, 1962-EXP, CAPCO, NO LISTING. *Courtesy of PCM Archives.*

(MARATHON) SUPER-M, 1962-EXP, CAPCO, NO LISTING. *Courtesy of PCM Archives.*

****Tower Oil Co.**

TOWER GASOLINE, 1940s, 13.5" GLASS, $1000-1400. *Courtesy of PCM Archives.*

****Aurora Gasoline Co.**

Left: SPEEDWAY GASOLINE, 1930s, 15" METAL, $500-750. *Courtesy of PCM Archives.*

SPEEDWAY 79 STRATOFUEL, 1938-1946, 13.5" GLASS, $750-1000. *Courtesy of PCM Archives.*

SPEEDWAY 79 (red background, blue lettering), 1955-1962, 13.5" GLASS, $400-650.
Courtesy of PCM Archives.

SPEEDWAY 79 (blue background, red lettering), 1955-1962, 13.5" GLASS, $400-650.
Courtesy of PCM Archives.

****Old Dutch Refining Co.**

OLD DUTCH ECONOMY, 1935-1940, GILL, $1400-1800. *Courtesy of Lonnie Hop.*

AROSPEED, 1940-1946, GLASS OVAL, $750-1000.
Courtesy of Lonnie Hop.

OLD DUTCH, 1940-1946, GILL, $1400-1800.
Courtesy of Lonnie Hop.

****Republic Oil Co.**

REPUBLIC SUPERIOR GASOLINE, 1946-1958, 13.5" GLASS, $300-550.
Courtesy of PCM Archives.

Left: REPUBLIC, 1922-1946, OPC SHIELD, $2000-2500.
Right: REPUBLIC MOTOR GASOLINE, 1930s, 15" METAL, $500-750.
Courtesy of PCM Archives.

REPUBLIC POWER PAK' D, 1958-1970, CAPCO, $175-275. *Courtesy of PCM Archives.*

REPUBLIC HI-T, 1958-1970, CAPCO, $150-250. *Courtesy of PCM Archives.*

REPUBLIC HI-WAY, 1946-1950, CAPCO, $175-275. *Courtesy of Bruce Bednar.*

HI-WAY, 1950-1958, CAPCO, $150-250. *Courtesy of PCM Archives.*

HI-WAY, 1958-1970, CAPCO, $150-250. *Courtesy of PCM Archives.*

SUPER HI-WAY, 1958-1970, CAPCO, NO LISTING. *Courtesy of Ronny Harrah.*

**Bonded

BONDED 98 in laurel wreath, 1955-1968, 13.5" GLASS, W, $750-1000. Courtesy of PCM Archives.

**Consolidated

CONSOLIDATED ETHYL, 1946-1960, OVAL GLASS, $600-850. *Courtesy of PCM Archives.*

**Road Chief

ROAD CHIEF "RULES THE ROAD", 1960s, CAPCO, $175-275. *Courtesy of PCM Archives.*

**Imperial Refineries

IMPERIAL W/LARGE OLD SHIELD, 1920-1929, 16.5" METAL, $1250-1750. *Courtesy of PCM Archives.*

IMPERIAL OF MICHIGAN, 1946-1970, CAPCO, $150-250. *Courtesy of PCM Archives.*

Left: IMPERIAL ETHYL (EC), 1946-1965, 13.5" GLASS, W, $200-350.
Right: IMPERIAL ETHYL (EGC), 1940-1946, 13.5" GLASS, W, $200-350.
Courtesy of Bruce Bednar.

IMPERIAL STOVE OIL, 1950-1965, CAPCO, $200-350. *Courtesy of PCM Archives.*

**Stephen

STEPHEN, 1940s, 13.5" GLASS, $200-350. *Courtesy of PCM Archives.*

**Rock Island Refining

ROCK ISLAND GASOLINE, 1930-1940, 15" METAL, $2800-3500. *Courtesy of PCM Archives.*

ROCK ISLAND NUNBETTER, 1930-1940, 15" METAL, $600-900. *Courtesy of PCM Archives.*

ROCK ISLAND ANTI-KNOCK, 1930-1940, 15" METAL, $600-950. *Courtesy of PCM Archives.*

ROCK ISLAND GASOLINE (early rock), 1940-1946, RED CAPCO, $1000-1400. *Courtesy of PCM Archives.*

ROCK ISLAND GASOLINE (early rock), 1940-1946, RED CAPCO, $1000-1400. *Courtesy of PCM Archives.*

ROCK ISLAND W/ROCK (red/white), 1946-1960, CAPCO, $600-900. *Courtesy of PCM Archives.*

ROCK ISLAND W/ROCK (red/white/blue), 1946-1960, CAPCO, $600-900. *Courtesy of PCM Archives.*

ROCK ISLAND W/ROCK (red/yellow), 1946-1960, CAPCO, $750-1100. *Courtesy of PCM Archives.*

****Colonial/Progressive**

COLONIAL, 1950-1965, 13.5" GLASS, $200-350. *Courtesy of PCM Archives.*

PROGRESSIVE, 1950-1965, 13.5" GLASS, $200-350. *Courtesy of PCM Archives.*

O-70, 1950s, 13.5" GLASS, $200-400. *Courtesy of PCM Archives.*

****Tulsa Oil Co.**

TULSA 95, 1950-1970, CAPCO, $450-700. *Courtesy of PCM Archives.*

TULSA HI-TEST, 1950-1970, CAPCO, $450-700. *Courtesy of PCM Archives.*

**United

UNITED HI-TEST W/ARROW, 1946-1965, 13.5" GLASS, $300-550. *Courtesy of PCM Archives.*

UNITED J-J W/ARROW, 1946-1955, 13.5" GLASS, $400-650. *Courtesy of PCM Archives.*

UNITED FUEL OIL W/FLAME, 1946-1965, 13.5" GLASS, $600-850. *Courtesy of PCM Archives.*

UNITED ETHYL RED/BLACK, 1960-1965, 13.5" GLASS, $200-400. *Courtesy of PCM Archives.*

UNITED, 1965-1980, CAPCO, $125-200. *Courtesy of PCM Archives.*

UNITED REGULAR, 1965-1980, CAPCO, $125-225. *Courtesy of PCM Archives.*

UNITED ETHYL, 1965-1980, CAPCO, $150-250. *Courtesy of PCM Archives.*

UNITED SUPER ETHYL, 1965-1980, CAPCO, $150-275. *Courtesy of PCM Archives.*

UNITED UNLEADED, 1975-1980, CAPCO, $125-250. *Courtesy of PCM Archives.*

**Naph-Sol Refining Co.

INDEPENDENT ZEPHYR GASOLINE, 1936-1940, 13.5" GLASS, $200-400. *Courtesy of PCM Archives.*

ZEPHYR ETHYL, 1940-1957, 13.5" GLASS, $200-425. *Courtesy of PCM Archives.*

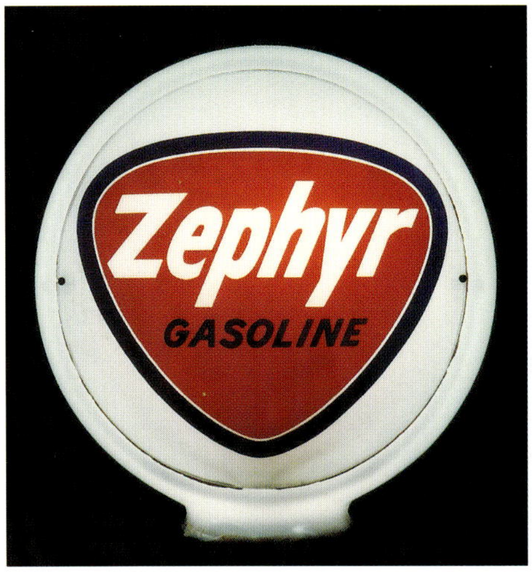

ZEPHYR W/SHIELD, 1957-1970, CAPCO, $275-475. *Courtesy of PCM Archives.*

(ZEPHYR) BLU-FLAME FUEL OIL, 1960s, CAPCO, $200-350. *Courtesy of Edwin Schwartz.*

**Osceola Refining Co.

OSCEOLA REGULAR, 1960s, PLASTIC, $250-400. *Courtesy of PCM Archives.*

Martin & Affiliates

**Martin Oil Company

MARTIN'S SUPER REGULAR w/tank car, 1940-1955, 13.5" LENS FOR GLASS BODY, $1600-2400. *Courtesy of PCM Archives.*

MARTIN'S SAFETY KEROSENE, 1940-1955, 13.5" LENS FOR GLASS BODY, $400-650. *Courtesy of PCM Archives.*

MARTIN SUPER REGULAR, 1955-1970, 13.5" LENS FOR GLASS BODY, $250-450. *Courtesy of PCM Archives.*

MARTIN PREMIUM ETHYL (EC), 1955-1970, 13.5" LENS FOR GLASS BODY, $250-450. *Courtesy of PCM Archives.*

****J.D. STREET and CO.**

ZEPHYR HIGH OCTANE, 1954-1960, OVAL CAPCO, $225-325. *Courtesy of PCM Archives.*

ZEPHYR ETHYL, 1954-1960, OVAL CAPCO, $225-325. *Courtesy of PCM Archives.*

Left: ZEPHYR ETHYL, 1939-1954, GILL BODY GLOBE, $250-425.
Right: ZEPHYR, 1939-1954, GILL BODY GLOBE, $250-400.
RED RIPPLE bodies: Add $1000 each. *Courtesy of R.V. Witherspoon.*

ZEPHYR, 1960-1975, OVAL CAPCO, $200-300. *Courtesy of PCM Archives.*

**Martin Oil Service

MARTIN PURPLE MARTIN ETHYL, 1955-1962, 13.5" LENS FOR GLASS BODY, NO LISTING. *Courtesy of PCM Archives.*

PURPLE MARTIN ETHYL. Glass, N.L. *Courtesy of PCM Archives.*

MARTIN SUPER REGULAR, 1962-1970, CAPCO, $200-300. *Courtesy of PCM Archives.*

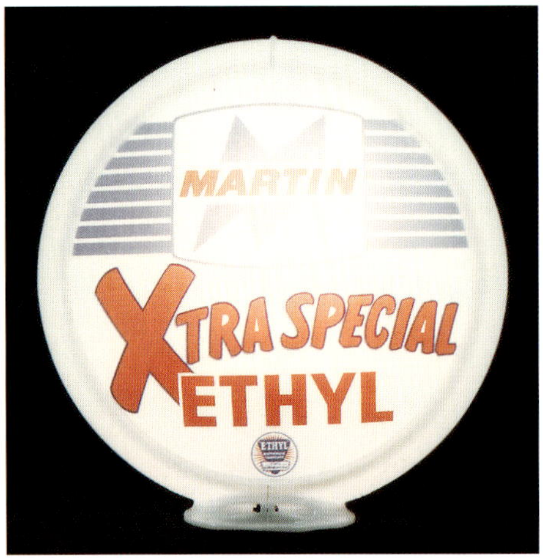

MARTIN EXTRA SPECIAL ETHYL, 1962-1970, CAPCO, $250-350. *Courtesy of PCM Archives.*

MARTIN PURPLE MARTIN ETHYL, 1962-1970, CAPCO, $350-600. *Courtesy of PCM Archives.*

MARTIN DIESEL #1, 1962-1970, CAPCO, $200-300. *Courtesy of PCM Archives.*

MARTIN RANGE OIL, 1962-1970, CAPCO, $300-550. POWER PLUS "THE BETTER GASOLINE". *Courtesy of PCM Archives.*

MIDWEST & AFFILIATES

**Midwest Oil Co.

MIDWEST ACE HIGH (w/airplane), 1925-1930, 15" METAL BODY GLOBE, $4500-7000. *Courtesy of Ace Feek.*

MIDWEST AVIO GASOLINE, 1920s, 15" METAL BODY GLOBE, $800-1250. *Courtesy of PCM Archives.*

ACE HIGH GASOLINE, 1940s,
13.5" LENS FOR GLASS BODY, $600-950.
Courtesy of PCM Archives.

ACE HIGH GASOLINE, 1940s, GILL BODY GLOBE, $400-650. *Courtesy of PCM Archives.*

ACE HIGH ETHYL w/attendant, 1940s, 13.5" LENS FOR GLASS BODY, NO LISTING.
Courtesy of PCM Archives.

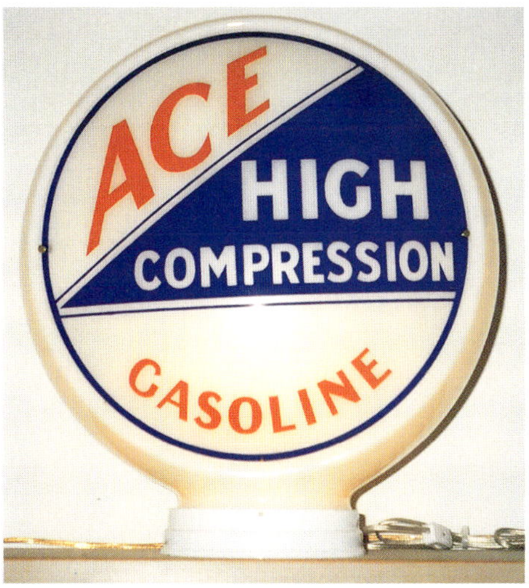

ACE HIGH COMPRESSION GASOLINE, 1930s,
13.5" LENS FOR GLASS BODY, $300-500.
Courtesy of Ace Feek.

Left: MIDWEST (points right), 1958-1970, OVAL CAPCO, $225-375.
Right: MIDWEST (points left), 1958-1970, OVAL CAPCO, $200-350.
Courtesy of PCM Archives.

AVIO QUALITY GASOLINE, 1946-1958,
13.5" LENS FOR GLASS BODY, $300-500.
Courtesy of PCM Archives.

**H.K. Stahl & Co.

POWERFLASH INDEPENDENT GASOLINE, 1930s, GILL BODY GLOBE, $700-1000, SHAMROCK (clover w/red band), SWENEY SKY HAWK GASOLINE.
Courtesy of PCM Archives.

TROPHY - OUR PREMIUM, 1935-1958,
GILL BODY GLOBE, $400-650.
Courtesy of PCM Archives.

TROPHY POWER FLASH, 1950-1958, GILL BODY GLOBE, $400-650. *Courtesy of PCM Archives.*

MOBIL & AFFILIATES
****Standard Oil Company of New York/SOCONY**

SOCONY MOTOR GASOLINE, 1915-1920, ONE PIECE ETCHED GLOBE, $2500-3800. *Courtesy of PCM Archives.*

SOCONY MOTOR GASOLINE (milk glass lens), 1915-1926, 13.675" METAL, $1800-2400. *Courtesy of PCM Archives.*

SOCONY MOTOR GASOLINE (porcelain lens), 1915-1926, 14" METAL SPC, $2000-2500. *Courtesy of PCM Archives.*

SOCONY red lettering w/white lines, 1926-1934, $300-550. *Courtesy of PCM Archives.*

SOCONY SPECIAL milk glass lens, 1922-1926, 15" METAL BODY GLOBE, $900-1400. *Courtesy of PCM Archives.*

SOCONY ETHYL (EGC), 1926-1934, $325-575. *Courtesy of PCM Archives.*

SOCONY AVIATION GASOLINE, 1920s, $1600-2200. *Courtesy of PCM Archives.*

SOCONY MOTOR OILS sm w/4" base, 1915-1920, ONE PIECE ETCHED GLOBE, $1300-1800. *Courtesy of PCM Archives.*

**Magnolia Petroleum

MAGNOLIA GASOLINE w/ seven-leaf flower, 1920-1926, $1700-2200. *Courtesy of PCM Archives.*

MAGNOLIA GASOLINE w/five-leaf flower, 1920-1926, $1700-2200. *Courtesy of PCM Archives.*

MAGNOLIA MAXIMUM MILEAGE, 1926-1934, $1700-2200. *Courtesy of PCM Archives.*

MAGNOLIA ANTI-KNOCK, 1922-1926, $1600-2100. *Courtesy of PCM Archives.*

MAGNOLIA ETHYL (EGC), 1926-1934, $750-1000. *Courtesy of PCM Archives.*

**General Petroleum Corporation

GENERAL VIOLET RAY GASOLINE, 1928-1932, 15" METAL BODY GLOBE, $3000-4000. *Courtesy of PCM Archives.*

GENERAL ETHYL, 1926-1934, 15" METAL BODY GLOBE, $2000-2500. *Courtesy of PCM Archives.*

**White Eagle Oil and Refining

WHITE EAGLE GASOLINE, 1918-1920, ONE PIECE ETCHED GLOBE, $7500-10000. *Courtesy of PCM Archives.*

WHITE EAGLE GASOLINE AND OILS, 1920-1925, 15" METAL BODY GLOBE, $1700-2200. *Courtesy of PCM Archives.*

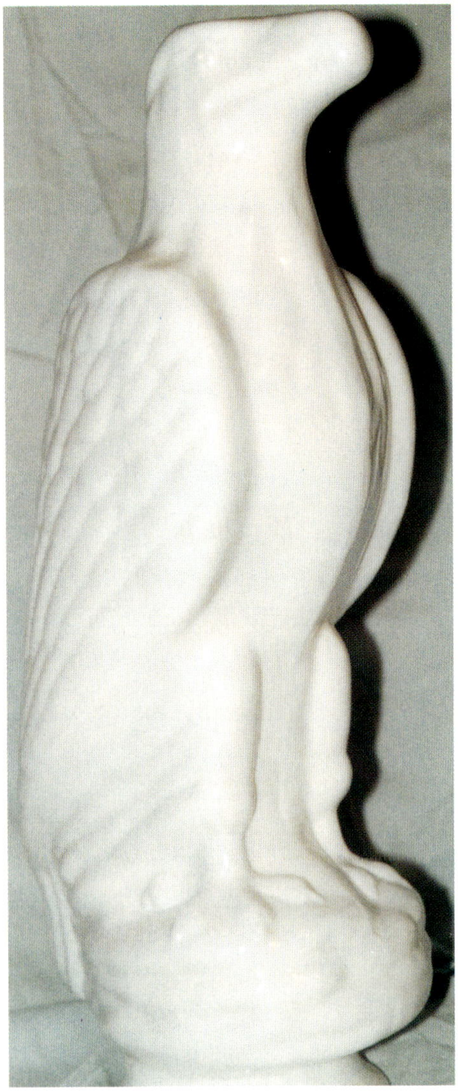

WHITE EAGLE blunt nose, 1924-1932, EAGLE, $1300-1600. *Courtesy of Larry Starkweather.*

WHITE EAGLE pointed nose, good detail, 1932, EAGLE, $1800-2400. *Courtesy of Larry Starkweather.*

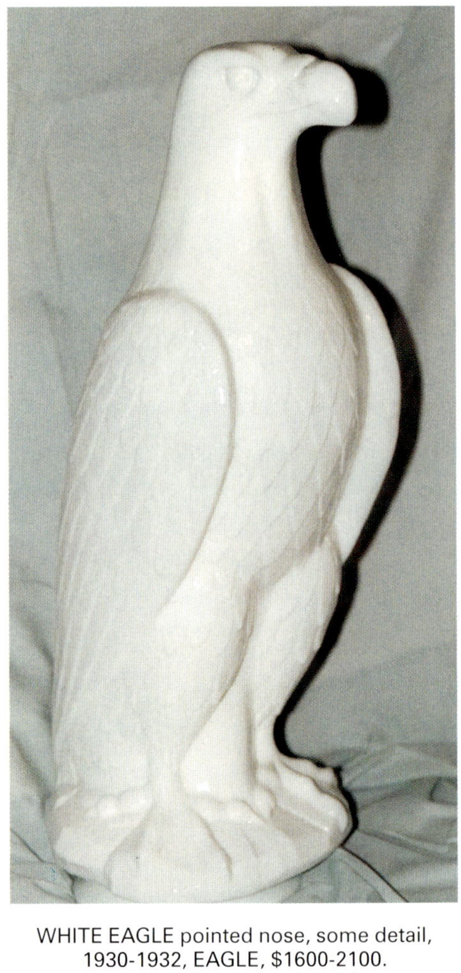

WHITE EAGLE pointed nose, some detail, 1930-1932, EAGLE, $1600-2100. *Courtesy of Larry Starkweather.*

****Vacuum Oil Company**

MOBILGAS - blue lettering w/ red line, 1932-1933, $300-550. *Courtesy of PCM Archives.*

MOBILGAS - blue lettering w/ red line, 1932-1933, 13.5″ LENS FOR GLASS BODY, $250-375. *Courtesy of PCM Archives.*

MOBILGAS ETHYL, 1928-1933, $325-575. *Courtesy of PCM Archives.*

MOBILGAS ETHYL (black ethyl), 1930-1933, 13.5″ LENS FOR GLASS BODY, $250-375. *Courtesy of PCM Archives.*

GARGOYLE MOBILOIL (small; raised border), 1920-1935, ONE PIECE CAST GLOBE OVAL, $1750-2250. *Courtesy of PCM Archives.*

GARGOYLE MOBILOIL (small; recessed border), 1920-1935, ONE PIECE CAST GLOBE OVAL, $1750-2250. *Courtesy of PCM Archives.*

****Lubrite Refining Co.**

LUBRITE, 1925-1929, 15″ METAL BODY GLOBE, $600-900. *Courtesy of PCM Archives.*

LUBRITE SKY HIGH, 1925-1929, 15″ METAL BODY GLOBE, $650-950. *Courtesy of PCM Archives.*

GARGOYLE MOBILOIL, 1930s, $1800-2500. *Courtesy of PCM Archives.*

**Wadhams Oil Corporation

W LAMP POST GLOBE, 1910s,
ONE PIECE ETCHED GLOBE, $900-1200.
Courtesy of PCM Archives.

WADHAMS 370 w/can, 1920-1922,
15" METAL BODY GLOBE, NO LISTING.
Courtesy of PCM Archives.

WADHAMS SPECIALO, 1920-1922,
15" METAL BODY GLOBE, $1700-2250.
Courtesy of PCM Archives.

WADHAMS 370, 1922-1925, 15" METAL BODY GLOBE, $800-1100. *Courtesy of PCM Archives.*

WADHAMS "INDEPENDENT" IOMA, 1926-1929, 15" METAL BODY GLOBE, $1200-1600.
Courtesy of PCM Archives.

WADHAMS ETHYL (EGC), 1926-1932,
15" METAL BODY GLOBE, $750-950.
Courtesy of PCM Archives.

WADAMS METRO "W", 1929-1932,
15" METAL BODY GLOBE, $600-900.
Courtesy of PCM Archives.

WADAMS METRO "W", 1932-1940,
13.5" LENS FOR GLASS BODY, $300-500.
Courtesy of PCM Archives.

WADHAMS GIANT, 1932-1940,
13.5" LENS FOR GLASS BODY, $300-550.
Courtesy of PCM Archives.

WADHAMS ETHYL, 1932-1940,
13.5" LENS FOR GLASS BODY, $300-550.
Courtesy of PCM Archives.

**White Star Refining Co.

WHITE STAR GASOLINE, 1918-1920,
15" METAL BODY GLOBE, $900-1250.
Courtesy of PCM Archives.

WHITE STAR GASOLINE, 1930-1934,
15" METAL BODY GLOBE, $900-1250.
Courtesy of PCM Archives.

**Independent Oil Co. of Pennsylvania

INDEPENDENT (W/3 MEN), 1925-1934,
15" METAL BODY GLOBE, $3500-5000.
Courtesy of PCM Archives.

INDEPENDENT GASOLINE (W/3 MEN) great detail, 1925-1934, 15" METAL BODY GLOBE, NO LISTING. *Courtesy of PCM Archives.*

**Metro

METRO GASOLINE/green lines, 1925-1933,
15" METAL BODY GLOBE, $400-650.
Courtesy of PCM Archives.

METRO GAS GARGOYLE MOBILOIL, 1933-1935, 15" METAL BODY GLOBE, $2200-2700.
Courtesy of PCM Archives.

Left: STAROLINE AND WHITE STAR ETHYL, 1926-1930, 15" METAL BODY GLOBE, $750-1000.
Center: WHITE STAR ETHYL, 1926-1930, 15" METAL BODY GLOBE, $650-900.
Right: WHITE STAR GASOLINE, 1926-1930, 15" METAL BODY GLOBE, $850-1200.
Courtesy of PCM Archives.

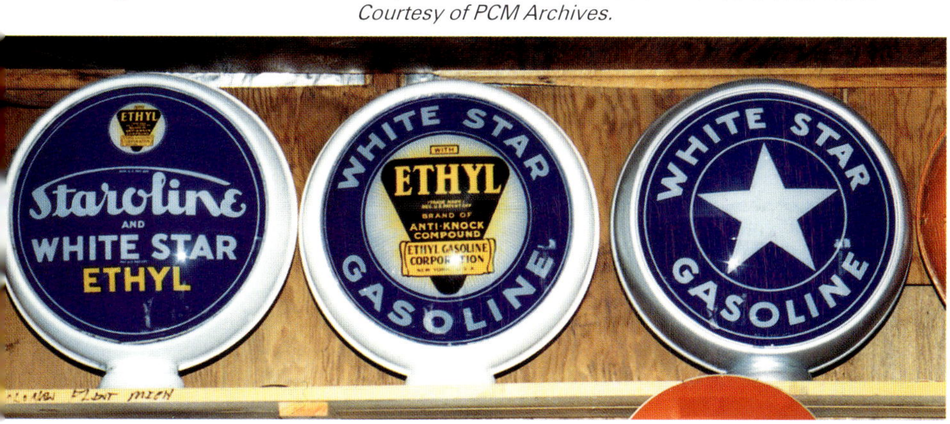

METRO red/green on white/red, 1940-1955, 15″ METAL BODY GLOBE, $400-650. *Courtesy of PCM Archives.*

METRO red/green on white/red, 1940-1955, 13.5″ LENS FOR GLASS BODY, NO LISTING. *Courtesy of PCM Archives.*

METRO red/green on white/red, 1940-1955, CAPCO, $175-275. *Courtesy of PCM Archives.*

****Gilmore Oil Co.**

GILMORE "ROAR WITH GILMORE", 1925-1942, 15″ METAL BODY GLOBE, $7000-9000. *Courtesy of PCM Archives.*

GILMORE BLU-GREEN GASOLINE, 1925-1933, 15″ METAL BODY GLOBE, $8000-11000. *Courtesy of PCM Archives.*

GILMORE WITH ETHYL (script letters), 1926-1942, 15″ METAL BODY GLOBE, $2000-3500. *Courtesy of PCM Archives.*

GILMORE FLEET W/FLAG, 1933-1942, 15″ METAL BODY GLOBE, NO LISTING. *Courtesy of PCM Archives.*

GILMORE RED LION W/TETRAETHYL, 1933-1942, 15″ METAL BODY GLOBE, $8000-10000. *Courtesy of PCM Archives.*

****Clipper Oil Company**

CLIPPER, 1940s, 15″ METAL BODY GLOBE, $3000-4500. *Courtesy of PCM Archives.*

CLIPPER, 1940s, OVAL CAPCO, $1100-1600.
Courtesy of PCM Archives.

****Mobilgas/SOCONY-Vacuum**

MOBILGAS SPECIAL blue letters, 1936-1962, 15" METAL BODY GLOBE, $500-750.
Courtesy of PCM Archives.

MOBILGAS SPECIAL blue letters, 1955-1962, CAPCO, $200-400. *Courtesy of PCM Archives.*

MOBILGAS with horse blue letters, 1934-1962, 15" METAL BODY GLOBE, $500-750.
Courtesy of PCM Archives.

MOBILGAS SPECIAL blue letters, 1936-1962, 13.5" LENS FOR GLASS BODY, $300-500.
Courtesy of PCM Archives.

MOBILGAS/MOBILGAS SPECIAL dual, 1955-1962, CAPCO, NO LISTING.
Courtesy of PCM Archives.

MOBILGAS with horse blue letters, 1934-1962, 13.5" LENS FOR GLASS BODY, $300-500.
Courtesy of PCM Archives.

MOBILGAS SPECIAL blue letters, 1936-1962, BALCRANK, $400-650.
Courtesy of PCM Archives.

MOBILFUEL DIESEL blue letters, 1935-1962, $800-1100. *Courtesy of PCM Archives.*

MOBILFUEL DIESEL blue letters, 1935-1962, 13.5" LENS FOR GLASS BODY, NO LISTING. *Courtesy of PCM Archives.*

MOBILGAS MARINE with non-outlined horse, 1930s, $1000-1500. *Courtesy of PCM Archives.*

MOBILGAS MARINE with horse, 1950-1962, 15" METAL BODY GLOBE, $1000-1500. *Courtesy of PCM Archives.*

MOBILGAS MARINE WHITE with horse, 1950-1962, 15" METAL BODY GLOBE, $1200-1750. *Courtesy of PCM Archives.*

AERO MOBILGAS with horse, 1935-1950, 15" METAL BODY GLOBE, $1800-2400. *Courtesy of PCM Archives.*

MOBIL/KEROSENE with horse, 1950-1966, 5" METAL BODY GLOBE, $750-1000. *Courtesy of PCM Archives.*

MOBILOIL (can shaped globe), 1947-1955, OPP, NO LISTING. *Courtesy of PCM Archives.*

MOBIL REGULAR, 1962-1966, CAPCO, $200-300. *Courtesy of PCM Archives.*

MOBIL PREMIUM, 1962-1966, CAPCO, $200-300. *Courtesy of PCM Archives.*

AMOCO & PAN-AM

The American Oil Company was founded in Baltimore in 1910, refining and marketing kerosene. In 1915 American Oil introduced a gasoline/ benzol blend called "AMOCO-GAS." Without crude oil production of its own and forced to buy on the open market, American Oil operated with the possibility of shortage at any time. This was remedied in 1923 when a supply agreement was signed with Pan American Petroleum, based in New Orleans, giving 50% interest in American Oil to Pan-Am in return for the assurance of a crude oil supply. Pan American (primarily a production company with trace marketing in New England, California, and in the process of expanding into several southern states) had begun its operations in 1916; however, by the time of the 1923 merger Pan-Am was already an international company. In 1925 Standard Oil of Indiana purchased a controlling interest in Pan-Am, in the wake of Pan-Am's involvement in the "Teapot Dome" Scandal. In 1933 Pan-Am, now a subsidiary of Standard of Indiana, purchased the remaining 50% of American Oil and converted the Pan-Am marketing facilities in Georgia and Florida to the Amoco brand, retaining the Pan-Am name in Alabama, Louisiana, Mississippi and Tennessee. Arkansas and Kentucky were added at a later date, while the Amoco brand name was used in all east coast states from Maine to Florida as well as West Virginia and Ohio. Standard bought the remaining interest in Pan-Am in 1954 and in 1956 the Pan-Am name was replaced by Amoco in the six southern states. About the same time, the Amoco brand had been introduced in Texas. While the eastern seaboard states and Texas retained the old red and black Amoco oval, marketing in the former Pan-Am states was under an Amoco torch and oval logo similar to the ones used by Standard and affiliates since 1946. In 1961, the entire Amoco marketing area was converted to the "American" brand name and pumps were re-imaged. At this time all globes were discontinued. See the "STANDARD" heading for Standard and Utoco globes.

AMERICAN OIL COMPANY—BALTIMORE, MARYLAND: (see Amoco introduction for American Oil history.)

AMOCO GAS 5 1915-1932 15" Metal H Cream $750-1000
AMOCO GAS 5 1915-1932 16.5" Metal H Cream $750-1000
Deep red upper and lower sections with black band horizontal across the middle. White lines top and bottom and white "AMOCO-GAS" on black band.

AMOCO 3 1932-1961 15" Metal H\L Cream $300-500
AMOCO 3 1940-1961 Gill $275-450
AMOCO 3 1940-1961 13.5" Glass $250-400
Red globe face with white outlined black band across center. White "AMOCO" on center band.

AMERICAN GAS 3 1926-1961 15" Metal H\L Yellow $300-475
AMERICAN GAS 3 1940-1961 Gill $250-400
AMERICAN GAS 3 1940-1961 13.5" Glass $250-400
Bright blue outer ring with large red dot in the center. White "AMERICAN" around top of blue ring, "GAS" around bottom.

AMERICAN STRATE 5 1922-1938 15" Metal H\L Red $750-1000
White globe face with red "AMERICAN" arched around top and red "STRATE" arched around bottom. Small red and blue shield in center with blue "GASOLINE" in outlined box across shield.

AMERICAN ETHYL 6 1933-1934 15" Metal H\L Yellow $650-850
White globe face with thick red outline ring. Red outlined yellow band arched across upper globe face. Black "AMERICAN" arched in yellow band. Ethyl (EGC) logo on lower globe face below band. Used in limited markets only.

AMERICAN ETHYL 7 1934-1935 15" Metal H\L Yellow NL (**N**o **L**isting)
This globe appears in a photo taken at an Amoco station in West Virginia in 1934. Colors cannot be determined but globe has dark colored outer ring around light colored center. Ethyl (EGC) shield in center. "AMERICAN" in upper half of ring, "GASOLINE" in lower half. Ring is split by light colored line at center on either side. Used in limited markets only.

ARTEX 5 1940-1961 16.5" Metal H Red $300-475
ARTEX 4 1940-1961 15" Metal H\L Red $275-450
ARTEX 4 1940-1961 Gill $275-425
White globe face with large red script "Artex" at angle across globe face. Lettering has bright blue drop shade.

(AMOCO) DIESEL FUEL 4 1946-1961 15" Metal H\L $275-400
Very plain blue "DIESEL" over "FUEL" on white face

****PAN-AM SOUTHERN CORPORATION—NEW ORLEANS, LOUISIANA**: Founded in 1916, Pan American Petroleum and Transport Company, as it was then called, rapidly became one of the largest international petroleum production and trading companies of the day. To secure crude oil reserves and production capacity, Standard purchased the major portion of Pan American in the spring of 1925, reorganizing its holdings as Pan American Eastern Petroleum Company. With the purchase also came Pan-Am's 50% interest in American Oil Company. In the United States Pan Am marketed largely through its subsidiary, Mexican Petroleum, with operations in New England and in California. With the purchase by Standard, the California operations were spun off as an independent, Pan American Petroleum, which was later sold to Richfield (See Richfield section for more information). During the 1920s Mexican Petroleum marketed gasoline branded Pan-Am in all of New York and New England, and established markets in the southeast beginning in 1925 that would form the traditional Pan-Am territory. After purchasing control of Pan American, the former Mexican Petroleum Company marketing in New England and in Georgia and Florida was transferred to American Oil, with the Amoco brand name replacing Pan-Am, and Standard organized a marketing subsidiary, Pan Am Petroleum Company, to continue marketing in Tennessee, Alabama, Mississippi and Louisiana under the Pan-Am brand name. This operation became Pan-Am Southern Corporation in 1937. In 1954 Standard finally purchased the last remaining outstanding shares of Pan Am and American Oil and began the process of merging the two operations. American's regular grade motor fuel, American Gas, was introduced at Pan Am stations in late 1955 and in 1956 the Pan-Am brand name was replaced by Amoco. In December 1960, all Standard of Indiana domestic marketing was consolidated under the American Oil Company name, and American signs replaced Amoco (and Utoco) signs in the south, east, and far west. See Standard section for more information.

PAN-AM 6 1920s OPE NL
Plain globe with etched "PAN-AM" lettering across center of globe face. Lettering is believed to be green.

PAN-AM GASOLINE CAST SQUARE 6 1923-1925 OPC NL
Rounded diamond shape one piece cast globe. No other details known.

PAN-AM GASOLINE/ MOTOR OIL 5 1925-1928 15" Metal H Red $650-900
Red outlined cream color face with green band across center and red ringed sunburst in center. Cream color "PAN-AM" on green band and green "GASOLINE" around top of globe face, green "MOTOR OILS" around bottom of globe face.

PAN-AM 5 1928-1929 15" Metal H Red $600-850
Red outlined cream color face with green band across center and red ringed sunburst in center. Cream color "PAN-AM" on green band.

PAN-AM 5 1929-1939 15" Metal H Red $500-750
Red outlined white globe face with blue band across center and red ringed sunburst in center. White color "PAN-AM" on blue band.

PAN-AM ETHYL 5 1929-1939 15" Metal H Red $650-900
White globe face with blue outline around edge and center circle. Large Ethyl (EGC) logo in center. Red "PAN-AM" around top, "GASOLINE" around bottom.

BLUE 6 1925-1929 15" Metal H Red NL
White globe face with blue outline ring. Blue outlined blue sunburst in center. Blue lines forming white band superimposed across center sunburst with large blue "BLUE" on band.

PAN-AM ORANGE 4 1929-1939 15" Metal H Red $700-950
White globe face with black band across center. White "PAN-AM" on black band. Orange sunburst with outline in center, orange "ORANGE" arched across top and orange "GASOLINE" arched around bottom.

PAN-AM ECONOMY 4 1939-1947 15" Metal H Red $650-850
White face with blue band across center. Red sunburst in center. White "PAN-AM" on blue band, red "ECONOMY" arched above sunburst, "GASOLINE" below sunburst.

PAN-AM QUALITY 4 1939-1947 15" Metal H Red $650-850
White face with blue band across center. Red sunburst in center. White "PAN-AM" on blue band, red "QUALITY" arched above sunburst, "GASOLINE" below sunburst.

PAN-AM PREMIUM 4 1939-1947 15" Metal H Red

$700-900
White face with blue band across center. Red sunburst in center. White "PAN-AM" on blue band, red "PREMIUM" arched above sunburst, "GASOLINE" below sunburst.

PAN-AM 5 1945-1947 Banded Glass $400-600
White face with blue band across center. Red sunburst in center. White "PAN-AM" on blue band.

PAN-AM
TORCH & Oval 4 1947-1956 12.5" Glass W $350-550
Pan-Am "Torch and Oval" logo on solid white face.

AMERICAN GAS 4 1955-1956 12.5" Glass W $250-400
Bright blue outer ring with large red dot in the center. White "AMERICAN" around top of blue ring, "GAS" around bottom.

AMOCO 3 1956-1961 12.5" Glass W $250-400
Red upper face and blue lower face with white band across center. White band has red line at top, blue line at bottom and "AMOCO" in blue.

AMERICAN 3 1956-1961 12.5" Glass W $250-400
Red upper face and blue lower face with white band across center. White band has red line at top, blue line at bottom and "AMERICAN" in blue.

(AMOCO)
DIESEL W/LINES 4 1956-1961 12.5" Glass W $250-350
White face with blue lines above and below center. Blue "DIESEL" across center inside lines.

CAPITOL OIL COMPANY — RICHMOND, VIRGINIA: Capitol was a Richmond based regional marketer that was purchased by American Oil in their first marketplace expansion into Virginia in 1926.

CAPITOL 6 1920-1926 15" Metal NL
White globe face with large outline drawing of capitol dome in center of globe face. Red stylized lettering "CAPITOL" across center of globe face.

**ASAMERA & AFFILIATES

****ASAMERA OIL CORPORATION — DENVER, COLORADO:** Asamera was a Canadian oil production company that invested in various marketing assets in the Rocky Mountain region of the United States in the 1970s. By the 1980s marketing assets had been sold to a former corporate executive and the operation is now known as Gallagher Oil Company, operating out of Fort Collins, Colorado.
>>No globes displaying the Asamera name are known.

****ORIENTAL REFINING COMPANY — DENVER, COLORADO:** Possibly the remnants of the former Dallas, Texas, based Oriental Oil Company, Oriental Refining Company was a Denver refiner-marketer from the 1930s until the 1970s when they were purchased by Canadian conglomerate Asamera to become Asamera Oil Company. Much of the marketing in that era was conducted under the Gasamat and Gasmart brand names. In the 1980s Asamera sold off their United States marketing assets to a former Asamera executive by the name of Gallagher. Gallagher Oil Company continues to operate the Gasamat stations today.

ORIENTAL
POLY OCTANE 5 1930s 13.5" Glass N $300-500
White globe face with black and two tone blue octagon on large green circle in center. Blue outline ring and borders of white center band. Blue "ORIENTAL" in upper white border, "COLORADOMADE" on lower with dark blue "Poly Octane" on center band.

ORIENTAL
REGULAR 3 1950s Capco $150-200
White globe face with blue "ORIENTAL" arched around top, "GASOLINE" around bottom, "REGULAR" across center.

ORIENTAL ETHYL 3 1950s Capco $150-200
White globe face with blue "ORIENTAL" arched around top, "GASOLINE" around bottom, "ETHYL" across center.

ORIENTAL
BRONZE 4 1950s Capco $150-200
White globe face with blue "ORIENTAL" arched around top, "GASOLINE" around bottom, "BRONZE" across center.

****ANDY-DANDY OIL CO. — FT. COLLINS, COLORADO:** In the mid-1950s, Ft. Collins, Colorado, entrepreneur Mr. Vern West began manufacturing a coin operated gas pump as a modification of existing Tokheim pumps. He named his company and his creation "Automatic Devices." In 1955 Vern formed a partnership with Mr. Gene Eaton, a local body-shop owner and the "Andy-Dandy Oil Company" was born, taking the "A" & "D" initials of the original "Automatic Devices." The partnership dissolved in 1958, but Vern West continued to operate the few (about 10) Andy Dandy stations in the Ft Collins area before selling out to Pat Griffin, the owner of the Gasamat operation.

ANDY-DANDY
OIL CO 5 1956-1964 Yellow Ripple $2500-4000
Red globe face with black & white "attendant" with pump hose in center of globe face. Silver dollar head on attendant. White "ANDY/DANDY" to left of figure, "OIL CO." to right.

****GASAMAT — FT. COLLINS, COLORADO:** Founded in Ft Collins, Colorado, in 1957 when the local Phillips 66 jobber became involved in experiments with coin operated gas pumps. By the 1960s Gasamat had developed into a chain of drive-through self service stations with live-in attendants. The company expanded into convenience stores and abandoned the drive-through operation in the 1970s. They operate around 50 stations today.

GASATERIA 5 1957 Oval Capco $200-350
White oval globe with red elongated hexagon covering most of globe face. Yellow logotype "Gasateria" on red hexagon.

GASAMAT 5 1957-1965 Oval Capco $200-350
White oval globe with red elongated hexagon covering most of globe face. Yellow logotype "Gasamat" on red hexagon.

GASAMAT
"SERVE SELF" 5 1965-1970 Capco $200-350
White globe face with red lines forming white shield on globe face. Red "Gasamat" logotype on upper shield over red "SERVE/SELF

GASMART 5 1957-1965 Oval Capco $200-350
White oval globe with red elongated hexagon covering most of globe face. Yellow logotype "Gas Mart" on red hexagon.

GASMART
"SERVE SELF" 5 1965-1970 Capco $200-350
White globe face with red lines forming white shield on globe face. Red "Gas Mart" logotype on upper shield over red "SERVE/SELF

ASHLAND & AFFILIATES

****ASHLAND OIL and REFINING — ASHLAND, KENTUCKY:** Ashland Oil was founded in 1924 as a refining division of the Swiss Oil Company of Lexington, Kentucky, which had numerous ties to National Refining. Ashland began marketing gasoline under the brand name "Pepper," with the gasoline dyed red and green as well as clear to market the famous Red Pepper, Green Pepper and White Pepper products. Stations under the Pepper brand name were operated in an area of about 250 miles around the Ashland, Kentucky, headquarters. Branded gasoline marketing was all but discontinued during World War II, and after the war when retail gasoline sales resumed, the brand name Ashland was chosen, Ashland Oil having replaced the corporate name Swiss Oil in about 1935. Ashland has grown through the many mergers listed below, with branded Ashland outlets found within about a 200 mile radius of Ashland, Kentucky, and with company owned and operated stations, mostly under the SuperAmerica brand, throughout the central and eastern United States.

****PEPPER:** Brand name used by Ashland Oil from 1924-1946.

RED PEPPER 7 1924-1928 OPE NL
Description not available.

RED PEPPER
AVIATION
w/pepper 7 1924-1928 OPE NL
Description not available

RED PEPPER
ETHYL 6 1924-1928 OPE NL
Description not available.

WHITE PEPPER 5 1924-1928 OPE W $1300-1800
White globe with black outline ring around yellow globe face. White outlined black expanding/contracting lettering "WHITE/PEPPER" on yellow face.

WHITE PEPPER 4 1928-1935 15" Metal H\L Red $800-1250

Upper half dark green, lower white. Dark green outline in lower half, "White" in white above. Dark green outlined red letters on lower half "Pepper."

GREEN PEPPER
ANTIKNOCK 5 1928-1935 15" Metal H\L Red $900-1500
White globe face with thin red outline. Orange outlined green "GREEN" over green outlined orange "PEPPER" on upper globe face. Orange line under "PEPPER", with orange outlined green "ANTI-KNOCK" below line on lower globe face.

RED PEPPER
BETTER ETHYL 5 1928-1935 15" Metal H\L Red $900-1400
White globe face with red "RED PEPPER" arched around top, and red "BETTER ETHYL"(in quotes) arched around bottom. Black outlined Ethyl (EGC) logo in center with blue rays out from logo.

SUPER PEPPER 5 1935-1946 15" Metal $800-1050
SUPER PEPPER 4 1935-1946 13.5" Glass W $500-750
White globe face with green outlined italic "SUPER" with speed lines over block "PEPPER". Glass globe has green outline.

RED PEPPER
ETHYL (EGC) 4 1935-1946 15" Metal L Red $900-1400

RED PEPPER
ETHYL (EGC) 6 1935-1946 13.5" Glass W NL
White globe face with black outlined red "RED/PEPPER" on upper face, large Ethyl (EGC) logo on lower face.

TRUCK PEPPER 6 1935-1946 13.5" Glass W NL
Red outlined globe face with white upper half and green lower half. Thin red line separates white area from green. Black outlined red "TRUCK" in upper white area, black outlined white "PEPPER" in lower green area.

****ASHLAND**: The Ashland trademark was emphasized more and more after the company was consolidated in 1936. Stations were considered "Ashland" stations that sold "Pepper" gasoline, as opposed to "Pepper" stations. The Ashland gasoline brand names replaced all the Pepper brand names in 1946. The image was revised in 1960, when red, white and light blue replaced red, white and green as marketing colors. Ashland changed its image in 1968, and again in 1995, in an effort to attract a larger network of distributors. The company introduced a completely new marketing image that remains in use to this day. Aluminum pump toppers displaying the 1968 and later 1995 logos are reminiscent of gas pump globes. These remain in use at some locations today.

ASHLAND 6 1946-1950 13.5" Glass W $300-450
White globe face with large red "Ashland" logotype across globe face. Black drop shade from stylized lettering. Does not have any other lettering on the globe face.

ASHLAND
FLYING
OCTANES 3 1946-1960 13.5" Glass W $275-425

ASHLAND
FLYING
OCTANES 2 1946-1960 Capco $200-300
White globe face with large red "Ashland" logotype across globe face. Black drop shade from stylized lettering. Black outlined green "FLYING/OCTANES" below logotype on lower globe face.

ASHLAND
FLYING
OCTANES 5 1950s Capco $225-350
Clear globe face with large red "Ashland" logotype across globe face. Black drop shade from stylized lettering. Black outlined green "FLYING/OCTANES" below logotype on lower globe face.

ASHLAND
ETHYL (EC) 4 1946-1960 13.5" Glass W $300-450

ASHLAND
ETHYL (EC) 3 1946-1960 Capco $225-325
White globe face with large red "Ashland" logotype across globe face. Black drop shade from stylized lettering. Black outlined green "FLYING/OCTANES" below logotype on lower globe face, to left of black circle outlining Ethyl (EC) logo.

ASHLAND DIESEL 4 1946-1960 13.5" Glass W $300-450
ASHLAND DIESEL 4 1946-1960 Capco $225-325
White globe face with large red "Ashland" logotype across globe face. Black drop shade from stylized lettering. Black outlined green "DIESEL" below logotype on lower globe face.

ASHLAND
FUEL OIL 5 1946-1960 13.5" Glass W $325-475

ASHLAND
FUEL OIL 5 1946-1960 Capco $250-350
White globe face with large red "Ashland" logotype across globe face. Black drop shade from stylized lettering. Black outlined green "FUEL OIL" below logotype on lower globe face.

ASHLAND
KEROSENE 5 1946-1960 13.5" Glass W $300-450

ASHLAND
KEROSENE 4 1946-1960 Capco $225-325
White globe face with large red "Ashland" logotype across globe face. Black drop shade from stylized lettering. Black outlined green "KEROSENE" below logotype on lower globe face.

ASHLAND GO MIX
OUTBOARD FUEL 6 1955-1960 Capco $450-650
Pale blue globe face with white band across center. White elongated octagon Ashland logo in upper blue area. Red "GO-MIX" with black speedboat and line in center white area and white "OUTBOARD/FUEL" in lower blue area.

ASHLAND
FLYING
OCTANES BLUE 2 1960-1968 Capco $200-300
White globe face with large red "Ashland" logotype across globe face. Black drop shade from stylized lettering. Black outlined *light blue* "FLYING/OCTANES" below logotype on lower globe face.

ASHLAND A-PLUS 1 1960-1968 Capco $150-225
White globe face with red outline ring. Large red "A" covers most of globe face, with small "ASHLAND" at top center o f "A" and "PLUS" on crossbar.

ASHLAND DIESEL 4 1960-1968 Capco $200-300
White globe face with large red "Ashland" logotype across globe face. Black drop shade from stylized lettering. Black outlined *light blue* "DIESEL" below logotype on lower globe face.

ASHLAND
FUEL OIL 5 1960-1968 Capco $225-325
White globe face with large red "Ashland" logotype across globe face. Black drop shade from stylized lettering. Black outlined *light blue* "FUEL OIL" below logotype on lower globe face.

ASHLAND
KEROSENE 4 1960-1968 Capco $200-300
White globe face with large red "Ashland" logotype across globe face. Black drop shade from stylized lettering. Black outlined *light blue* "KEROSENE" below logotype on lower globe face.

VALVOLINE
GO MIX 3 1960-1968 Capco $400-600
Pale blue globe face with white band across center. White bow tie "VALVOLINE" in upper blue area, Red "GO-MIX" with black speedboat and line in center white area and white "OUTBOARD/FUEL" in lower blue area. Duplicated from Valvoline section.

****TRI-STATE—KENOVA, WEST VIRGINIA**: Tri-State, an independent refiner serving much the same territory as Ashland did in its earliest days, merged with Ashland in 1930. The Tri-State gasoline brands were retained until the early 1960s as an Ashland secondary brand.

TRI STATE
GASOLINE 6 1928-1932 OPB $1400-1800
White globe face with green outline ring and green outline around white center circle. Green "TRI-STATE" arched around top, "GASOLINE" around bottom with large green triangle in center circle. White interlocking "TS" in green triangle.

TRI STATE
SUPER MOTOR 4 1930s 13.5" Glass W $300-500
White globe face with green outline ring. Black outlined bold red lettering "TRI-/STATE" over bold green italic "SUPER/MOTOR" with speed lines. Lettering "TRI-" is rounded on top.

TRI-STATE
ETHYL (EGC) 4 1930s 13.5" Glass W $250-400
White globe face with red outline. Black outlined bold red lettering "TRI-/STATE" at top of globe face with large Ethyl (EGC) logo on lower face. Lettering "TRI-" is rounded on top.

TRI-STATE 3 1946-1960 13.5" Glass W $225-375
White globe face with green outline. Black outlined thin red lettering "TRI-/STATE" on center of globe face.

TRI-STATE

ETHYL (EC) 3 1946-1960 13.5" Glass W $250-400
White globe face with green outline. Black outlined thin red lettering "TRI-/STATE" on upper globe face with small outlined Ethyl (EC) logo below.

****NATIONAL REFINING — CLEVELAND, OHIO**: The National Refining Company was a large refiner/marketer based in Cleveland, Ohio, and founded in 1882. Marketing was concentrated throughout the Eastern and Central states with a subsidary in Canada that was spun off as a separate company in the 1930s and survived until its 1964 merger with Shell. National marketed the famous White Rose Gasoline and for many years their boy and slate logo was one of the best recognized trademarks in the country. After World War II, National was greatly weakened in the marketplace and the major portion of the company merged with Ashland in 1950. National White Rose stations were rebranded Ashland about 1957. Unique in the gasoline business, the White Rose brand name was franchised to Mid-West Refineries of Muskegon, Michigan, in about 1946. Although Mid-West merged with Leonard in 1955, the White Rose name remained in use in Michigan through the early 1970s.

WHITE ROSE
ENARCO GAS 5 1915-1918 OPB W $1600-2000
White globe with orange outline ring and orange ring around white center circle. Orange "WHITE ROSE" arched at top between inner and outer rings, "GASOLINE" at bottom, orange script "Enarco" above "MOTOR OIL" in center circle.

WHITE ROSE
ENARCO GAS 5 1915-1918 15" Metal $650-900
White globe face with orange outline ring and orange ring around white center circle. Orange "WHITE ROSE" arched at top between inner and outer rings, "GASOLINE" at bottom, orange script "En-ar-co" above "MOTOR OIL" in center circle.

WHITE ROSE
GASOLINE
BOY/SLATE 4 1920-1930 15" Metal $2200-2800
White globe face with multi-color boy holding red/yellow bordered black slate. White lettering on slate "WHITE ROSE/GASOLINE", small lettering around slate border.

WHITE ROSE
GASOLINE NO
KNOCK BOY/
SLATE FLAT
MILK GLASS 6 1920-1930 15" Metal HP $2200-2800
Flat milk glass lenses. White globe face with multi-color boy holding red/yellow bordered black slate. White lettering on slate "WHITE ROSE/GASOLINE", small lettering around slate border.

WHITE ROSE
NO KNOCK
BOY/SLATE 5 1922-1930 15" Metal $2300-2900
White globe face with multi-color boy holding red/yellow bordered black slate. White lettering on slate "WHITE ROSE/NO KNOCK", small lettering around slate border.

WHITE ROSE
ETHYL BOY/
SLATE 5 1922-1930 15" Metal $2300-2900
White globe face with multi-color boy holding red/yellow bordered black slate. White lettering on slate "WHITE ROSE/ETHYL", small lettering around slate border.

ROYAL GASOLINE 4 1925-1928 15" Metal $350-550
White globe face with purple outline ring and purple 1/3 width band across center. White "ROYAL/GASOLINE" on center purple band.

ROYAL GASOLINE 3 1925-1928 15" Metal $375-575
White globe face with purple outline ring and purple 1/3 width band across center. White "ROYAL/GASOLINE" on center purple band, small "THE NATIONAL REFINING/COMPANY" just below purple band.

WHITE ROSE
GASOLINE
BOY/SLATE 4 1926-1930 OPB $2400-3000

WHITE ROSE
GASOLINE
BOY/SLATE 4 1926-1930 OPP $1800-2400
White globe with multi-color boy holding red/yellow bordered black slate. White lettering on slate "WHITE ROSE/GASOLINE", small lettering around slate border.

WHITE ROSE
GASOLINE
BOY/SLATE 4 1926-1930 OP Baked on 3PC Body $1600-2000
White globe with multi-color boy holding red/yellow bordered black slate. White lettering on slate "WHITE ROSE/GASOLINE", small lettering around slate border.

WHITE ROSE
NO KNOCK
BOY/SLATE 5 1926-1930 OPB $2500-3200

WHITE ROSE
NO KNOCK
BOY/SLATE 5 1926-1930 OPP $1500-2400
White globe with multi-color boy holding red/yellow bordered black slate. White lettering on slate "WHITE ROSE/NO KNOCK", small lettering around slate border.

WHITE ROSE
ETHYL BOY/
SLATE 5 1925-1930 OPB $2600-3600
White globe with multi-color boy holding red/yellow bordered black slate. White lettering on slate "WHITE ROSE/ETHYL", small lettering around slate border.

WHITE ROSE
GASOLINE
BOY/SLATE 4 1928-1935 13.5" Glass W $1200-1600

WHITE ROSE
GASOLINE
BOY/SLATE 6 1928-1935 Balcrank $1300-1700
White globe face with multi-color boy holding red/yellow bordered black slate. White lettering on slate "WHITE ROSE/GASOLINE", small lettering around slate border.

WHITE ROSE
ETHYL GASOLINE5 1928-1935 13.5" Glass W $300-550
White globe face with black outline ring and black ring around white center circle. Red "WHITE ROSE" arched around top, "ETHYL" around bottom with Ethyl (EGC) logo in center circle.

ROYAL GASOLINE-
THE NAT.REF.CO. 5 1928-1935 14" Gill $300-500

ROYAL GASOLINE-
THE NAT.REF.CO. 3 1928-1935 14" Wide Gill $250-450

ROYAL/GASOLINE-
THE NAT.REF.CO. 3 1928-1935 13.5" Glass W $250-400
White globe face with purple outline ring and purple 1/3 width band across center. White "ROYAL/GASOLINE" on center purple band, small "THE NATIONAL REFINING/COMPANY" just below purple band.

WHITE/BOY &
SLATE/ROSE 3 1936-1940 13.5" Glass W $500-850
Green globe face with 1/3 width cream band across center. Cream "WHITE" in upper green area, "ROSE" in lower green area with boy and slate logo in center. "EN-AR-CO" on slate.

WHITE/ETHYL/
ROSE 5 1936-1940 13.5" Glass W $250-375
Green globe face with 1/3 width cream band across center. Cream "WHITE" in upper green area, "ROSE" in lower green area with outlined circle in center with small Ethyl logo in circle.

ROYAL/GASOLINE-
THE NAT.REF.CO. 4 1936-1940 13.5" Glass W $250-375
White globe face with green outline ring and 1/3 width band across center of globe face. White "ROYAL/GASOLINE" on center band, small green "THE NATIONAL REFINING/COMPANY" just below green band.

SUPER ROYAL 3 1936-1940 13.5" Glass W $250-375
Green globe face with 1/3 width cream colored band across center of globe face. Large green "SUPER/ROYAL" on cream band.

NATIONAL/WHITE
ROSE/*** 3 1940-1955 13.5" Glass W $250-375
Green globe face with cream colored box with rounded sides in center. Cream "NATIONAL" in upper green area, red "WHITE/ROSE" in center box, three white stars on lower green area.

NATIONAL/WHITE
ROSE/ETHYL 4 1940-1955 13.5" Glass W $250-375
Same as above with "ETHYL" replacing stars in lower green area.

NATIONAL/
ROYAL/*** 4 1940-1950 13.5" Glass W $250-375
Same as above with large green "ROYAL" replacing "WHITE ROSE" in center box.

NATIONAL/SUPER
ROYAL/*** 3 1940-1950 13.5" Glass W $250-375

Same as above with large green "SUPER/ROYAL" replacing "WHITE ROSE" in center box.

AROUND THE WORLD 5 1950-1955 13.5" Glass N $3000-4500
White globe face with multicolor world at top of globe face. Black outlined red banner across world with white "AROUND THE WORLD" on banner. Truck and 1950s cars riding on banner with airplane above. Very detailed. Black "GASOLINE" above two yellow arched lines on lower globe face.

****WHITE ROSE:** In the early 1950s the following globes were used in some select southern U.S. markets. It is unclear if the White Rose brand had been franchised to someone else at that time, but these markedly different globes would lead you to believe so.

WHITE/BOY & SLATE/ROSE 6 1936-1940 13.5" Glass W NL
Teal globe face with 1/3 width cream band across center. Red "WHITE" in upper teal area, "ROSE" in lower teal area with boy and slate logo in center. "EN-AR-CO" on slate.

WHITE/ETHYL/ ROSE 6 1936-1940 13.5" Glass W NL
Teal globe face with 1/3 width cream band across center. Red "WHITE" in upper teal area, "ROSE" in lower teal area outlined circle in center with small Ethyl(EC) logo in circle.

****WHITE ROSE/MID-WEST REFINING — GRAND RAPIDS, MICHIGAN:** The White Rose brand name was licensed to Mid-West Refineries in 1946, and although Mid-West merged with Leonard Refineries in 1955, the White Rose name continued in use in Michigan until 1970.

WHITE ROSE 3 1946-1970 Capco $200-300
Green globe face with cream colored box with rounded sides in center. Red "WHITE/ROSE" in center box.

NEW/WHITE ROSE/ETHYL 4 1946-1955 Capco $200-300
Same as above with cream "NEW" in upper green area and "ETHYL" in lower green area.

ETHYL/WHITE ROSE/ETHYL 4 1955-1970 Capco $225-325
Same as above with "ETHYL" replacing "NEW" in upper green area, as well as remaining in the lower green area.

****WHITE ROSE — CANADIAN OIL COMPANIES:** See listings for Canadian Oil Companies White Rose in the Shell section.

****AETNA OIL — LOUISVILLE, KENTUCKY:** Aetna Oil, founded in 1923, marketed through service stations in central and western Kentucky and southern Indiana. Aetna merged with Ashland in 1950. The Aetna brand was replaced by Ashland in the early 1960s.

AETNA GOLD BACKGROUND W/MOUNTAIN 5 1930s 15" Metal NL
Description not available.

AETNA MOTOR GAS W/MOUNTAIN 5 1930s 13.5" Glass NL
Red globe face with thick red/thin white outline. Red "mountain" on lower face with white lettering "MOTOR/GASOLINE". White upper area (background) with red "AETNA" across top of globe face.

AETNA BENZOL GAS 6 1930s 13.5" Glass $450-700
White globe face with red outline ring and double red outline around red mountain in center circle. Red "AETNA" above mountain, white "PETROLEUM PRODUCTS" below. Black "THE BETTER" arched around top, "MOTOR FUEL" around bottom with red "BENZOL-GAS" with black dropshade diagonal across globe face.

AETNA TOP QUALITY MOTOR OILS 5 1930s 13.5" Glass $350-500
Similar layout as above with black "TOP QUALITY" arched around top, "PARAFFIN BASE" around bottom and red/black "MOTOR OILS" diagonal across globe face.

AETNA XL LUBRICATED ANTI-KNOCK 6 1930s 13.5" Glass $400-650
White globe face with black outline ring. Grey/black/white circle design in center with large "XL" in center of circle. Red "AETNA" arched around top, "LUBRICATED-ANTI-KNOCK" arched around bottom.

AETNA ETHYL (EGC) 4 1930s 13.5" Glass $300-425
White globe face with thick red outline ring. Red "AETNA" arched around top with large Ethyl (EGC) logo below.

AETNA LAB FUEL 6 1930s 13.5" Glass NL
Description not available.

AETNA ON WHITE 2 1940s 13.5" Glass $200-350

AETNA ON WHITE 3 1940s Capco $175-275
White globe face with green outline ring. Red "AETNA" logotype lettering with green dropshade across globe face.

AETNA ON GREEN 3 1950s 13.5" Glass $225-375
Medium green globe face with black/white outlined yellow oval across center. Red "AETNA" logotype lettering with green dropshade across oval.

AETNA ON LIME GREEN 3 1950s Capco $200-300
Lime green globe face with black/white outlined yellow oval across center. Red "AETNA" logotype lettering with green dropshade across oval.

(AETNA) MOTO-ZIP 4 1940s 13.5" Glass $300-475

(AETNA) MOTO-ZIP 4 1950s Capco $250-350
Medium green globe face with black/white outlined yellow oval across center. Dark green "MOTO-ZIP" logotype lettering across oval.

AETNA A-PLUS 5 1960-1961 Capco $200-275
White globe face with red outline ring. Large red "A" covers most of globe face, with small "AETNA" logotype at top center o f "A" and "PLUS" on crossbar.

****FRONTIER REFINING — BUFFALO, NEW YORK:** Founded in Buffalo in 1931, Frontier marketed gasoline in western New York and Pennsylvania. Frontier merged with Ashland in April 1950 and Frontier stations were rebranded Ashland about 1960.

FRONTIER DOUBLE REFINED (BURST) 4 1935-1950 13.5" Glass N $250-400
Blue globe face with large white burst covering most of globe face. Red "FRONTIER" across center of burst with blue "DOUBLE" above and "REFINED" below.

FRONTIER DOUBLE REFINED (BOWTIE) 4 1950-1955 13.5" Glass N $250-400
White globe face with large blue outlined red "FRONTIER" in bowtie lettering across globe face. Blue "DOUBLE/REFINED" below "FRONTIER".

FRONTIER ETHYL (EC) 6 1950-1955 13.5" Glass N $200-350
White globe face with large blue outlined red "FRONTIER" across globe face. Red "ETHYL" below "FRONTIER". Small yellow circle on upper globe face with Ethyl (EC) logo with black rays on circle.

FRONTIER DOUBLE REFINED 3 1955-1960 13.5" Glass N $200-350
White globe face with large blue outlined red "FRONTIER" across globe face. Blue "DOUBLE/REFINED" below "FRONTIER".

FRONTIER ETHYL 3 1955-1960 13.5" Glass N $225-375
White globe face with large blue outlined red "FRONTIER" across globe face. Red "ETHYL" below "FRONTIER".

****FREEDOM-VALVOLINE — OIL CITY, PENNSYLVANIA:** Freedom Valvoline was created from the merger of Freedom Oil Works of Freedom, Pennsylvania, and Valvoline Oil Company of Binghampton, New York, in 1944. The company was in the process of selling off its fuel marketing operations to concentrate on lubricants manufacturing when they merged with Ashland in 1950. The Valvoline brand remains Ashland's premier lubricants brand worldwide today.

****VALVOLINE OIL COMPANY — BINGHAMPTION, NEW YORK:** Valvoline (founded as the Continuous Oil Refining Company in Binghampton, New York, in 1866) introduced the famous trademark "Valvoline" in 1873. So well known were the company's Valvoline lubricants that the company name was changed to the Valvoline Oil Co. in 1902. In the early 1920s Valvoline began marketing gasoline in New York and Pennsylvania and continued to do so until the 1944 merger with Freedom Oil Co. of Freedom, Pennsylvania. Freedom had also marketed gasoline since the 1920s and the few remaining Valvoline stations were rebranded Freedom after World War II. Freedom was in the process of discontinuing gasoline marketing, selling it's stations to Esso, when the company merged with Ashland Oil in early 1950. Since that time the primary product has been Valvoline Oils, marketed by all the Ashland retail network and throughout the world.

VALVOLINE GASOLINE & MOTOR OIL 4 1920s OPE W/N $2000-2500

White globe with no outline ring around face. Green/white "Corliss valve" logo in center. Green "VALVOLINE" in bowtie lettering across valve. Red "GASOLINE" arched around top, "MOTOR OIL" around bottom. Small green "TRADEMARK" above valve logo.

VALVOLINE GASOLINE 6 1920s OPE N NL

White globe with green globe face. White outline "Corliss valve" logo in center with white "VALVOLINE" in bowtie lettering across valve. White "GASOLINE" below "VALVOLINE".

VALVOLINE PENNSYLVANIA MOTOR OILS 5 1920s OPB W $1600-2200

White globe with green/white "Corliss valve" logo in center of face. Green "VALVOLINE" in bowtie lettering across center. Green "PENNSYLVANIA" arched around top, "MOTOR OILS" around bottom.

VALVOLINE 100% PURE PENNA MOTOR OILS 5 1920s OPB W $1600-2200

White globe with green/white "Corliss valve" logo in center of face. Green "VALVOLINE" in bowtie lettering across center. Green "100% PURE PENNA" arched around top, "MOTOR OILS" around bottom.

VALVOLINE GASOLINE METAL FACE 5 1910s 15" Metal $800-1300

Green/white outlined green metal litho globe face with white outlined Corliss valve logo in ceter of globe face. Green outlined white bowtie lettering "VALVOLINE" across center with white "OILS" with red inline arched around top, "GASOLINE" around bottom.

VALVOLINE GASOLINE ETCHED FACE 5 1920s 15" Metal $1800-2500

Etched white globe face with green/white "Corliss valve" logo in center. Green "VALVOLINE" in bowtie lettering across valve. Red "GASOLINE" arched around top, "MOTOR OIL" around bottom.

VALVOLINE GASOLINE 5 1930s 15" Metal $1100-1600

Green globe face with white outline around green/white "Corliss valve" logo. Large white bowtie "VALVOLINE" across logo on center of globe above white "GASOLINE".

VALVOLINE RED STAR 5 1920s 15" Metal $1000-1500

White globe face with orange outline ring. Orange outline around white/green "Corliss valve" logo. Green bowtie "VALVOLINE" across valve on center of globe, green "RED STAR" arched around top, "GASOLINE" around bottom.

VALVOLINE AERO 5 1920s 15" Metal $1200-1700

White globe face with orange outline ring. Orange outline around white/green "Corliss valve" logo. Green bowtie "VALVOLINE" across valve on center of globe, green "AERO" arched around top, "GASOLINE" around bottom.

VALVOLINE 5 1930s Banded Glass $500-850
VALVOLINE 5 1930s 13.5" Glass $450-800
VALVOLINE 5 1930s Capco $350-650

White globe face with green outline ring. Green outlined white "Corliss valve" logo with red outlined green bowtie lettering "VALVOLINE" across "valve" on center of globe face. Small green "TRADE MARK/ REGISTERED 1873" above valve logo.

VALVOLINE ETHYL 5 1930s Banded Glass $550-875
VALVOLINE ETHYL 5 1930s 13.5" Glass $500-850
VALVOLINE ETHYL 5 1930s Capco $400-675

White globe face with green outline ring. Green outlined white "Corliss valve" logo with red outlined green bowtie lettering "VALVOLINE" across "valve" on center of globe face. Small green "TRADE MARK/ REGISTERED 1873" above valve logo. Black outlined white circle on upper globe face with small Ethyl (EGC) logo with yellow rays in circle. Red "ETHYL across bottom of globe face below valve.

MAGNET 5 1930s 13.5" Glass $750-1200

Magnet is believed to be the brand name of the motor-grade gasoline offered by Valvoline in the 1930s. Black outlined white globe face with large black outlined horseshoe magnet in center of globe face. Red outlined black "MAGNET" across center, with small black "TRADE" to upper left of magnet image, "MARK" to right.

VALVOLINE MOTOR OILS 5 1930s 13.5" Glass $500-800

White globe face with green outline ring. Green outlined white "Corliss valve" logo with red outlined green bowtie lettering "VALVOLINE" across "valve" on center of globe face. Red "PENNSYLVANIA" arched around top, "MOTOR OIL" around bottom with small red "100% PURE" above small green "TRADEMARK" above valve logo.

VALVOLINE GO MIX 3 1960-1968 Capco $400-650

Pale blue globe face with white band across center. White bowtie "VALVOLINE" in upper blue area, Red "GO-MIX" with black speedboat and line in center white area and white "OUTBOARD/FUEL" in lower blue area.

****FREEDOM OIL WORKS — FREEDOM, PENNSYLVANIA:**

FREEDOM GASOLINE 5 1910s 15" Metal $850-1200

Yellow globe face with blue "FREEDOM" arched around top, "GASOLINE" around bottom with four dark blue keystones joined at the bottom to a center square to form a cross. With yellow "W" in center square, yellow lettering "FOWCO" runs horizontal and verticle on cross.

FREEDOM MOTOR 6 1920s OPE N NL

White globe with blue outline ring around white globe face. Blue "FREEDOM" in elongated typestyle above "MOTOR" in normal block lettering.

FREEDOM ETHYL 6 1926-1930 OPE N NL

White globe face with outline ring. Large Ethyl (EGC) logo in center with blue "FREEDOM" arched around top, "GASOLINE" around bottom.

FREEDOM 4 1930s 15" Metal $650-900

White globe face with four blue keystones joined at the bottom to a center square to form a cross. With white "E" in center square, white lettering "FREEDOM" runs vertical and horizontal on cross.

FREEDOM ETHYL (EGC) 7 1930s 15" Metal NL

White globe face with outline ring. Large Ethyl (EGC) logo in center with blue "FREEDOM" arched around top, "GASOLINE" around bottom.

FREEDOM 6 1930s Yellow Glass Hull NL

Special yellow glass body. White globe face with four dark blue keystones joined at the bottom to a center square to form a cross. With white "E" in center square, white lettering "FREEDOM" runs vertical and horizontal on cross.

FORT PITT GASOLINE 6 1920s-1930s 15" Metal NL

Description not available.

****INDEPENDENT QUALITY SERVICE:** Independent Quality Service was an Ashland secondary brand operated for independent, unbranded dealers from the mid-1940s through the 1960s.

IQS 3 1940s 13.5" Glass W $300-450

White globe face with blue outer ring. White "INDEPENDENT" arched around top, "QUALITY SERVICE" arched around bottom with red reverse bowtie "I.Q.S." in center.

IQS ETHYL 4 1940s 13.5" Glass W $325-475

White globe face with blue outer ring. White "INDEPENDENT" arched around top, "QUALITY SERVICE" arched around bottom with red I.Q.S./ETHYL in center.

IQS ETHYL 4 1940s 13.5" Glass W $325-475

White globe face with blue outer ring. White "INDEPENDENT" arched around top, "QUALITY SERVICE" arched around bottom with black I.Q.S./ETHYL in center.

****PAYLESS - DANCE — NEW ALBANY, INDIANA:** Payless is the 1952 creation of James Thornton, operating discount service stations "from Canada to the Gulf Coast" as their slogan claimed. In the middle 1960s Payless acquired the marketing of Louisville's Dance Oil Company—Mr. Thornton having married Mr. Dance's daughter, Bonnie. Ashland Oil, long a supplier to both Dance and Payless, bought the operation in 1968. A few Payless outlets remain today, although most have been converted to SuperAmerica. James Thornton formed Louisville based Thornton Oil Corporation after selling Payless, and continues to operate Thornton stations in six states today.

****DANCE:**

DANCE "92" A
WHALE OF A BUY5 1940s Capco $200-350
White globe face with red "DANCE" arched around top, "A WHALE OF A BUY" around bottom and large yellow "92" in center.

DANCE
"100" ETHYL 5 1940s Capco $200-350
Same as DANCE 92, above, with yellow "100" and small Ethyl (EC) logo in center replacing "92".

DANCE PINK
GOLDEN
POWER PACK 4 1960s Capco $1600-2200
Pink globe face with large white "D" with red dropshade covering most of face. Black/white halftone figure of ballerina superimposed over "D" with white band across lower globe face. Gold "GOLDEN POWER PACK" on band.

DANCE BLUE
GENUINE ETHYL 4 1960s Capco $1600-2200
Light blue globe face with same design as above. Red "GENUINE ETHYL" on lower band.

DANCE GREEN
SUPER HI-TEST 5 1960s Capco $1800-2400
Green globe face with same design as above. "SUPER Hi-TEST" on lower band.

****PAYLESS:**

PAYLESS WHALE
OF A BUY 6 1952-1960 Capco NL
White globe face with large black line drawing of a whale in blue water across center of globe face. Blue outlined white text balloon across top of globe face with red "WHALE OF A BUY" in balloon. Red "PAYLESS" in formed letters to fit below whale.

HI TEST POWER
PACK 6 1940s 13.5" Glass $200-350
Red outlined white globe face with large red "HI-TEST" across upper globe face above dark blue "POWER/PACK".

PAYLESS ROAD
TESTED REGULAR 4 1960-1965 Oval Capco
$250-400
White globe face with deep violet outline. Red/white/violet textured "PAYLESS" across top of globe face with yellow outlined blue "REGULAR" across lower globe face with blue freeform "ROAD TESTED" across center.

PAYLESS EXTRA
SPECIAL 4 1960-1965 Oval Capco $250-400
White globe face with deep violet outline. Red/white/violet textured "PAYLESS" across top of globe face with yellow outlined blue "EXTRA/SPECIAL" in freeform lettering below.

PAYLESS GOLDEN
THUNDERBIRD
SUPER ETHYL 6 1960-1965 Oval Capco $400-650
White globe face with deep violet outline. Red/white/violet textured "PAYLESS" across top of globe face with blue outlined yellow script "Golden Thunderbird" across center of globe face above yellow outlined violet "SUPER ETHYL".

PAYLESS NO.2
DIESEL CLEAN
FILTERED 6 1960-1965 Oval Capco $400-650
White globe face with green/black/white textured "PAYLESS" over black outlined green "No 2 DIESEL" over much smaller green "CLEAN/FILTERED".

PAYLESS
KEROSENE
HEATER OIL 6 1960-1965 Oval Capco $400-650
White globe face with green/black/white texture "PAYLESS" over green "KEROSENE over black outlined green "HEATER OIL".

****SUPERAMERICA — MINNEAPOLIS, MINNESOTA:** Northwestern Refining Co., the St. Paul Park, Minnesota, refining unit of the Erickson family, began operating the SuperAmerica gasoline/convenience store (c-store) units about 1962. In 1970 SuperAmerica, a marginal c-store operation, was sold to Ashland Oil and by the middle 1970s began to expand out of its north-central base. Many of the Ashland affiliate brand locations, closed during the gasoline shortages of 1973 and 1978, were rebuilt as SuperAmerica convenience stores and the operation today is considered the model of efficient and customer service oriented gasoline c-store chains.

SA 5 1962-1965 Plastic $150-225
White plastic trapezoid shaped halves assembled back to back with large red trapezoid shield on globe face. White lettering "SA" on red shield above white "SUPERAMERICA".

SA (red logo plastic
pump topper) 5 1965-1985 Plastic $75-125
Unusual shaped multi-piece rectangular globe. Looking down from the top, the globe is hour-glass shaped with a flat waistline. Along each side of the waistline is the white on red shield "SA" logo for SuperAmerica. On each side in the flared area are black numbers indicating the pump position number. A black plastic top covers the internal area, allowing access to light bulb inside. Many are installed non-illuminated.

SA (red/blue logo
plastic pump
topper) 4 1985-Pres Plastic $75-125
Unusual shaped multi-piece rectangular globe. Looking down from the top, the globe is hour-glass shaped with a flat waistline. Along each side of the waistline is the red "S" / blue "A" logo for SuperAmerica. Three blue lines run across the bottom of the globe face. On each side in the flared area are black numbers indicating the pump position number. A black plastic top covers the internal area, allowing access to light bulb inside. Many are installed non-illuminated.

****OSKEY BROS. PETROLEUM — ST. PAUL, MINNESOTA:** Oskey Bros. has operated the North Star brand stations in the north-central states since the 1930s. Ashland acquired the chain about 1980 and has since closed many locations or converted them to SuperAmerica. The company has no connection with Canada's North Star Oil Company.

NORTH STAR
E-TANE (ethyl) 4 1955-1965 13.5" Glass $300-500
Blue outlined white globe face with large blue star in center. Red "NORTH STAR" arched around top, "E-TANE" around bottom.

NORTH STAR
R-TANE (regular) 4 1955-1965 13.5" Glass $300-500
Blue outlined white globe face with large blue star in center. Red "NORTH STAR" arched around top, "R-TANE" around bottom.

NORTH STAR
DIESEL 5 1955-1965 13.5" Glass $300-500
Blue outlined white globe face with large blue star in center. Red "NORTH STAR" arched around top, "DIESEL" around bottom.

NORTH STAR
MARINE 5 1955-1965 13.5" Glass NL
Blue outlined white globe face with large blue star in center. Red "NORTH STAR" arched around top, "MARINE" around bottom.

NORTH STAR Oval 4 1965-1980 2pc Plastic Oval $200-300
Oval formed plastic faces glued to plastic band forming body. Blue ring around white area at left extending to blue area at right split diagonal. Large red "N" on white area, large white star on blue area.

NORTH STAR Oval 3 1965-1980 1pc Plastic Oval $200-300
Oval deep formed plastic faces glued back-to-back with blue ring around white area at left extending to blue area at right split diagonal. Large red "N" on white area, large white star on blue area.

****RED HEAD OIL CO.—WOOSTER, OHIO**: Red Head (operating discount gasoline stations with service bays, unique in the industry) was founded in Wooster, Ohio, in the early 1930s as Fair Price Stations, Inc., rebranding to Red Head in the mid-1930s. Following World War II, the chain greatly expanded and eventually operated stations in Ohio, Pennsylvania and West Virginia when they were bought out by Ashland in about 1970. Many have since been converted to SuperAmerica, but some Red Head stations remain today.

HI-OCTANE
RED HEAD
GASOLINE 5 1935-1946 13.5" Glass W NL
White globe face with black/white/red line drawing of red-headed boy in center of globe face. Red "RED" to left of boy, "HEAD" to right with red "GASOLINE" arched below and black "HI-OCTANE" arched around top fo globe face.

RED HEAD 4 1946-1965 13.5" Glass W $300-450
White globe face with red/black lines forming band across center. "HI-OCTANE" arched around top, "GASOLINE" around bottom, red "RED HEAD" across center band.

RED HEAD ETHYL 4 1946-1965 13.5" Glass W $325-475
White globe face with black outline ring and black outline around center white circle. Red "RED HEAD" logotype lettering arched around top, "GASOLINE" around bottom with Ethyl (EC) logo in center.

RED HEAD
STOVE
GASOLINE 6 1946-1965 13.5" Glass W $350-500
White globe face with red/black lines forming band across center. "STOVE" arched around top, "GASOLINE" around bottom, red "RED HEAD" across center band.

****SOUTHERN OIL OF NY—HORSEHEADS, NEW YORK:** Founded in 1923, Southern Oil operated the Rotary brand stations in New York and Pennsylvania from the mid-1920s until the 1970s when the operation was acquired by Ashland Oil.

ROTARY 5 1920s 15" Metal $800-1250
Blue outlined white globe face with red "rotating circle" design in center. Blue outlined white bowtie band across center of globe face with blue "ROTARY" in bowtie lettering on band. Blue "SOUTHERN OIL CO." arched around top, "OF NEW YORK, Inc." around bottom.

ROTALINE HI COMPRESSION
ANTI-KNOCK
GASOLINE 5 1920s 15" Metal $900-1400
Black outlined yellow lobe face with black outlined red circle in center. Black outlined red "ROTALINE" arched around top, "GASOLINE" around bottm with black outlined yellow triangle superimposed over black triangle in red center circle. Black "HIGH/COMPRESSION/ANTI-KNOCK/MOTOR/FUEL" in center triangle.

ROTARY BENZOL
GASOLINE 5 1920s 15" Metal $1000-1600
Yellow outer ring with black outline around edge and around center. Red center with yellow triangle on top of black triangle in center. Black lettering on yellow triangle "BENZOL/BLEND/ANTI-KNOCK/MOTOR/FUEL/?". Red "ROTARY" arched at top in yellow ring, "GASOLINE" below.

ROTARY w/rotor 4 1930s 15" Metal $750-1100
White globe face with red/blue outline rings. Red one-line "spiral" on center of globe face broken by large blue lettering "ROTARY" across center.

ROTALINE 5 1930s 15" Metal $500-800
White globe face with red "ROTARLINE" across center. Black "HIGH" arched around top, "COMPRESSION" around bottom.

ROTA-ZOL
BLEND "?" 4 1930s 15" Metal $500-750
White globe face with white border area then blue/red outline rings. Large blue question mark (?) at top center of glbe face above red "ROTA-ZOL". Small blue triangle with blue outlined white triangle on top of it at bottom of globe face. Red "BLEND" in white triangle.

ROTARY ETHYL
(EGC) 4 1930s 15" Metal $550-800
Yellow outer ring with black outline around edge and around center. White center with large Ethyl (EGC) logo and blue rays. Black "ROTARY" arched at top in yellow ring, "GASOLINE" below.

ROTARY 3 1946-1960 Gill $350-550
White globe face with red/blue outline rings. Red one-line "spiral" on center of globe face broken by large blue lettering "ROTARY" across center.

ROTALINE 3 1946-1960 Gill $350-550
White globe face with blue/red outline rings. Blue one-line "spiral" on center of globe face broken by large red lettering "Rotaline" across center.

ROTA-ZOL
BLEND "?" 4 1946-1960 Gill $300-500
White globe face with blue/red outline rings. Large blue question mark (?) at top center of globe face above red "ROTA-ZOL". Small blue triangle with blue outlined white triangle on top of it at bottom of globe face. Red "BLEND" in white triangle.

ROTA-ZOL
BLEND 4 1946-1960 Gill $275-450
Same as above without question mark.

ROTARY
KEROSENE 5 1946-1960 Gill $300-500
White globe face with green outline ring. Small green "ROTARY" at top of globe above "KEROSENE" in green half-bowtie lettering. Two short green lines below.

BENZOLA 6 1940s Gill NL
Description not available.

****HOOD OIL CO.—NORCROSS, GEORGIA**: Hood's was originally a jobber operation known as Hood-Harden Oil Company, distributing Shell products. The company introduced the discount brand "HOODS" following World War II, and rebranded Fleet about 1960 and acquired by Ashland about 1967. In the 1980s Fleet was sold to the son of the founder and continues in operation today.

HOOD'S FLO-POWER
REGULAR 4 1950s Red Capco $600-1000
Light blue globe face with red/white rocket diagonal across globe face. Black "HOOD'S" arched across top, "REGULAR" across bottom with italic red "FLO-POWER" across center with black speed lines.

HOOD'S FLO-POWER
SUPER 4 1950s Red Capco $600-1000
Light blue globe face with red/white rocket diagonal across globe face. Black "HOOD'S" arched across top, "SUPER" across bottom with italic red "FLO-POWER" across center with black speed lines.

HOOD'S FLO-POWER
ETHYL 4 1950s Red Capco $600-1000
Light blue globe face with red/white rocket diagonal across globe face. Black "HOOD'S" arched across top, "ETHYL" across bottom with italic red "FLO-POWER" across center with black speed lines.

****TRESLER OIL — CINCINNATTI, OHIO**: Tresler was founded in Cincinnati in 1922 and operated retail service stations in Ohio, Indiana, Tennessee and Kentucky until 1985 when they were purchased by Ashland. All have since been converted to SuperAmerica.

TRESLER
BENZOL GAS 6 1926-1930 OPE RW NL
White globe with black "TRESLER" arched around top, "GAS" around bottom with red "GENUINE" above black "BENZOL" in center of globe face.

COMET GASOLINE 7 1920s 15" Metal NL
This globe is known only in old photographs. Description not available.

TRESLER ETHYL
GASOLINE 7 1930s 15" Metal NL
This globe is known only in old photographs. White globe face with dark colored "TRESLER" arched around top, "GASOLINE" around bottom with large Ethyl (EGC) logo inside center circle.

TRESLER
BENZOL GAS 6 1930s 13.5" Glass $500-800
White globe face with red outline ring and large red center circle. Blue "TRESLER" arched around top, "BENZOL GAS" around bottom.

TRESLER ETHYL
GASOLINE 7 1930s 13.5" Glass NL
This globe is known only in old photographs. White globe face with dark colored "TRESLER" arched around top, "GASOLINE" around bottom with large Ethyl (EGC) logo inside center circle.

COMET (w/comet) 7 1938-1946 13.5" Glass NL

This globe is known only in an old photograph. It appears to be a dark color lens with the word "COMET" in a comet across the center of the globe face.

COMET 7 1946-1955 13.5" Glass NL

COMET 5 1950-1955 Capco $200-325

Blue globe face with red outlined white oval in center. Red "COMET" across center of globe face.

COMET ETHYL 7 1946-1955 13.5" Glass $350-550

COMET ETHYL 5 1950-1955 Capco $225-350

White globe face with blue outlined white oval in center. Red "COMET" over blue "ETHYL" on oval.

COMET DIESEL 6 1946-1955 13.5" Glass $375-600

COMET DIESEL 5 1950-1955 Capco $225-350

Blue globe face with red outlined white oval across center of globe face. Large red "COMET/DIESEL" in oval.

REGULAR TRESLER COMET 4 1955-1964 Capco $250-400

White globe face with blue outlined white oval on lower globe face. Red "COMET" with speed lines in oval. Blue trapezoid on top of oval with white "TRESLER", red "REGULAR" at top of globe face.

SILVER COMET 4 1955-1964 Capco $850-1250

White globe face with blue bird with upstretched wings covering most of globe face. Black outlined white lettering "SILVER" at top of globe face, red "COMET" arched with speed lines across center of globe face.

TRESLER COMET REGULAR 5 1964-1970 Capco $250-400

White globe face with large red outlined square Tresler Comet logo in center consisting of interlocking blue "T" and red "C". Blue band across below square with white "REGULAR" in band.

SUPREMIUM 5 1964-1970 Capco $225-325

White globe face with red outline. Thick blue outlines around with band across center. Red italic "SUPREMIUM" on white band.

**** CINCINNATI OIL WORKS CO. — CINCINNATI, OHIO**: One of the nations first distributors of petroleum products, Cincinnati Oil Works was founded in 1841 to refine and distribute the small quantities of kerosene that could be refined from crude oil skimmed from creeks and found in natural seepages. In the early years of this century, Cincinnati Oil Works began marketing gasoline, selling their fuel under the brand name "White Rose" and actually pre-dating the use of that trademark by the better known National Refining Company. For many years Cincinnati Oil Works continued to market their White Rose gasoline through a small number of stations in the Cincinnati area. In 1956 Tresler Oil Company purchased the operation and stations were branded Tresler Comet stations at that time.

WHITE ROSE FILTERED GAS 6 1910s OPE W NL

White globe with red "WHITE ROSE" arched around top, "GASOLINE" around bottom with black "FILTERED" across center.

WHITE ROSE 4 1920s-40s 15" Metal Red $450-700

White globe face with red circle around white center circle. Blue "WHITE ROSE" in bowtie lettering across center.

ZIP ROSE HIGH COMPRESSION 6 1920s-1940s 15" Metal $600-900

Red globe face with white outline ring and small white circle in center. Large blue "ZIP/ROSE" diagonal across upper globe face over smaller blue "HIGH/COMPRESSION".

****UNION OIL CO. — CINCINNATI, OHIO**: An Ohio marketer with ties to Tresler Oil Company.

UNION OIL CO. UNION PRODUCTS 4 1930s 15" Metal $500-850

White globe face with blue/red outline rings. Blue band diagonal across center with red outlined white "UNION" on band. Blue "UNION OIL CO." arched around top, "PRODUCTS" around bottom.

UNION OIL CO. UNION PRODUCTS 4 1940s 13.5" Glass W $375-600

White globe face with blue/red outline rings. Blue band diagonal across center with red outlined white "UNION" on band. Blue "UNION OIL CO." arched around top, "PRODUCTS" around bottom.

UNION OIL CO. HI-TEST PRODUCTS 5 1940s 13.5" Glass W $400-600

White globe face with blue/red outline rings. Blue band diagonal across center with red outlined white "HI-TEST" on band. Blue "UNION OIL CO." arched around top, "PRODUCTS" around bottom.

UNION OIL CO. FARMCO PRODUCTS 5 1940s 13.5" Glass W $400-600

White globe face with blue/red outline rings. Blue band diagonal across center with red outlined white "FARMCO" on band. Blue "UNION OIL CO." arched around top, "PRODUCTS" around bottom.

UNION GASOLINE MOTOR OIL 5 1950s 13.5" Glass W $400-600

Blue globe face with white band across center with blue "UNION" on band. White "GASOLINE." arched around top, "MOTOR OIL" around bottom.

****PYROIL CHEMICAL CO. — LA CROSSE, WISCONSIN**: Pyroil marketed a chemical additive for gasoline. Stations adding this lubricant to their products sold it under the Pyroil Lubrication Gasoline trademark. Pyroil was bought out by Ashland in 1988 and continues to market automotive chemicals today.

PYROIL LUBRICATING GASOLINE 5 1930s 15" Metal $600-850

PYROIL LUBRICATING GASOLINE 4 1930s 13.5" Glass W $350-650

PYROIL LUBRICATING GASOLINE 4 1930s 14" or 13.5" Banded Glass $500-750

White globe face with red outline and horizontal red lines forming a circle in center of globe face. Red outlined black "PYROIL" across center of globe face with black "LUBRICATING" arched around top and "GASOLINE" around bottom.

CAMPECO PYROIL LUBRICATING GASOLINE 6 1930s 18" Neon Metal NL

White globe face with red outline and horizontal red lines forming a circle in center of globe face. Red outlined black "PYROIL" across center of globe face with black "CAMPECO" logotype arched around top and "GASOLINE" around bottom. Chicago based Camden Petroleum was a Pyroil franchise.

ATLANTIC & AFFILIATES

****THE ATLANTIC REFINING CO. — PHILADELPHIA, PENNSYLVANIA**: The Atlantic Refining Company was founded in Philadelphia in 1866, pulled together from earlier interests of the various founders and united as a production and refining company. In 1874 Atlantic sold out to Standard Oil and, until the breakup of Standard in 1911, functioned as a refining division. Atlantic began marketing in 1915, after the breakup of Standard, and throughout most of its history was in constant search for crude oil to keep its refineries and service stations operating. In 1965 Atlantic, then operating service stations all along the eastern seaboard, began merger talks with Richfield Oil Corporation of California, a west coast refiner/marketer that traces its roots back to the turn of the century. Atlantic merged with Richfield in 1966 and from 1966 until 1970 marketed under the Atlantic name in the east and Richfield in the west. Upon the com;pany's 1969 merger with Sinclair a new identity was needed, resulting in the introduction of the Arco brand name. Arco replaced both the Atlantic and Richfield brands as well as some of the Sinclair brand. By the 1980s, Sinclair had been sold off and was operating as a successful independent and Arco had abandoned marketing in the east, where the name Atlantic was reintroduced by the independent marketing company that had purchased the former Arco marketing. In 1988, the new Atlantic was sold to Sunoco and stations operate under the Atlantic name in the northeast today, although it has been announced that they will be gradually changed to Sunoco. See separate Richfield and Sinclair listings for more information on these affiliated companies.

ATLANTIC
POLARINE
GASOLINE 5 1915-1920 OPE Chimney $2200-3500
White globe with red edged red/white/black Atlantic "crossed arrows" logo in center. Blue "ATLANTIC" arched around top, GASOLINE" around bottom with small black "POLARINE" between logo and "GASOLINE".

ATLANTIC
POLARINE
GASOLINE 5 1915-1920 OPE $2000-3000
White globe with red edged red/white/black Atlantic "crossed arrows" logo in center. Blue "ATLANTIC" arched around top, GASOLINE" around bottom.

ATLANTIC GASOLINE
MILK GLASS
ETCHED FACES 5 1918-1920 15" Metal L/H Red $2000-2800
White globe face with red edged red/white/black Atlantic "crossed arrows" logo in center. Blue "ATLANTIC" arched around top, "GASOLINE" around bottom with small blue "POLARINE" between logo and "GASOLINE".

ATLANTIC
GASOLINE
blue lettering 4 1918-1933 16.5" Metal L/H Red $600-900

ATLANTIC
GASOLINE
blue lettering 4 1918-1933 15" Metal L/H Red $600-900
White globe face with red/white/black Atlantic "crossed arrows" logo in center on irregular red field. Blue "ATLANTIC" arched around top, "GASOLINE" around bottom. Porcelain sign attachment "WHITE FLASH" add $300-$500.

ATLANTIC
GASOLINE
CAST FACES 6 1918-1920 16.5" Metal L/H Red NL

ATLANTIC
GASOLINE
CAST FACES 5 1918-1920 15" Metal L/H Red $2000-2800
Cast detail globe face. White painted globe face with red/white/black Atlantic "crossed arrows" logo in center on irregular red field. Blue "ATLANTIC" arched around top, "GASOLINE" around bottom.

ATLANTIC
MOTOR OIL 6 1920s 16.5" Metal L/H Red NL
White globe face with red/white/black Atlantic "crossed arrows" logo in center on irregular red field. Blue "ATLANTIC" arched around top, "MOTOR OIL" around bottom.

ATLANTIC
ETHYL (EGC)
BLUE BURST 4 1927-1936 16.5" Metal L/H Red $550-800

ATLANTIC
ETHYL (EGC)
BLUE BURST 4 1927-1936 15" Metal L/H Red $550-800
White globe face with blue outlines around edge and center. Blue "ATLANTIC" arched around top, "GASOLINE" around bottom with large Ethyl logo in center with blue rays extending out to blue outline around center.

ATLANTIC ETHYL
(EGC) YELLOW
BURST 4 1927-1936 16.5" Metal L/H Red $550-800

ATLANTIC ETHYL
(EGC) YELLOW
BURST 4 1927-1936 15" Metal L/H Red $550-800
Same as above with yellow rays extending out from Ethyl logo.

ATLANTIC 68-70 4 1925-1928 16.5" Metal L/H Red $500-750
ATLANTIC 68-70 5 1925-1928 15" Metal L/H Red $500-750
White globe face with blue outlines around edge and center. Red "ATLANTIC" arched around top, "GASOLINE" around bottom with blue "68-70" in center.

ATLANTIC CAPITOL 4 1928-1935 16.5" Metal L/H Red
$500-750
ATLANTIC CAPITOL 4 1928-1935 15" Metal L/H Red
$500-750
White globe face with red/blue squares positioned above center and blue/red squares below center in checkerboard style. White "ATLANTIC" on upper squares, "GASOLINE" on lower squares with black "CAPITOL" across center of globe face.

ATLANTIC
RAYOLITE
KEROSENE 5 1935-1940 16.5" Metal L/H Red $700-1200

ATLANTIC
RAYOLITE
KEROSENE 5 1935-1940 15" Metal L/H Red $700-1200
Green outlined white outer ring around globe face with green "ATLANTIC" arched in ring at top, "KEROSENE" at bottom. Bright yellow circle in center of globe face. Green band across center with yellow distinctive lettering "Rayolight" on green band. "Rayolight" underlined connecting lower portions of "y" and "g".

ATLANTIC
WHITE FLASH 4 1932-1953 16.5" Metal L/H Red $450-700

ATLANTIC
WHITE FLASH 4 1932-1953 15" Metal L/H Red $450-700
Narrow white outer ring around globe face. Center of face is divided into quarters with upper left and lower right red and upper right-lower left blue. White "ATLANTIC" arched around edge of upper left red quarter. White "WHITE" with large "dot" above "I" above centerline of globe face, white "FLASH" below centerline, very distinctive lettering.

ATLANTIC HI-ARC 4 1936-1953 16.5" Metal L/H Red $500-750
ATLANTIC HI-ARC 4 1936-1953 15" Metal L/H Red $500-750
White globe face with blue band arched across center. Double red bars arched upper left and lower right of band. Blue lettering "ATLANTIC" across top of globe face, white "HI-ARC" on blue band.

(ATLANTIC)
DIESEL FUEL 4 1935-1953 16.5" Metal L/H Red $250-450

(ATLANTIC)
DIESEL FUEL 4 1935-1953 15" Metal L/H Red $250-450
White globe face with bright blue lettering "DIESEL/FUEL" center of globe face. No company identification.

ATLANTIC
WHITE FLASH 3 1932-1953 Gill $275-450
Narrow white outer ring around globe face. Center of face is divided into quarters with upper left-lower right red and upper right-lower left blue. White "ATLANTIC" arched around edge of upper left red quarter. White "WHITE" with large "dot" above "I" above centerline of globe face, white "FLASH" below centerline, very distinctive lettering.

ATLANTIC HI-ARC 3 1936-1953 Gill $275-450
White globe face with blue band arched across center. Double red bars arched upper left and lower right of band. Blue lettering "ATLANTIC" across top of globe face, white "HI-ARC" on blue band.

ATLANTIC
CROSSED
ARROWS 6 1925-1935 Gill NL
White globe face with red/white/black Atlantic crossed arrows logo in center on irregularly shaped red field. This lens fits Gill bodies but was originally used as a light-up portion of a porcelain identification sign.

ATLANTIC 2 1953-1966 Gill $200-325
ATLANTIC 5 1953-1966 Capco $150-200
White globe face with Atlantic red rectangle logo in center.

ATLANTIC
PREMIUM 3 1953-1957 Gill $275-450

ATLANTIC
PREMIUM 5 1953-1957 16.5" Metal L/H Red $650-850
White globe face with Atlantic red rectangle logo and blue mast-arm style pole tilted on center of globe face with blue "PREMIUM" below sign.

ATLANTIC
IMPERIAL 3 1957-1966 Gill $275-450

ATLANTIC
IMPERIAL 5 1964-1966 Capco $250-425
White globe face with large red/gold/blue Imperial shield on globe face. Imperial shield has gold mirrored outline.

ATLANTIC
IMPERIAL
W/O MIRROR 5 1964-1966 Capco $200-400
Same as above with plain yellow-gold outline around shield.

ATLANTIC
NEW LOGO 4 1966-1970 Gill $250-375
White globe face with red/white/blue Atlantic hexagon logo on globe face.

(ATLANTIC)

DIESEL FUEL 4 1936-1970 Gill $200-325
White globe face with bright blue lettering "DIESEL/FUEL" center of globe face. No company identification.

(ATLANTIC)
KEROSENE 5 1940-1970 Gill $225-350
White globe face with blue "KEROSENE" across center. Double red bars at upper left and lower right of "KEROSENE". No company identification.

****RED "C" OIL COMPANY—BALTIMORE, MARYLAND:** Formed in 1870 as a kerosene refiner and marketer, the company was continued as a marketing firm purchased by Atlantic Refining in the mid-1920s.

RED "C" 5 1920s 15" Metal $300-450
White globe face with blue outline ring set in from the edge of the lens. Large blue outlined red "C" in center of globe face.

CHAMPLIN & AFFILATES

****CHAMPLIN OIL AND REFINING — ENID, OKLAHOMA:** Champlin Oil and Refining was an Enid, Oklahoma, based refiner/marketer founded in Mr. H. H. Champlin in 1916. The first Champlin service station was opened in 1923 at 221 East Broadway in Enid, Oklahoma. Expansion in the 1920s and 1930s was solid. in 1938 the company acquired Eason Oil, which they would operate as a secondary brand into the 1950s. The company remained an independent until 1954, when the company was merged with a petroleum production conglomerate known as "The Chicago Corporation." Two years later The Chicago Corporation was renamed Champlin Oil and Refining Company. In 1969 Champlin became a unit of the Union Pacific Railroad. Marketing was concentrated throughout the mid-continent, eventually expanding to 19 states from Tennessee to Colorado. In 1981 operations shifted from Enid, Oklahoma, to Fort Worth, Texas, and in 1984 retail marketing was sold off to American Petrofina. The Fina brand replaced Champlin at some 170 company stations as well as being offered to Champlin jobbers. The production and refining divisions survive as an Irving, Texas, based subsidiary of Citgo Petroleum which is owned by Petros de Venezuela.

CHAMPLIN
GASOLINE 4 1923-1930 OPE N $1200-1800
White globe with blue "CHAMPLIN" arched around top and "GASOLINE" around bottom with red "Iron Cross" logo in center with lettering "USE/CHAMPLIN/OILS".

CHAMPLIN
GASOLINE 5 1923-1930 OPE RW $1200-1800
White globe with blue "CHAMPLIN" arched around top and "GASOLINE" around bottom with red "Iron Cross" logo in center with lettering "USE/CHAMPLIN/OILS".

CHAMPLIN
GASOLINE
INVERTED 6 1923-1930 OPE N NL
Same as above with entire design inverted. Used as a light fixture globe.

CHAMPLIN
PRESTO 6 1927-1935 OPE N NL
White globe with blue "CHAMPLIN" arched around top and "GASOLINE" around bottom with red "PREST-O-/ANTI-KNOCK" in center.

CHAMPLIN
GASOLINE 5 1925-1935 15" Metal Red $450-850

CHAMPLIN
GASOLINE 3 1930-1936 13.5" Glass W $275-450

CHAMPLIN
GASOLINE 4 1930-1936 Banded Glass $300-475
White globe face with blue outer ring and blue "CHAMPLIN" arched at top, "GASOLINE" at bottom with red Champlin "Iron Cross" logo with notched ends on blue center crossbar. White "CHAMPLIN" on center crossbar with white "USE" above and "OILS" below.

CHAMPLIN
ETHYL (EGC) 3 1930-1946 13.5" Glass W $275-450

CHAMPLIN
ETHYL (EGC) 3 1930-1946 Banded Glass $300-475
White globe face with blue "CHAMPLIN" arched at top and "GASOLINE" at bottom with Ethyl (EGC) logo in center.

CHAMPLIN
PRESTO 4 1930-1956 13.5" Glass W $275-450

CHAMPLIN
PRESTO 4 1930-1956 Banded Glass $300-475
White globe face with blue outer ring and blue outlined white bowtie shaped center band. Blue "CHAMPLIN" arched at top, "GASOLINE" at bottom with red "PRESTO" in bowtie lettering in center band.

CHAMPLIN 1 1956-1962 Capco $175-275
Blue outlined white globe face with large Champlin "Iron Cross" logo, blue crossbar on cross has rounded instead of notched ends. White lettering "CHAMPLIN" on crossbar, no "use" or "oils" on cross as on earlier version of this logo.

CHAMPLIN
OBLONG LOGO 3 1962-1970 Capco $150-250
White globe face with blue/red outline oblong 1960s Champlin logo with blue "CHAMPLIN" across center.

CHAMPLIN
OBLONG LOGO
Plastic Lens 2 1962-1970 Capco $175-275
Plastic lens. White globe face with blue/red outline oblong 1960s Champlin logo with blue "CHAMPLIN" across center.

CHAMPLIN
OBLONG ALL
BLUE LOGO 5 1962-1970 Capco $150-250
White globe face with blue outline oblong 1960s Champlin logo with blue "CHAMPLIN" across center.

****HARBOR — TULSA, OKLAHOMA:** Harbor is a Tulsa based Champlin secondary brand from the 1960s.**

HARBOR 4 1960s Capco $250-400
White globe face with nautical flag covering most of globe face. Flag has three horizontal stripes red/white/black with black "HARBOR" on center white stripe.

HARBOR
Plastic Lens 4 1960s Capco $250-400
Plastic lens. White globe face with nautical flag covering most of globe face. Flag has three horizontal stripes red/white/black with black "HARBOR" on center white stripe.

****EASON OIL CO.—ENID, OKLAHOMA:** Eason Oil Co. was an Oklahoma based refiner/marketer, one of the early participants in the Sovereign Service marketing program. By the early 1950s, they were based in Enid, Oklahoma, and eventually merged with Champlin in the mid-1950s.

EASON
GASOLINE 5 1930s-40s 15" Metal $650-900
Green globe face with white/green outline ring around white center circle. Large black outlined red "E" in center circle. Black outlined white "EASON" arched in green outer ring at top, "GASOLINE" at bottom.

EASON ETHYL 5 1930s-40s 15" Metal $650-850
Red globe face with white/red outline ring around white center circle. Large Ethyl (EGC) logo in center circle. Black outlined white "EASON" arched in red outer ring at top, "GASOLINE" at bottom.

EASON
GASOLINE 5 1940s Gill $450-750
Green globe face with white/green outline ring around white center circle. Large black outlined red "E" in center circle. Black outlined white "EASON" arched in green outer ring at top, "GASOLINE" at bottom.

EASON
DICTATOR 6 1940s Gill $450-750
Green outlined white globe face with green Eason logo with red "E" at top center of globe face. Three red lines across below logo above green "DICTATOR" over smaller "UNLIMINTED POWER". Red "Gasoline" below with three green lines across lower globe face.

EASON
DICTATOR 6 1946-59 Green Capco $350-600
Green outlined white globe face with green Eason logo with red "E" at top center of globe face. Three red lines across below logo above green "DICTATOR" over smaller "UNLIMINTED POWER". Red "Gasoline" below with three green lines across lower globe face.

****GOLDEN RULE OIL COMPANY—KANSAS CITY, MISSOURI:** This firm marketed under the Blue Ribbon brand name until the company was purchased by Champlin in 1935.

BLUE RIBBON
GASOLINE 6 1930s-1940s 15" Metal NL
White globe face with blue ribbon vertical in center of globe face with gold fringe at top and white lettering "BLUE" above vertical "RIBBON" on ribbon. Blue "BLUE RIBBON" arched around top, red "GASOLINE" around bottom.

CHEVRON & AFFILIATES
—STANDARD OIL CO. OF

CALIFORNIA

****STANDARD OIL COMPANY OF CALIFORNIA**: Standard Oil of California, known in the industry as "Socal," is the west's largest gasoline marketer. Tracing its origins to the 1879 founding of the Pacific Coast Oil Company, Standard Oil—which had maintained a San Francisco marketing office since 1879—purchased the Pacific Coast Oil Co. in 1900 and reorganized it as Standard Oil of California.

Standard of California is credited with running one of the first retail gasoline stations, opened at the company's bulk plant in Seattle, Washington, in 1907. Always an innovator, Standard was soon operating hundreds of retail stations throughout the west. By the mid-1930s, company stations were under the brand "Standard Stations Inc." and dealer operated locations were "Standard Oil Dealers." In 1945 the Chevron name was introduced for Standard gasolines. Independent dealers in the Standard western territory, Standard Texas southern territory and Calso (CALifornia Standard Oil) territory became "Chevron Gas Stations." Company owned and operated stations remained "Standard" stations, selling Chevron and Chevron Supreme gasolines. In 1961 Standard of California bought Standard of Kentucky and from 1961 on Chevron gasolines were sold at Standard Oil stations in the southeast. About 1977 the Standard name came down from all but a very small number of service stations and was replaced by the Chevron name. Standard of Kentucky globes can be found under the listing for "Standard Oil."

****STANDARD STATIONS, INC. and STANDARD OIL/CHEVRON DEALERS**:

RED CROWN
GASOLINE
pat. 1915 4 1918-1930 15" Metal HP Red $1200-1800
White globe face with blue outline ring. Blue center circle with white burst covering most of circle. Red California style crown on burst in center. Blue outlined red lettering "RED CROWN" arched around top, "GASOLINE" around bottom.

RED CROWN
ETHYL on
"chevron" 5 1926-1930 15" Metal HP Red $750-1250
Red globe face with blue/white/red outlined chevron shaped shield covering most of globe face. Small blue "RED CROWN" at top in shield above larger "ETHYL" above small "GASOLINE". Small Ethyl (EGC) logo with blue rays in circle at bottom.

STANDARD
GASOLINE
(W/CROWN) 3 1930-1946 15" Metal HP Red $900-1400
Blue globe face with red outline ring. Red outline around center white circle. Small red California style crown on center circle. White "STANDARD" arched around top, "GASOLINE" around bottom.

STANDARD
ETHYL (EGC) 4 1930-1940 15" Metal HP Red $750-1250
Red globe face with blue/white/red outlined chevron shaped shield covering most of globe face. Small blue "STANDARD" at top in shield above larger "ETHYL" above small "GASOLINE". Small Ethyl (EGC) logo with blue rays in circle at bottom.

STANDARD FLIGHT
(w/Blue Arrow) 3 1934-1936 15" Metal HP Red $650-950
Blue outline ring around white outer ring around dark blue center circle. White lettering "Flight" at angle across upper area of center circle with white outlined light blue "arrow" with speed lines passing left-to-right across globe face. White "GASOLINE" arched around bottom of center circle.

STANDARD FLIGHT
(W/Red Arrow) 3 1936-1946 15" Metal HP Red $650-900
Blue outline ring around white outer ring around dark blue center circle. White lettering "Flight" at angle across upper area of center circle with red "arrow" with speed lines passing left-to-right across globe face. White "GASOLINE" arched around bottom of center circle.

STANDARD' S
SUPREME
WITH ETHYL 4 1940-1946 15" Metal HP Red $700-1000
White globe face with geartooth edge and red outline area. Blue "STANDARD'S" arched around top over small Ethyl (EGC) logo in circle. Blue outlined red italic "SUPREME" just under Ethyl logo with small blue "(WITH ETHYL)" below "SUPREME". Blue "GASOLINE" arched around bottom.

CHEVRON
GASOLINE 4 1946-1957 15" Metal HP Red $900-1600
Red ring around blue globe face. White band across blue area above center. Red "CHEVRON" with blue dropshade and winged "V" on white band, extending into outer ring, red "GASOLINE" on blue area just below white band.

CHEVRON
SUPREME 3 1946-1957 15" Metal HP Red $900-1600
Red ring around white globe face. Red "CHEVRON" with blue dropshade and winged "V" just above center, extending into outer ring, above blue "SUPREME". Small red italic "GASOLINE" below Supreme.

CHEVRON
GASOLINE 4 1950-1957 15" Metal HP Red $850-1500
Red ring around blue globe face. White band across blue area above center. Red "CHEVRON" with NO dropshade and winged "V" on white band, NOT extending into outer ring, red "GASOLINE" on blue area just below white band.

CHEVRON
SUPREME 4 1950-1957 15" Metal HP Red $850-1500
Red ring around blue globe face. White band across blue area above center. Red "CHEVRON" with NO dropshade and winged "V" on white band, NOT extending into outer ring, red "GASOLINE" on blue area just below white band.

****THE CALIFORNIA COMPANY — ROCKY MTN. MARKETING**: In the late 1930s, Standard of California expanded its marketing into the Rocky Mountain area. Unable to use the Standard name in this area due to its use by Standard of Indiana, gasoline was marketed under the "Calso" (CALifornia Standard Oil) name. In 1946 the Calso stations, all dealer/jobber operated, were rebranded with the new "Chevron Gas Station" logo. Chevron globes, listed above, were used at all Standard Stations in the west and Chevron stations in the Rocky Mountain area.

CALSO W/GREEN
& BLUE LINES 3 1935-1938 15" Metal HP Blue $600-850
White globe face with blue outline ring. Blue "CALSO" across top of globe face above smaller blue "GASOLINE". Pale green, white, and blue lines horizontal below lettering.

CALSO W/GREEN &
BLUE LINES
UNSURPASSED 3 1938-1946 15" Metal HP Blue $650-900
White globe face with blue outline ring. Blue "CALSO" across top of globe face above smaller blue "GASOLINE". Pale green, white, and blue lines horizontal below lettering. Small blue "UNSURPASSED" below lines.

CALSO ETHYL
(on chevron) 5 1935-1938 15" Metal HP Blue $650-900
Green globe face with blue/white/green outlined chevron shaped shield covering most of globe face. Small blue "CALSO" at top in shield above larger "ETHYL" above small "GASOLINE". Small Ethyl (EGC) logo with blue rays in circle at bottom.

CALSO'S SUPREME
w/ ETHYL (EGC) 3 1938-1946 15" Metal HP Blue $750-1250
White globe face with geartooth edge and green outline area. Blue "CALSO'S" arched around top over small Ethyl (EGC) logo in circle. Blue outlined green italic "SUPREME" just under Ethyl logo with small blue "(WITH ETHYL)" below "SUPREME". Blue "GASOLINE" arched around bottom.

(CALSO) FLIGHT
(w/Green Arrow) 4 1938-1946 15" Metal HP Blue $650-900
Blue outline ring around white outer ring around dark blue center circle. White lettering "Flight" at angle across upper area of center

circle with green "arrow" with speed lines passing left-to-right across globe face. White "GASOLINE" arched around bottom of center circle.

**CALIFORNIA OIL COMPANY — NORTHEAST MARKETING: In 1945-46 Standard of California purchased the terminal facilities of Monogram Oil Co. and set up "The California Oil Company" in Perth Amboy, New Jersey as an East Coast marketing subsidiary. As in the Rockies, stations were branded Calso and were found from Maine to Virginia. In 1959, Calso Stations were rebranded Chevron. Chevron stations could be found in small numbers throughout the northeastern United States until Chevron purchased Gulf in 1984 and sold off both the Gulf and Chevron marketing in that area in the mid-1980s.

CALSO
GASOLINE 4 1946-1949 13.5" Glass W $300-500
Green globe face with white horizontal band above center. Blue "CALSO" on white band with "L" extending into upper green area. Blue lettering "GASOLINE/PRODUCT OF/THE CALIFORNIA OIL COMPANY" on lower green area.

CALSO
SUPREME 4 1946-1949 13.5" Glass W $300-500
Green globe face with white horizontal band below center. Blue "CALSO" in upper green area with red "SUPREME" with blue dropshade on white band. Blue "A CALIFORNIA/GASOLINE" on lower green area.

CALSO
RED/WHITE 3 1949-1959 Gill $300-500

CALSO
RED WHITE 3 1949-1959 Capco $250-350
Red globe face with white outer ring. Large white "CALSO" with extended "L" covering most of globe face. White "GASOLINE" below "CALSO". Small black "THE CALIFORNIA OIL COMPANY" below "GASOLINE".

CALSO
SUPREME 3 1949-1959 Gill $300-500

CALSO
SUPREME 3 1949-1959 Capco $250-350
Red globe face with white outer ring. Large white "CALSO" with extended "L" covering most of globe face. White "SUPREME" above black "GASOLINE" below "CALSO". Small black "THE CALIFORNIA OIL COMPANY" below "GASOLINE".

CALSO RED Black 5 1949-1959 Gill $300-500
CALSO RED Black 3 1949-1959 Capco $250-350
Red globe face with white outer ring. Large white "CALSO" with extended "L" covering most of globe face. Black "GASOLINE" below "CALSO". Small black "THE CALIFORNIA OIL COMPANY" below "GASOLINE".

CHEVRON
GASOLINE 4 1959-1969 Gill $350-550

CHEVRON
GASOLINE 4 1959-1969 Capco $300-450
White globe face with large red "CHEVRON" with winged "V" above center. Italic blue "GASOLINE" above very small "CALIFORNIA OIL COMPANY" below.

CHEVRON
SUPREME 5 1959-1969 Gill $375-600

CHEVRON
SUPREME 5 1959-1969 Capco $300-450
White globe face with large red "CHEVRON" with winged "V" above center. Italic blue "SUPREME GASOLINE" above very small "CALIFORNIA OIL COMPANY" below.

**STANAVO: Formed in 1929 by Standard of Indiana, New Jersey, and California as a joint venture to research aviation petroleum needs and market those products through their individual marketing outlets. Stanavo first marketed aviation gasoline in 1931. With the withdrawl of Standard of New Jersey in 1937, the two other participants decided to discontinue the effort and Stanavo was dissolved in June 1938.

STANAVO 5 1931-1938 16.5" Metal NL
White globe face with outlined bird-like airplane image across center of globe face. Blue outlined red "STANAVO" across top with "AVIATION GASOLINE" to lower right below airplane.

**PDQ — CHEVRON SECONDARY BRAND:
PDQ w/ pelican 6 1940s 15" Metal NL (**No**Listing)
Description not available.

**SIGNAL: Stations operated as part of Signal Oil from 1932 until 1947 and as part of Standard Oil Company of California from 1947 until 1966. In 1966, the company was sold to Standard Oil of New Jersey and rebranded Enco.

SIGNAL
PURR-PULL 6 1932-1933 15" Metal NL
Yellow painted globe face with purple lettering "SIGNAL" with the "S" and "L" extending below the baseline of the other lettering. Purple band between the bases of the extended letters with yellow "PURR PULL" on band.

GO SIGNAL 4 1933-1948 15" Metal $2500-3800
Black globe face with orange outline ring. Large white stoplight with red/black/red signals and large flag on globe face. Black "GO" on stoplight flag. Large yellow "SIGNAL" below stoplight across bottom of globe face.

GO SIGNAL 4 1933-1948 15" Metal $2500-3800
Black globe face with orange outline ring. Large white stoplight with red/black/red signals and large flag on globe face. Black "GO" on stoplight flag. Small yellow "SIGNAL" below stoplight across bottom of globe face.

SIGNAL
PRODUCTS 5 1933-1948 15" Metal $4000-7500
Yellow globe face with orange outline ring and orange ring around black center circle. Stoplight, same as above without post, on center circle. Stoplight has side flag with black "GO" lettering. White outlined black "SIGNAL" arched around top, "PRODUCTS" around bottom.

CITIES SERVICE & AFFILIATES

**CITIES SERVICE OIL COMPANY — TULSA, OKLAHOMA: Cities Service was founded in 1910 as a public utility, operating municipal natural gas, lighting, ice, and other utility type services, hence the name "Cities Service." Venturing into petroleum marketing about 1914, Cities Service was a scattered collection of refining and marketing companies operating gas stations throughout the East and Midwest. By the mid-1930s, the various concerns were all marketing under the Cities Service brand name. In 1946, the company's black and white station colors gave way to green and white stations. In the 1950s many elaborate service stations were built but Cities Service continued to have an image problem. In 1965 the time-tested Cities Service brand name was replaced by Citgo and the entire marketing program was revitalized. Many new stations were added, only to close in record numbers during the 1970s gas shortages. In 1981 Southland Corporation, operator of the 7-Eleven convenience stores, and the government owned oil company in Venezuela purchased the marketing division, Citgo Petroleum, while Occidental Petroleum purchased the remaining divisions of Cities Service Oil Company. In the late 1980s, Southland, in bankruptcy, sold their share of Citgo to Venezuela and the company, with the assured supply of a government owned oil company, again began aggressively marketing. Citgo now supplies thousands of stations in 46 states and new locations are added every day.
>>NOTE: No "CITGO" globes known to exist.

CITIES SERVICE OILS
ONCE ALWAYS 6 1914-1918 OPE Round small $2500-3800
Globe with black "CITIES SERVICE OILS" arched around top. Black and white CS clover and triangle logo in center. Black outlined pointed rectangle box below logo with black "ONCE-ALWAYS" in box.

HIGH TEST/
KOOLMOTOR/
ANTI-KNOCK 5 1914-1918 OPE Round small $2500-3800
Green globe face with white outline ring. Green/white outlined white bar and circle band across center. Black "KOOLMOTOR" on white center band. Black "HIGH TEST ANTI-KNOCK" arched around top, "GASOLENE" around bottom.

CITIES SERVICE OILS
ONCE ALWAYS 3 1918-1936 16.5" Metal HP/LP Black or White $400-650

CITIES SERVICE OILS
ONCE ALWAYS 3 1918-1936 15" Metal HP/LP Black or White $400-650

CITIES SERVICE OILS

ONCE ALWAYS 5 1918-1936 11.5" Metal HP/LP Black or White $850-1300

White globe face with black "CITIES SERVICE OILS" arched around top. Black and white CS clover and triangle logo in center. Black outlined pointed rectangular box below logo with black "ONCE-ALWAYS" in box.

CITIES SERVICE OILS
ONCE ALWAYS 6 1918-1936 15" Metal H/LP Black/ White NL

White globe face with thin black outline and black "CITIES SERVICE OILS" arched around top. Black and white CS clover and triangle logo in center. Black outlined pointed rectangular box below logo with black "ONCE-ALWAYS" in box.

CITIES SERVICE OILS
AVIATION
(overpaint) 5 1930s 15" Metal H/LP Black/ White $700-1100

White globe face with black "CITIES SERVICE OILS" arched around top. Black and white CS clover and triangle logo in center. Black outlined pointed rectangular box below logo with black "ONCE-ALWAYS" in box. Red "AVIATION" painted on outside across lower globe face.

CITIES SERVICE OIL CO.
KOOLMOTOR 4 1920s 15" Metal H/LP Black/ White $800-1250

White globe face with black "CITIES SERVICE OIL CO." arched around top, "GASOLINE" around bottom and red "KOOLMOTOR" across center.

HITEST ANTI KNOCK/
KOOLMOTOR/
GASOLENE 4 1920-1925 15" Metal or 16.5" metal L/HP Black
$650-1000

Green globe face with white outline ring. Green/white outlined white bar and circle band across center. Black "KOOLMOTOR" on white center band. Black "HIGH TEST ANTI-KNOCK" arched around top, "GASOLENE" around bottom.

HIGH TEST/
KOOLMOTOR/
ANTI-KNOCK 4 1925-1934 15" Metal L/HP Black $750-1250

Green globe face with black outline ring. Green/white outlined white bar and circle band across center. Black "KOOLMOTOR" on white center band. White "HIGH-TEST" arched around top, "ANTI-KNOCK" around bottom.

HIGH TEST/
KOOLMOTOR/
ANTI-KNOCK 4 1926-1934 OP Clover $2200-3000

White globe with black outlined green globe face. Green/white outlined white bar and circle band across center. Black "KOOLMOTOR" on white center band. White "HIGH-TEST" arched around top, "ANTI-KNOCK" around bottom.

CITIES SERVICE OILS
ONCE ALWAYS 5 1926-1934 OP Clover $2300-3200

White globe with black clover outline ring. Black "CITIES SERVICE OILS" arched around top, black and white clover and triangle in center, and black outlined pointed rectangular box below logo. Black "ONCE ALWAYS" in box.

CITIES SERVICE OILS
(ETHYL DECAL) 5 1932-1934 OP Clover $2300-3300

White globe with black clover outline ring. Black "CITIES SERVICE OILS" arched around top, black and white clover and triangle in center, and black outlined pointed rectangular box below logo. Black "ONCE ALWAYS" in box. Large Ethyl (EGC) logo on decal covering clover logo in center.

CITIES SERVICE OILS
ONCE ALWAYS 4 1934-1936 13.5" Glass N $400-750

White globe face with black clover outline ring. Black "CITIES SERVICE OILS" arched around top, black and white clover and triangle logo in center, and black outlined pointed rectangular box below logo. Black "ONCE-ALWAYS" in box.

CITIES SERVICE OILS
KOOLMOTOR 4 1934-1936 13.5" Glass N $450-850

White globe face with black clover outline ring. Black "CITIES SERVICE OILS" arched around top, black and white clover and triangle logo in center, and black outlined green "KOOLMOTOR" below logo.

CITIES SERVICE OILS
ETHYL 5 1934-1936 13.5" Glass N $450-850

White globe face with black clover outline ring. Black "CITIES SERVICE OILS" arched around top, black and white clover and triangle logo in center, and black outlined orange "ETHYL" below logo.

HIGH TEST/
KOOLMOTOR/
ANTI-KNOCK 4 1932-1934 Glass Clover $1250-1750

Green globe face with black clover outline ring. Green/white outlined white bar and circle band across center of globe face. Black "KOOLMOTOR" across white center band. White "HIGH-TEST" arched around top, "ANTI-KNOCK" around bottom.

CITIES SERVICE OILS
ONCE ALWAYS 4 1932-1936 Glass Clover $900-1400

White globe face with black clover outline ring. Black "CITIES SERVICE OILS" arched around top, black and white clover and triangle logo in center, and black outlined pointed rectangular box below logo. Black "ONCE-ALWAYS" in box.

CITIES SERVICE (OILS)
[LOGO]
KOOLMOTOR 4 1934-1936 Glass Clover $1000-1600

White globe face with black clover outline ring. Black and white clover and triangle logo in center. Black "CITIES SERVICE" arched around top, black outlined yellow "KOOLMOTOR" arched around bottom.

CITIES SERVICE (OILS)
[LOGO] ETHYL 5 1932-1936 Glass Clover $950-1500

White globe face with black clover outline ring. Black and white clover and triangle logo in center. Black "CITIES SERVICE" arched around top, black outlined red "ETHYL" straight across bottom.

CITIES SERVICE 4 1936-1947 Glass Clover $850-1350
CITIES SERVICE 5 1936-1947 14" Hull Clover $1000-1600

White globe face with black clover shaped outline ring. Small black CS clover and triangle at top above black "CITIES/SERVICE".

CITIES SERVICE
KOOLMOTOR 4 1936-1947 Glass Clover $900-1500
CITIES SERVICE
KOOLMOTOR 5 1936-1947 14" Hull Clover $1000-1700

White globe face with black clover shaped outline ring. Small black CS clover and triangle at top above black "CITIES/SERVICE", black outlined yellow "KOOLMOTOR" below "CITIES SERVICE".

CITIES SERVICE
ETHYL (EGC) 5 1936-1947 Glass Clover $950-1600
CITIES SERVICE
ETHYL (EGC) 5 1936-1947 14" Hull Clover $1100-1750

White globe face with black clover shaped outline ring. Small black CS clover and triangle at top above black "CITIES/SERVICE", with small Ethyl (EGC) logo below "CITIES SERVICE". Black outlined red "ETHYL" superimposed across logo.

CITIES SERVICE 6 1936-1947 15" Metal $600-950

White globe face with black clover shaped outline ring. Small black CS clover and triangle at top above black "CITIES/SERVICE".

CITIES SERVICE
KOOLMOTOR 5 1936-1947 15" Metal $750-1100

White globe face with black clover shaped outline ring. Small black CS clover and triangle at top above black "CITIES/SERVICE", black outlined yellow "KOOLMOTOR" below "CITIES SERVICE".

CITIES SERVICE
ETHYL (EGC) 6 1936-1947 15" Metal NL

White globe face with black clover shaped outline ring. Small black CS clover and triangle at top above black "CITIES/SERVICE", with small Ethyl (EGC) logo below "CITIES SERVICE". Black outlined red "ETHYL" superimposed across logo.

CITIES SERVICE
REGULAR 5 1947-1953 15" Metal $650-1000

White globe face with green clover outline around edge. Green "CITIES SERVICE" over smaller "REGULAR" inside clover outline.

CITIES SERVICE
PREMIUM 5 1947-1953 15" Metal $650-1000

White globe face with orange clover outline around edge. Orange "CITIES/SERVICE" over smaller "PREMIUM" inside clover outline.

CITIES SERVICE
REGULAR 6 1947-1953 Green Capco NL

White globe face with green clover outline around edge. Green "CITIES SERVICE" over smaller "REGULAR" inside clover outline.

CITIES SERVICE
PREMIUM 6 1947-1953 Red Capco NL

White globe face with orange clover outline around edge. Orange "CITIES/SERVICE" over smaller "PREMIUM" inside clover outline.

CITIES SERVICE (green) 5 1953-1960 13.5" Glass W $250-425

CITIES SERVICE (green) 2 1953-1960 Capco $200-300

White globe face with green clover outline around edge. Large green "CITIES/SERVICE" in center.

CITIES SERVICE (red or orange) 5 1953-1960 13.5" Glass W $250-325

CITIES SERVICE (red or orange) 4 1953-1960 Capco $200-300

White globe face with red or orange clover outline around edge. Large red or orange "CITIES/SERVICE" in center. Both colors, red or orange, also reddish orange exist.

CITIES SERVICE (yellow) 5 1953-1960 13.5" Glass W $275-450

CITIES SERVICE (yellow) 4 1953-1960 Capco $250-350

White globe face with yellow clover outline around edge. Large yellow "CITIES/SERVICE" in center.

CITIES SERVICE CLOVER & TRIANGLE (window) 4 1932-1960 15" Metal $250-400

White globe face with thick green clover outline with triangle in center. This was used as a light-up sign. A metal globe band was framed into a service station building to form a window and these lenses were installed in the frame.

CITIES SERVICE CLOVER & TRIANGLE (window) 4 1932-1960 15" Metal $250-400 Per Lens

Red globe face with thick white clover outline with triangle in center. This was used as a light-up sign. A metal globe band was framed into a service station building to form a window and these lenses were installed in the frame.

CITIES SERVICE CLOVER & TRIANGLE (window) 4 1932-1960 15" Metal $250-400 Per Lens

Flat lens. Red globe face with thick white clover outline with triangle in center. This was used as a light-up sign. A metal globe band was framed into a service station building to form a window and these lenses were installed in the frame.

****CITGO:** The Citgo brand name was introduced in 1965 to replace the Cities Service brand. The Citgo brand remains in use at over 14,000 service stations throughout the United States today.
>>NO Citgo globes are known at this time.

****LINCOLN OIL CO.—FLINT, MICHIGAN:** Lincoln was a southern Michigan marketer, operating more than 50 stations when they were merged into Cities Service in 1930.

LINCOLN OILS 6 1920-1930 15" Metal $800-1250

Globe face is divided into three even bands, red on top, white in the middle and blue on the bottom. Large blue LOCO logo on center band with diagonal red arrow thru logo. White "LINCOLN" on red band, "OILS" on blue.

****ELWOOD OIL CO.—ELWOOD, INDIANA:** Indiana jobber brand converted to Cities Service in 1931.

RED PARROT GASOLINE W/RED PARROT 7 1923-1931 15" Metal NL

White globe face with "RED PARROT" arched around top, "GASOLINE" around bottom and detailed red parrot in center. Undated repros exist!

****CARTER HALL COMPANY — GLENS FALLS, NEW YORK:** New York distributor of Crew Levick products that was purchased by Cities Service in 1929.

CITIES SERVICE "CARTER HALL CO" 6 1932-1936 15" Metal NL

White globe face with black "CITIES SERVICE OILS" arched around top. Black and white CS clover and triangle logo in center. Black "CARTER HALL CO" across bottom of globe face.

****LOUISIANA OIL REFINING CO.—SHREVEPORT, LOUISIANA:** Founded in 1917, Louisiana Oil Refining was originally a Gulf Coast refiner/marketer. Loreco became a part of Cities Service in 1932.

LORECO GASOLINE/ MOTOR OIL 5 15" Metal $650-900

White globe face with white/red outline rings. Black "GASOLINE" arched around top, "MOTOR OIL" around bottom and red/white outlined red diamond in center with white "LORECO" on diamond.

LORECO ETHYL 3 15" Metal $400-750

White globe face with black/white outline rings. Black outline around white center circle. Red "LORECO" arched around top, "GAS" around bottom with large Ethyl (EGC) logo in center with blue rays.

LOREX MOTOR FUEL 4 15" Metal $600-950

White globe face with dark blue "LOREX" with light blue dropshade in decreasing height lettering across globe face. Red italic "Motor Fuel" below "LOREX".

****CREW-LEVICK COMPANY — PHILADELPHIA, PENNSYLVANIA:** Founded in Philadelphia in 1862, they became part of CS about 1914 and were the originators of the Cities Service "expanding circle" (commonly known as the cloverleaf) logo. They remained a Cities Service subsidiary for many years, until the Cities Service companies were reorganized in the postwar era.

CREW-LEVICK 6 1910s OPE Oval Chimney Capped NL

White oval chimney cast globe with black diamond design in center of globe face. White "CREW LEVICK" across diamond with red details.

CITIES SERVICE "CREW-LEVICK" 4 1920s 15" Metal Cast Face $1250-1800

Cast face with white globe face with black "CITIES SERVICE OILS" arched around top. Black and white CS clover and triangle logo in center. Black "CREW-LEVICK" around bottom.

CITIES SERVICE "CREW-LEVICK" 5 1932-1936 15" Metal $750-1100

White globe face with black "CITIES SERVICE OILS" arched around top. Black and white CS clover and triangle logo in center. Black "CREW-LEVICK" around bottom.

****WARNER-QUINLAN COMPANY—NEW YORK, NEW YORK:** Founded in New York City in 1903, Warner-Quinlan was marketing throughout New England when purchased by Cities Service in 1933.

WARNOCO 6 1920-1926 OPE NL

White globe with red "WARNOCO" arched around top of globe face. Small red "W" on lower globe face with blue detailed wings to either side.

SUPER X MOTOR FUEL 5 1926-1930 OPB N $900-1500

White globe face with red outline ring. Large black/white outlined red "X" on black diamond in center. Black/white outlined red "SUPER" arched around top, "MOTOR FUEL around bottom with black "REG.U.S." and "PAT.OFF." to either side of "X".

SUPER X HIGHTEST 6 1926-1930 OPB $900-1500

Yellow globe face with red outline ring. Large red "X" in center with smaller "HIGH" and "TEST" to either side. Red "SUPER" arched around top, "GASOLINE" around bottom.

MILEAGE W/OWL 4 1927-1930 15" Metal $2500-3750

White globe face with blue outline rings. Blue outline around white center circle. Red and blue perched owl and blue triangle design in center circle with banner across owl. "FOR WISE DRIVERS" on banner. Blue "MILEAGE" arched around top, "GASOLINE" around bottom.

MILEAGE PLUS GASOLINE 4 1927-1930 15" Metal $550-900

White globe face with outlined white center circle. "MILEAGE" arched around top, "GASOLINE" around bottom with large "PLUS" in center.

****SHOEMAKER COAL CO.—AUBURN, NEW YORK:** Purchased by Cities Service about 1925. Little is known about this company.

BLUE GRASS GASOLINE
"SHOEMAKER
COAL" CAST 6 1920s 15" Metal NL

White globe face with cast lettering. Blue "BLUEGRASS" arched around top, "GASOLINE" around bottom with reverse bowtie "SHOEMAKER/COALCORP" in center. Small blue "AUBURN, N.Y." at bottom.

CITIES SERVICE
"SHOEMAKER
COAL CO." 6 1935-1936 15" Metal NL

White globe face with black "CITIES SERVICE OILS" arched around top. Black and white CS clover and triangle logo in center. Black "SHOEMAKER COAL CORP" arched around bottom.

****A.R. NEWCOMBE OIL CO.—KINGSTON, NEW YORK:** New York based marketer purchased by Cities Service in 1930.

KEYSTONE
GASOLINE 5 1920s 12" Metal NL

Red outlined dark blue globe face with white/red outline around white center circle. Smaller red circle inside white center circle. White "KEYSTONE" arched around top, "GASOLINE" around bottom with small red "MOBIL-/OILS" at left in blue area, "VELVET/OILS" at right. Blue "A.R.NEWCOMBE" arched above red circle in white center circle, "OIL CORPORATION" below. White keystone in red center circle with blue "PENNA./PRODUCTS" and red star on keystone.

KEYSTONE
GASOLINE 6 1920s 15" Metal $650-900

White/red outlined blue globe face with white outline around red center circle. Large white keystone in center circle with white "KEYSTONE" arched around top, "GASOLINE" around bottom with small white star to either side.

****JENNEY MANUFACTURING CO. — BOSTON, MASSACHUSETTS:** Founded in Boston in 1812, Jenney Manufacturing was originally a mid-nineteenth century dealer in whale oils, coal, kerosene and related products. Jenney entered the New England gasoline market in the early years of the twentieth century. By 1920 Jenney had a network of elaborate service stations in Boston and the surrounding area. With close ties to Cities Service for many years, Jenney was merged into Cities Service about 1965 and the Citgo brand replaced Jenney.

JENNEY SERVICE
STATION 7 1910s OPE NL

White globe with "JENNEY/SERVICE/STATION" across globe face.

JENNEY AUTO
OILS/GASOLINE 6 1920s 15" Metal $3500-5000

White globe face with black outlined orange circle in center. Orange and black industrial scene in center circle with "BOSTON" near bottom of scene. White band diagonal across upper area of center circle. Black outlined orange "AUTO OILS" arched around top, "GASOLINE" around bottom with script black "Jenney" on diagonal band. White "MANUFACTURING CO" on "Jenney" underline.

JENNEY ETHYL
GASOLINE (EGC) 5 1920s 15" Metal $900-1400

Yellow outlined black globe face with yellow outline around white center circle. Yellow "JENNEY" arched around top, "GASOLINE" around bottom with large Ethyl (EGC) logo with blue rays in center circle.

JENNEY AERO
GASOLINE 5 1935-1940 15" Metal $3500-5500

Orange outline around globe face. Blue band across center covering most of globe face. Narrow black areas above and below. Black and orange airplane on blue center area with white "AERO" across airplane. Orange "JENNEY" in upper black area, "GASOLINE" below.

CUT PRICE GAS 5 1930-1940 15" Metal $300-500

White globe face with blue "CUT/PRICE/GAS" across center of globe face.

SOLVENIZED
JENNEY
HY POWER 5 1935-1940 Gill $2800-3800

White globe face with black outlined orange circle in center. Orange and black industrial scene in center circle with "BOSTON" near bottom of scene. White band diagonal across upper area of center circle. Black outlined orange "SOLVENIZED" arched around top, "HY-POWER" around bottom with script black "Jenney" on diagonal band. White "MANUFACTURING CO." on "Jenney" underline.

JENNY AERO
SOLVENIZED 4 1935-1940 Gill $3500-4500

White globe face with wide orange and narrow black outline rings. Black lines offset white center band. Orange and black airplane in center with black outlined blue "AERO" across airplane. Black outlined orange "JENNEY" at top, "SOLVENIZED" at bottom.

JENNY AERO 3 1940-1960 Gill $300-500

White globe face with thick blue and thin red outline rings. Large red "AERO" across center. Blue "JENNEY" at top, "GASOLINE" at bottom.

JENNY SUPER
AERO 4 1940-1960 Gill $300-500

White globe face with thick blue and thin red outline rings. Large blue "SUPER-AERO" across center. Red "JENNEY" across top, "GASOLINE" across bottom.

****LINDSAY-McMILLAN CO. — MILWAUKEE, WISCONSIN:** Founded in Milwaukee in 1907, the company went through several reorganizations before becoming Lindsay McMillan Co. in 1919. Known primarily as a lubricants manufacturer, the company was operating nearly 100 service stations when purchased by Cities Service in 1931.

DELCO MOTOR
OILS 5 1920s OPE NL

White globe with black script "Delco" diagonal across center of globe face. Red "LINDSAY McMILLAN CO's" arched around top, "MOTOR OILS" around bottom.

DELCO GASOLINE 5 1926-1930 OPB W $1400-2000

White globe with red outline ring around globe face. Black script "Delco" over black "GASOLINE" in center of globe face. Small IOMA thunderbird logo under "GASOLINE". Red "AS GOOD AS GAS CAN BE" arched around top, "LINDSAY McMILLAN CO" around bottom.

DELCO BENZOL
GAS 5 1926-1930 OPB W $1500-2100

White globe wtih red outline ring around globe face. Black script "Delco" over black "BENZOL" in center of globe face. Small red & black logo below "BENZOL". Red "BETTER THAN GASOLINE" arched around top, "LINDSAY McMILLAN CO" around bottom.

****WINONA OIL COMPANY - WINONA, MINNESOTA:** Founded as a lubricants manufacturer in 1910, Winona became a small Minnesota / Wisconsin marketer that operated stations in the 1910s and 1920s under the "POMIAC" brand.

POMIAC
GASOLINE 6 1920s OPE NL

Small globe with 4" base. White globe with black outline ring around face. Large black triangle with red thunderbird IOMA logo in center of globe face. White "INDEPENDENT" above bird, "OILS" below. Black "POMIAC" arched around top, "GASOLINE" around bottom.

POMIAC 6 1920s 15" Metal $400-600

White globe face with red "POMIAC" across center. Single red line above and below lettering.

****MILTON OIL CO. — SEDALIA, MISSOURI & PIONEER OIL CO.— CARTHAGE/JOPLIN, MISSOURI:** As best as can be determined, Milton Oil Co. of Sedalia, Missouri, introduced the "DIXCEL" brand sometime around 1930. Pioneer Oil of Carthage & Joplin, Missouri, was apparently a jobber for Dixcel that sometime after World War II acquired the brand from Milton Oil. Milton continued to operate under the "Savex" brand well into the 1970s & Pioneer rebranded from Dixcel to Pioneer sometime around 1960. The Dixcel / Pioneer operation was purchased by Cities Service in the early 1960s.

DIXCEL GASOLINE 6 1930s 15" Metal $650-950

Black outlined orange globe face with black outlined white half-circles on upper and lower globe face with orange band between. Orange outlined black "DIXCEL" formed to fit in upper half circle, "GASOLINE" in lower.

A TRIAL TELLS/
DIXCEL/
GASOLINE 6 1930s 15" Metal $700-1000
White globe face with large orange circle in center. Black band across globe face. White "DIXCEL" on black band with black "A Trial Tells" arched around top of orange area & black "GASOLINE" arched around bottom.

DIXCEL 6 1930s 13.5" Glass $300-500
Black outlined orange globe face with black band across center. Large white "DIXCEL" in band.

A TRIAL TELLS/
DIXCEL/
GASOLINE 4 1930s 13.5" Glass $350-550
White globe face with large orange circle in center. Black band across globe face. White "DIXCEL" on black band with black "A Trial Tells" arched around top of orange area & black "GASOLINE" arched around bottom.

PIONEER OIL CO/
DIXCEL/
GASOLINE 4 1940s 13.5" Glass $350-550
Orange globe face with black outline ring & black ring around white center circle. Black band across center with white "DIXCEL" on band. White "PIONEER OIL CO." arched around top, "GASOLINE" around bottom.

PIONEER OIL CO/
MOTOR/
GASOLINE 8 1940s 13.5" Glass $350-550
Orange globe face with black outline ring & black ring around white center circle. Black band across center with white "MOTOR" on band. White "PIONEER OIL CO." arched around top, "GASOLINE" around bottom.

PIONEER OIL CO
MOTOR
GASOLINE 8 1940s 13.5" Glass $350-550
Blue globe face with white center circle. Blue band across center with white "MOTOR" on band. White "PIONEER OIL CO." arched around top, "GASOLINE" around bottom.

MILTON OIL CO/
DIXCEL/
GASOLINE 4 1940s 13.5" Glass $300-500
Orange globe face with black outline ring & black ring around white center circle. Black band across center with white "DIXCEL" on band. White "MILTON OIL CO." arched around top, "GASOLINE" around bottom.

DIXCEL PREMIUM
ETHYL(EC) 4 1940s 13.5" Glass $350-550
Orange globe face with black outline ring & black outline around white center circle. White "DIXCEL" arched around top, "PREMIUM" around bottom with Ethyl (EC) logo in center.

****PIONEER**: As best as can be determined, the Dixcel operation rebranded to Pioneer in the 1950s.

PIONEER 4 1950s Capco $175-250
Orange globe face with black outline ring. White outlined black "PIONEER" across center of globe face.

PIONEER
REGULAR 4 1955-1965 Capco $175-250
Orange globe face with black outline ring. White outlined black "PIONEER" over black "REGULAR". Black underline under "REGULAR".

PIONEER
PREMIUM 4 1955-1965 Capco $175-250
Orange globe face with black outline ring. White outlined black "PIONEER" over black "PREMIUM". Black underline under "PREMIUM".

****MILTON-SAVEX—SEDALIA, MISSOURI**: Milton Oil Company became a discounter operation in the postwar years and operated stations under the Savex brand into the 1970s.

SAVE WITH SAVEX3 1946-1970 13.5" Glass N $250-400
White globe face with green band across center. White "SAVEX" on green band. Black "SAVE" arched around top over "WITH", black "GASOLINE" arched around bottom.

SAVE ETHYL 3 1946-1970 13.5" Glass N $250-400
Green globe face with white circle in center. White "SAVEX" arched around top, "PREMIUM" around bottom with lage Ethyl (EC) logo in center.

****VALLEY OIL CO.—MIDDLETOWN, CONNECTICUT**: Founded in 1913, Valley Oil Company was a small New England gasoline marketer that became involved in the IOMA organization in the 1920s and remained a member for many years. The company was purchased by Cities Service about 1960.

VALOCO 5 1930s Banded Glass $450-750
White globe face with orange outline ring and orange lines forming bar and circle design across center of globe face. Green "VALOCO" across center band with green "MOTOR" arched around top, "GASOLINE" around bottom. Green triangle/red thunderbird IOMA logo in upper circle area.

CLARK & AFFILIATES

****CLARK OIL AND REFINING CORP.—ST. LOUIS, MISSOURI**: Clark Oil Co. was begun as a one station operation by Mr. Emory Clark in 1933 in Milwaukee, Wisconsin. Mr. Clark had been involved in earlier petroleum marketing enterprises and was instrumental to the early success of the Independent Oil Men's Association. By the early 1930s, Clark had evolved into a chain operation, selling "discount" gas through company operated stations in a number of Wisconsin towns. Significant growth came after World War II, when the operation expanded to include stations from Pennsylvania to Kansas, totalling more than 1750 at the peak, including 122 stations acquired when Clark bought out the Owens Oil Co. of Bloomington, Illinois, in 1968. In 1981 Clark was purchased by St. Louis based Apex Oil and was again sold in 1988, this time to Canadian conglomerate Horsham Corporation. In the mid-1990s, Clark was again independent, and in a move to establish a firm base for future expansion, purchased the former Gulf Oil Company refinery facilities at Port Arthur, Texas. With the refinery purchase, Clark began the active solicitation of branded distributors in select Gulf Coast markets, the first time in history that Clark products were sold through stations that were not company owned and salary operated. Today, the Clark brand appears at over 900 stations in 12 states.

CLARK HIGH GRADE
GASOLINE 6 1930s 15" Metal Flat NL
White globe face with red "CLARK OIL CO." arched around top, "GASOLINE" around bottom with red "HIGH/GRADE" diagonally across center. Globe frame is flat metal as opposed to formed high or low profile band.

CLARK'S SUPER
GAS 5 1932-1946 13.5" Glass W $350-550
Black outlined orange globe face with black outlined white band across center. Black "CLARK" on white band. White outlined black script "Super" across upper orange area, "Gas" across lower orange area.

CLARK SUPER 100 4 1946-1956 13.5" Glass W $350-500

CLARK SUPER 100 4 1936-1956 Orange Capco $300-450
Orange globe face with black outlined white "CLARK" at the top of the globe face, italic black "SUPER" across the center and a stylized black and white "100" on the lower globe face.

(CLARK) FUEL OIL 4 1946-1956 13.5" Glass W $200-300
Orange globe face with black 'FUEL/OIL" across center.

CLARK 3 1956-1982 13.5" Glass W $225-325

CLARK 1 1956-1982 Capco $175-250
Orange globe face with black outline. White band across center of globe face with black "CLARK" lettering.

CLARK 6 1970s One Piece Plastic NL
Experimental, possibly a sales sample. One piece plastic (halves assembled back to back) with cast image of Clark logo. Orange globe face with black outline. White band across center of globe face with black "CLARK" lettering.

****TARGET - WESTERN OIL COMPANY—INDIANAPOLIS, INDIANA**: Target is a 1960s Clark secondary brand.**

TARGET 4 1960s 13.5" Glass W $250-375
White globe face with three red rings circling so as to simulate a

target. Red outlined white band across center with black italic "Target" lettering.

****OWENS OIL CO. — BLOOMINGTON, ILLINOIS:** Owens was a Bloomington, Illinois, based discount gasoline chain that operated 122 stations in 10 states when purchased by Clark in 1968. Despite the buyout, the Owens brand remained in use for many years after the buyout by Clark.

OCO PEP REGULAR
GAS 4 1950s Capco $250-400
White globe face with red outline ring. Red/white/black arched red band across upper globe face. White "OCO PEP" on red band. Red script "Regular" below band above black "GASOLINE".

OCO PEP ETHYL
(EC) 4 1950s Capco $250-400
Red globe face with black outline ring. Large white cloud in center of globe face with three sets of three "burst" lines extending in different directions from the cloud. Black "OCO/PEP" diagonal on cloud. Black outlined white circle at bottom center with Ethyl (EC) logo with yellow rays in circle.

COLONIAL & AFFILAITES
****COLONIAL OIL INDUSTRIES — SAVANNAH, GEORGIA:** Colonial Oil Industries was founded in Savannah, Georgia, in 1921 as the American Oil Company. From 1921 until 1928, gasoline was sold under the American brand; however, with the expansion of the Pure Oil Co. into Georgia, American became a branded Pure Oil jobber. In 1932, when Pan American Petroleum merged with the American Oil Co. (AMOCO), and replaced the Pan-Am brand in Georgia with the Amoco name, Colonial sold the rights to the American name and became Colonial. In 1950, Colonial sold the Pure branded operation to Pure and concentrated on a growing wholesale/terminal business. The Colonial gasoline brand, which dated from the mid-1930s, was revitalized in the mid-1950s in an effort to give their dealers a unified brand. In the early 1960s, Colonial began operating a discount service station chain, branded "Interstate" and selling Colonial gasoline. By the early 1980s they had replaced many Interstate stations with "Enmark" convenience stores, which they continue to operate today.

COLONIAL OIL
INDUSTRIES 6 1946-1956 13.5" Glass N NL
White globe face with early version of shield logo covering most of globe face. Shield has red outline with red upper band, white center and blue lower. Blue outlined white eagle at top of shield. White "COLONIAL" on upper red band, blue "OIL" on center band, red outlined white banner across lower blue band with red "INDUSTRIES" on banner, small white "INC." below.

COLONIAL EAGLE 4 1956-1970 13.5" Glass N $450-700
COLONIAL EAGLE 3 1956-1970 Capco $350-600
White globe face with red ring around outer edge. Large red/white/blue Colonial shield with eagle (with blue and white coloring) covers most of globe face. Blue "EAGLE" arched around top, "GASOLINE" arched around bottom.

COLONIAL ROYAL
EAGLE 3 1956-1970 13.5" Glass N $450-700
White globe face with red ring around outer edge. Large red/white/blue Colonial shield with eagle (with blue and white coloring) covers most of globe face. Blue "ROYAL EAGLE" arched around top, "GASOLINE" arched around bottom.

COLONIAL ROYAL
EAGLE 4 1956-1970 Capco $350-600

COLONIAL GOLDEN
EAGLE 4 1956-1962 13.5" Glass N $450-700

COLONIAL GOLDEN
EAGLE 4 1956-1962 Capco $350-650
White globe face with gold ring around outer edge. Large red/white/blue Colonial shield with eagle (with blue and gold coloring) covers most of globe face. Blue outlined gold "GOLDEN/EAGLE" arched around top, "GASOLINE" arched around bottom.

COLONIAL
KEROSENE 5 1956-1970 Capco $350-650
White globe face with red ring around outer edge. Large red/white/blue Colonial shield with eagle (with blue and white coloring) covers most of globe face. Blue "KEROSENE" arched around top, "KEROSENE" arched around bottom.

COLONIAL DIESEL 5 1956-1970 Capco $350-650
White globe face with red ring around outer edge. Large red/white/blue Colonial shield with eagle (with blue and white coloring) covers most of globe face. Blue "DIESEL" arched around top, "DIESEL" arched around bottom.

CONOCO & AFFILIATES
****CONTINENTAL OIL COMPANY — PONCA CITY, OKLAHOMA:** Continental Oil—founded in 1875 in Denver, Colorado—became an operating unit of the Standard Oil Trust, having been purchased by Standard in 1885. After the 1911 breakup of Standard, Continental was spun off as an independent company and by 1913 was involved in production, refining and marketing. The Continental name survived a 1929 takeover by Marland Oil, and over the course of time the Conoco brand name has appeared in nearly all states of the Southeast, Midwest, mountain and Pacific coast areas. Conoco expanded its scope worldwide during the 1950s, including the purchase of several marketing companies such as Western Oil, Kayo and Douglas, greatly expanding the company's marketing presence throughout the United States. Conoco has been a part of the giant chemical company Dupont since 1982 and markets in 37 states today.

CONOCO MINUTEMAN
GASOLINE 4 1913-1929 15" Metal HP Red $2500-3300
Yellow globe face with blue "CONOCO" arched around top, red "GASOLINE" arched around bottom with black silhouette of Continental soldier in center of globe face.

CONOCO MINUTEMAN
ETHYL 4 1927-1929 15" Metal HP Red $1500-2200
Yellow outer ring around globe face with blue "CONOCO" arched around top, red "GASOLINE" arched around bottom around white circle in center with large Ethyl (EGC) logo in center.

CONOCO TRIANGLE
W/GREEN
OUTLINE 4 1929-1950 15" Metal HP White $650-950
White globe face with green outlined red Conoco triangle logo in center.

CONOCO TRIANGLE
W/GREEN
OUTLINE 6 1929-1950 18" Neon Metal NL
White globe face with green outlined red Conoco triangle logo in center.

CONOCO ETHYL
(EGC) 4 1929-1941 15" Metal HP White $650-950
White globe face with green outline. Red "CONOCO" arched around top, "GASOLINE" around bottom with large Ethyl (EGC) logo in center.

CONOCO TRIANGLE
W/GREEN OUTLINE 3 1929-1950 13.5" Glass W
$500-750

CONOCO TRIANGLE
W/GREEN OUTLINE 3 1929-1950 Teal Ripple
$2200-3000

CONOCO TRIANGLE
W/GREEN OUTLINE 3 1933-1950 Green Capco
$250-400

CONOCO TRIANGLE
W/GREEN OUTLINE 3 1933-1950 Capco $200-300
White globe face with green outline. Green outlined red Conoco triangle logo in center. Please note that globes with green capco frames were used on pumps dispensing regular grade "Conoco Bronze" gasoline, while those with white capco frames were used on pumps dispensing motor grade "Conoco Demand" gasoline.

CONOCO ETHYL
(EGC) 4 1929-1941 13.5" Glass W $500-750

CONOCO ETHYL
(EGC) 4 1933-1941 Red Capco $250-400
White globe face with green outline. Red "CONOCO" arched around top, "GASOLINE" around bottom with large Ethyl (EGC) logo in center.

CONOCO ETHYL
(EC) 4 1941-1950 13.5" Glass W $500-750

CONOCO ETHYL

(EC) 4 1941-1950 Red Capco $225-375
Same as above with Ethyl(EC) logo.
CONOCO RED TRIANGLE 3 1950-1969 Green Capco $200-350
CONOCO RED TRIANGLE 3 1950-1969 Red Capco $200-325
White globe face with red and white Conoco triangle logo in center-no green.

****MARLAND OIL COMPANY—PONCA CITY, OKLAHOMA:** Marland traces its early history to the Oklahoma oil fields of the early 1910s. By 1920, marketing had begun under the famous Red Triangle logo and by 1922 there were Marland service stations in 11 states. In 1929, Marland purchased Continental but chose to retain the Continental Oil Co. name for the new corporation. The new Conoco logo was to be the famous Marland triangle rebranded with the name "CONOCO".

MARLAND GASOLINE 4 1920s OPE N R/W $1800-2600
White globe with black outline around globe face and black outlined red "MARLAND OILS" triangle in center with black "MARLAND" arched around top and "GASOLINE" around bottom.
MARLAND GASOLINE (small Marland Oils) 5 1920s OPE R/W $1800-2600
White globe with NO outline around globe face and black outlined red "MARLAND OILS" triangle in center with black "MARLAND" arched around top and "GASOLINE" around bottom.
MARLAND HI-TEST5 1920s OPE N R/W $2000-2800
White globe with black outline around globe face and black outlined red "MARLAND OILS" triangle in center with black "MARLAND" arched around top and "HIGH TEST" around bottom.
MARLAND GASOLINE 5 1920-1929 15" Metal $1200-1800
White globe face with black outline and black outlined red "MARLAND OILS" triangle with black "MARLAND" arched around top and "GASOLINE" arched around bottom.
MARLAND GASOLINE 4 1930s Gill $500-850
MARLAND GASOLINE 4 1927-1929 Balcrank $550-900
MARLAND GASOLINE 4 1930s 13.5" Glass W $500-800
Mint green globe face with dark blue outline. Blue outlined red "MARLAND OILS" triangle in center with blue "MARLAND" arched around top and "GASOLINE" arched around bottom.
MARLAND GASOLINE red letters 5 1927-1929 Balcrank $550-900
Mint green globe face with dark blue outline. Blue outlined red "MARLAND OILS" triangle in center with red "MARLAND" arched around top and "GASOLINE" arched around bottom.
MARLAND HI-TEST4 1930s Gill $550-900
MARLAND HI-TEST5 1927-1929 Balcrank $600-1000
MARLAND HI-TEST5 1930s 13.5" Glass W $550-800
Mint green globe face with dark blue outline. Blue outlined red "MARLAND OILS" triangle in center with blue "MARLAND" arched around top and "HI TEST" arched around bottom.
(MARLAND) AVIATION GASOLINE 5 1927-1929 13.5" Glass W $3500-5000
White globe face with black outline ring. Black outlined dark green "AVIATION" arched around top, "GASOLINE" around bottom with red/white/black detailed airplane and cloud scene in center.

****NATIONAL OIL CO. — RICHMOND, VIRGINIA:** Founded in 1896, National was a mid-Atlantic regional marketer operating stations in Virginia and the Carolinas until purchased by Conoco in 1937. Much of the marketing was converted to Conoco or sold to Cities Service at that time.

NATIONAL 7 1920s 15" Metal NL
From sketcy details in an old photo. Red globe face with black outlined white "bar and circle" band across center of globe face. Red outlined black "NATIONAL" across band.

****WESTERN OIL AND FUEL — MINNEAPOLIS, MINNESOTA:** Western Oil and Fuel was a Minneapolis based marketing operation founded in 1926 that introduced the famous Husky brand in the mid-1930s. This company operated the Husky stations into the late 1950s. In 1938, Husky Refining Company was established by Western Oil and Fuel as a separate operation in Cody, Wyoming, although Western Oil and Fuel continued to use the Husky brand name until at least 1954. Some dealers are known to have used the name as late as 1960, although Conoco had purchased Western by that time and was in the process or reorganizing the company. Conoco bought out Western in 1958 and, at about that time, introduced "Western" as another gasoline trademark. By 1970, Conoco had converted all the Mileage and DS locations to the Conoco brand, but the Western stations, modeled after the competing Holiday and SuperAmerica convenience stores, remained "Western" up through the late 1970s when they were rebranded "JET."

HUSKY "QUALITY PETROLEUM PRODUCTS" 5 1935-1940 13.5" Glass $1700-2300
Globe face has blue outline ring around upper white area. Lower 1/3 is dark blue, separated from upper area by red line. Black/white/red husky dog head with extreme detail with orange/yellow "northern lights" design on upper globe face. Yellow "HUSKY" in lower blue area. Black "QUALITY PETROLEUM PRODUCTS" arched above dog.
HUSKY w/ northern lights 6 1940-1960 12.5" Glass NL
HUSKY w/ northern lights 4 1940-1960 13.5" Glass $1800-2400
HUSKY w/ northern lights 4 1940-1960 Gill $1700-2500
Globe face has blue outline ring around upper white area. Lower 1/3 is dark blue. Blue/gray/white/red husky dog head with good detail with orange/yellow "northern lights" design on upper globe face. Yellow "HUSKY" in lower blue area.
HUSKY ETHYL (EGC) 6 1935-1940 Gill NL
Red globe face with black outline around white center circle. Inside circle is white with lower 1/3 blue. Blue/gray/white/red husky dog head with good detail and orange/yellow "northern lights" design in upper 2/3 of center circle. Small black outlined white circle superimposed over lower blue area, extending into red outer area. Ethyl (EGC) logo with yellow rays in small circle.
>>These Husky globes as well as later Husky Oil globes appear in Husky section in this book.

****WESTERN:** Brand introduced in 1959 and used at locations that were similar in design to convenience stores then being operated by competitive brands Holiday and SuperAmerica. The Western name remained in use at these stations until the late 1970s, when the Jet brand was introduced in this country for Conoco's convenience store outlets.

WESTERN 4 1960-1969 Shaped Plastic $200-300
Red and white plastic trapezoidal globe in shape of 1960s Western logo. Removable formed plastic lens with blue "Western" logotype across center.

****ZENITH GAS POOL — DULUTH, MINNESOTA:** Zenith Gas Pool was an independent marketer that introduced the Mileage trademark in the 1920s. By the mid-1930s, Zenith had become part of Western Oil and Fuel and Mileage was serving as a brand name for the company's motor grade gasoline. In 1954, when interest in the Husky brand name shifted to Husky Oil Company of Cody, Wyoming, Western adopted Mileage as its primary trademark until the Western brand was introduced. In the 1960s, Mileage was paired more and more with the other Western brand, DS, and about 1970 Conoco began conversion of all Mileage and DS stations to the Conoco brand.

MILEAGE GASOLINE 5 1930s 15" Metal $1800-2300
White globe face with black outline ring set in from edge of lens. Large red line drawing of spoke wheel in center of globe face with red outlined black script "Mileage" logotype across wheel. Red "GASOLINE" across on lower globe face.
MILEAGE NOT A NOX 5 1930s 15" Metal $600-900
White globe face with band across center. Script "Mileage" and de-

sign at top. White "NOT-A-NOX" on black center band, "GASOLINE" arched below.

MILEAGE (w/ wheel) 5 1940-1949 Gill $600-900
White globe face with large orange line drawing of spoke wheel in center of globe face with orange outlined black script "Mileage" logotype across wheel.

MILEAGE 3 1949-1960 Gill $200-350
Bright blue globe face with white band across center. Red "Mileage" on white band.

SUPER MILEAGE 3 1949-1960 Gill $225-375
Same as above, italic "SUPER" added above "age" in Mileage.

MILEAGE 4 1960-1969 Shaped Plastic $200-300
Red and white plastic trapazoidal globe in shape of 1960s Mileage logo. Removable formed plastic lens with blue "Mileage" across center.

MILEAGE 4 1960-1969 Shaped Plastic $200-300
Red and white plastic trapazoidal globe in shape of 1960s Mileage logo. Red outer area with white oval across center with blue "Mileage" across center.

****DIRECT SERVICE OIL CO. — MINNEAPOLIS, MINNESOTA:** Direct Service Oil Co. was a Minneapolis marketer, known to be in operation in the early 1930s, that was purchased by Western Oil and Fuel in 1956. Stations retained the "DS" trademark until replaced by Conoco in 1970.

DS 7 1940s 13.5" Glass $200-350
White Globe face with large red outlined black circle in center. Red "DS" superimposed over circle.

MAJOR 5 1940s 13.5" Glass $250-400
White globe face with blue outline ring. Red banner folded across center with black outlined white "MAJOR" on banner.

SUPER MAJOR EHTYL 5 1940s 13.5" Glass $250-450
Yellow globe face with yellow outline ring around white lower half. Black outlined white "SUPER" on upper yellow area above black "Major" at color split. Small Ethyl (EC) logo on lower globe face.

DS 4 1960-1969 Shaped Plastic $200-300
Red and white plastic trapazoidal globe in shape of 1960s DS logo. Removable formed plastic lens with blue "DS" across center.

****MALCO REFINING — ROSWELL, NEW MEXICO:** Malco, founded in 1931, was a Roswell, New Mexico, refiner/marketer operating stations in New Mexico under the Malco and Numex brands and under the Valley brand in Arizona. Conoco purchased Malco in 1959 but was forced to sell on anti-trust grounds in 1964. Malco operated as an independent into the 1970s.

MALCO W/ NM LOGO 4 1940s 13.5" Glass $350-550
Yellow globe face with blue outlined white circle in center. Four blue lines radiating out from circle on each side. Blue outlined red "MALCO" on center white circle.

MALCO ETHYL 7 1940s 13.5" Glass NL
Yellow globe face with blue outlined white circle in center. Large Ethyl (EGC) logo in center circle. Red "MALCO" arched around top, "GASOLINE" around bottom.

MALCO w/bird 4 1950s 13.5" Glass $750-1250
MALCO w/bird 4 1950s Capco $600-950
Yellow/green outlined white globe face with yellow outlined green bird logo at top of globe face. Yellow outlined black "MALCO" across center of globe face with large green "REG.T.M." just below "MALCO".

****VALLEY OIL COMPANY — ROSWELL, NEW MEXICO and MESA, ARIZONA:** This was a Malco secondary brand from the 1950s.

VALLEY 5 1950s Capco $225-375
White globe face with blue mountain and valley scene with blue/white lines above and below. Yellow outlined black "VALLEY" across center of globe face.

****NEW MEXICO OIL COMPANY — ALBEQUERQUE, NEW MEXICO:** Founded in 1927 as a Powerine subsidiary, in the postwar era this company's Numex brand became a Malco secondary operation.

NUMEX 5 1950s 13.5" Glass $300-450
White globe face with dark lettering across center. Two parallel red line above and below "NUMEX".

NUMEX ETHYL 5 1950s 13.5" Glass $350-500
White globe face with dark lettering across center. Two parallel red line above and below "NUMEX". Small Ethyl (EGC) logo on lower globe face.

****DOUGLAS OIL CO.—LOS ANGELES, CALIFORNIA:** Douglas Oil dates back to 1927 and merged with Krieger in 1946. In 1961, Conoco bought Douglas and operated the nearly 300 Douglas stations until 1971 when they were sold off and rebranded, mostly Texaco.

DOUGLAS W/ HEART 5 1950s Capco $450-750
White globe face with Douglas "red heart and blue wing" logo on globe face. No wording appears on this globe.

****KRIEGLER:** This company was a California independent that merged with Douglas in 1946.

KRIEGER "KRIEG-O-LENE" 6 1930s 16.5" Metal NL
White globe face with large yellow outlined black diamond covering most of globe face. Yellow band across center of diamond with black "KRIEG-O-LENE" on band. Orange "KRIEGER" at top with yellow underline and lettering "FIRST GRADE". Orange "GASOLINE" over "QULAITY/at a/PRICE" on bottom of diamond.

****F.P. KENDALL and CO. — CHATTANOOGA, TENNESSEE:** F. P .Kendall and Company opened their first Kayo station in Chattanooga, Tennessee, in 1940. The chain expanded throughout the South and had approximately 170 stations in seven states when purchased by Conoco in 1959. Kayo was operated as a separate brand throughout the 1960s and 1970s. Around 1980 the nearly 900 Kayo stations were rebranded "JET". Conoco began replacing the Jet brand in 1990.
>>No Kayo brand globes known at this time.

CO-OPERATIVE ASSOCIATIONS

****FARMLAND INDUSTRIES — KANSAS CITY, MISSORI:** Founded in 1929 as Consumers Co-operative Association, the company originally marketed gasoline under the "Union Certified" trademark. After 1931, gasolines were simply branded Co-Op. The operation became known as Consumers Co-Operative Association in 1936 and as Farmland Industries in 1966. Today Farmland, commonly refered to as "Double Circle" in reference to their logo, supplies products to farm co-ops throughout the Midwest. They continue to market through co-ops under the "CO-OP" brand name and at retail under the "AMPRIDE" brand.

UNION CERTIFIED 5 1929-1931 Gill $300-500
Blue outlined orange globe face with large white outlined "UNION" arched across top, "CERTIFIED" across bottom.

USE/CO-OP/ PRODUCTS (ON Shield) 3 1930s Gill $300-500
White globe face with red "16 point" border around Co-op logo. White ring around edge of logo with red center. Black "USE" on upper ring, "PRODUCTS" on lower ring, and large black "CO-OP" on diagonal band across center.

USE/CO-OP/ PRODUCTS 5 1930s 13.5" Glass $250-450
USE/CO-OP/ PRODUCTS 5 1930s Gill $300-500
White globe face with red "7 point" border around Co-op logo. White ring around edge of logo with red center. Black "USE" on upper ring, "PRODUCTS" on lower ring, and large black "CO-OP" on diagonal band across center.

USE/CO-OP ETHYL (EGC) 5 1930s 13.5" Glass $275-475
USE/CO-OP ETHYL (EGC) 5 1930s Gill $300-550
White globe face with red "7 point" border around Co-op logo. White ring around edge of logo with red center. Black "USE" on upper ring, large black "CO-OP" on diagonal band across center, and black out-

lined white circle offset to right on lower globe face, covering lower section of outer ring, with Ethyl (EGC) logo with yellow rays in center.

FARMCO
(ACROSS Shield) 3 1930s Gill $300-550
White globe face with red "16 point" border around Co-op logo. White ring around edge of logo with red center. Black "FARMCO" on horizontal band across center.

CO-OP WHITE
ON RED 2 1940s 13.5" Glass $225-350

CO-OP WHITE
ON RED 2 1940s Gill $225-375
Red globe face with black/white outline and white "CO-OP" with black dropshade.

CO-OP RED
ON WHITE 3 1940s 13.5" Glass $225-350

CO-OP RED
ON WHITE 3 1940s Gill $225-400
White globe face with red outline and red "CO-OP" across center.

CO-OP GREEN
ON WHITE 4 1940s Gill $225-375
White globe face with green outline and green "CO-OP" across center.

CO-OP RED
ON BLUE 4 1940s Gill $275-450
Red globe face with blue outline and blue "CO-OP" across center.

CO-OP DOUBLE
CIRCLE 3 1962-1985 Plastic $200-325
White plastic formed faces assembled back-to-back with aluminum ring. Formed face of globe is in shape of interlocking circles with red outline on left circle and blue on right circle with black "CO-OP" across center.

FALCON: Falcon was a postwar attempt at retail markets by Consumers Co-Operative Association (CCA).

FALCON
GASOLINE 6 1946-1955 Gill NL
White globe face with detailed multicolored falcon at center of globe face. Red "FALCON" logotype lettering arched aound top, "GASOLINE" around bottom.

FALCON 4 1955-1970 Green Capco $800-1250
White globe face with detailed brown/yellow falcon in flight covering most of globe face. Green band across center of globe face with white "FALCON".

FALCON ETHYL 5 1955-1970 Red Capco $800-1250
White globe face with detailed brown/yellow falcon in flight covering most of globe face. Red band across center of globe face with white "FALCON", red "ETHYL" at bottom.

UNION OIL COMPANY: Co-operative in existence in North Dakota in the 1940s.

UNION OIL CO
CO-OP 5 1940-1955 Gill $300-450
White globe face with red "16 point" Co-Op seal with white center circle on globe face. White outlined black "UNION" arched around top of shield, "OIL CO." around bottom with black "CO-OP" across center of seal.

UNION OIL CO
FARMCO 5 1930s Gill $300-450
White globe face with red "16 point" Co-Op seal with white center circle on globe face. White outlined black "UNION" arched around top of shield, "OIL CO." around bottom with black "FARMCO" across center of seal.

FARMERS UNION CENTRAL EXCHANGE — ST. PAUL, MINNESOTA: Farmers Union Central Exchange was founded in 1931. The Exchange is a combination of several farm co-operative organizations that operate across the North Central states. The CENEX brand name was introduced about 1974 to establish a common identity for the various companies. In 1988, Farmers Union Central Exchange merged with Land O' Lakes and all petroleum products are marketed under the CENEX brand name today.

FARMERS UNION
w/plow & shield 5 1920s 15" Metal $2000-3000
Yellow outlined dark green globe face with yellow/red outlined yellow shield in center with dark green "FARMERS/UNION" over plow. Interlocked F/U in red circle at top of shield. Yellow outlined red "GASOLINE" arched around bottom.

FARMERS UNION
COOPERATIVE
STRENGTH 6 1932-1934 13.5" Glass NL

FARMERS UNION
COOPERATIVE
STRENGTH 6 1932-1934 Gill NL
Mint green globe face with dark green outline around red six point shield. Yellow outlined green banner with irregular edges across center of globe face. Green circle at top of shield with interlocking yellow "FU" monogram on circle, mint green plow on bottom of shield with yellow "FARMERS/UNION" on banner. Red CO-OPERATIVE STRENGTH" arched around bottom of globe face.

FARMERS UNION
GASOLINE 4 1934-1938 13.5" Glass $900-1500
Yellow outer ring around white center of globe face. Dark green "FARMERS UNION" arched around top of globe face, "GASOLINE" at bottom with red/yellow/black Farmers Union shield with plow in center. Black "CO-OPERATIVE STRENGTH" arched under shield.

FARMERS UNION
GASOLINE W/
CO-OP seal 3 1938-1952 13.5" Glass $450-750

FARMERS UNION
GASOLINE W/
CO-OP seal 4 1938-1952 Gill $500-800
Yellow outer ring around white center of globe face. Dark green "FARMERS UNION" arched around top of globe face, "GASOLINE" at bottom with red/green "outline" co-op 7-point shield in center with green "NATIONAL" at top, "CO-OPERATIVES" at bottom, black "CO-OP" across center.

FARMERS UNION
ETHYL 4 1942-1952 13.5" Glass $450-750

FARMERS UNION
ETHYL 7 1942-1952 Gill $500-800
White globe face with wide green outlined yellow band arched across upper face. Red outlined green "FARMERS/UNION" on yellow band. Ethyl (EC) logo on lower globe face below band.

FARMERS UNION
FULL POWER 5 1938-1952 Gill $500-750
Red outlined white globe face with black outlined red bowtie shaped band across center of globe face. Yellow outlined green "FARMERS" in upper white area and "UNION" in lower with black outlined white "FULL POWER" across center.

(CENEX) CO-OP
Shield 4 1952-1972 Plastic $250-450
Formed plastic shield halves riveted back-to-back. Black border area around yellow shield area of globe face with red "V" shaped section at top. Black "CO-OP" diagonal across center.

(CENEX) SUPER
CO-OP Shield 5 1952-1972 Plastic $250-450
Formed plastic shield halves riveted back-to-back. Black border area around yellow shield area of globe face with red "V" shaped section at top. Black "CO-OP" diagonal across center and yellow "SUPER" on upper red "V" area.

(CENEX) DIESEL
CO-OP Shield 5 1952-1972 Plastic $300-500
Formed plastic shield halves riveted back-to-back. Black border area around yellow shield area of globe face with red "V" shaped section at top. Black "CO-OP" diagonal across center and yellow "DIESEL" on upper red "V" area.

MIDLAND COOPERATIVES — ST. PAUL, MINNESOTA: Midland Cooperatives, founded in 1926, operated farmers co-op service stations across the upper Midwest for many years. Midland sold out to Land O' Lakes in 1985 and Land O' Lakes merged with Farmers Union Central Exchange (CENEX) in 1988. All gasoline has been sold under the Cenex brand since that time.

USE CO-OP
GASOLINE
(red/yellow) 5 1926-1930 15" Metal $1800-2500
White/red outlined yellow globe face with red outline around white center circle. Green outlined red "CO-OP" with green line drawing of shaking hands across center of globe face. Green outlined red "USE" arched around top, "GASOLINE" around bottom.

USE CO-OP
GASOLINE 5 1930s 13.5" Glass $750-1250
Green outlined yellow-orange globe face with green outlined white center circle. Green/white outlined green "CO-OP" with shaking hands logo across center. White outlined green "USE" arched around top, "GASOLINE" around bottom.

HI TEST/CO-OP/
ANTI NOK 5 1930s 13.5" Glass $800-1300
Red outlined orange ring around outer globe face. White circle in center. Black outllined red "HI-TEST" arched around top, "ANTI-NOK" around bottom with black outlined red "CO OP" across center with green outlined "shaking hands" extending from both "O's" in CO-OP into center circle.

CO-OP COMPETITIVE
GASOLINE 5 1930s 13.5" Glass $750-1250
Yellow-orange / green outlined white globe face with green/white outlined yellow-orange band diagonal across center. White outlined green "COMPETITIVE" on band with green "CO-OP" with shaking hands logo in upper white area, "GASOLINE" in lower white area.

MIDLAND/CO-OP/
WHOLESALE 4 1940-1955 Gill $800-1250
Green outlined orange ring around outer globe face. White circle in center. White outlined green "MIDLAND" arched around top, "WHOLESALE" around bottom with white outlined green "CO-OP" across center with green outlined "shaking hands" extending from both "0s" in CO-OP into center circle.

CO-OP SUPER
OCTANE
GASOLINE 5 1940-1955 Gill $275-450
Green outlined orange ring around outer globe face. White circle in center. White outlined green "CO-OP" arched around top, "GASOLINE" around bottom with green script "Super/Octane" diagonal in center.

MIDLAND "M" 3 1955-1960 13.5" Glass $150-300
White globe face with black outlined Midland "M" logo with dots and red rectangle below "M" with white "MIDLAND" lettering.

MIDLAND "M"
NO BOX 3 1960-1987 Capco $125-225
White globe face with large black "M" with red triangle and dots only. No border or lettering.

****FARMERS EDUCATIONAL and CO-OPERATIVE STATE UNION — OMAHA, NEBRASKA**: Formed in the 1920s, Farmers Educational and Co-Operative State Union sold products under the Statex brand name after World War II. The Union then sold products under the Felco brand name from the late 1950s until the 1970s, when they merged with the Farmers Union Central Exchange and the facilities were rebranded Cenex.

FARMERS UNION
W/PLOW SCENE 5 1920s 15" Metal $2200-3500
White globe face with red/white co-op logo in center. Red lettering arched at top of logo "FARMERS EDUCATIONAL AND COOPERATIVE STATE UNION", "OF NEBRASKA" at bottom, farm field scene with plow in center. Blue "FARMERS" arched around top of globe face, "UNION" around bottom.

FARMERS UNION
HIGH OCTANE 4 1930s 15" Metal $500-850
White globe face with blue "FARMERS" lettering arched at top, "UNION" at bottom (identical to above) with red outline around center white circle and red "HIGH/OCTANE" on circle.

FARMERS UNION
CO-OPERATIVE
HIGH OCTANE 4 1930s 13.5" Glass $250-400
White globe face with red outline ring and red lines form irregular bar and circle bowtie band across center of globe face. Black "FARMERS" arched around top, "UNION" around bottom with black outlined red "CO-OPERATIVE" in bowtie lettering across center. Black "High" in center circle at top, "Octane" arched below.

STATEX 5 1940-1958 13.5" Glass N $225-375
Dark Green globe face with white "STATEX" across center.

STATEX HI OCTANE 3 1950-1958 13.5" Glass N $250-400
Light green globe face with irregular green logo at top with white "STATEX" on logo. Dark green "GASOLINE" below logo and dark green band across bottom with white "HI-OCTANE".

STATEX PREMIUM 3 1950-1958 13.5" Glass N $250-400
Light green globe face with irregular green logo at top with white "STATEX" on logo. Dark green "GASOLINE" below logo and dark green band across bottom with white "PREMIUM".

FELCO 4 1958-1970 Capco $150-250
White globe face with green diamond logo with rounded corners in center of globe face. White "f" on logo and black "felco" below logo.

****MISSOURI FARM BUREAU — JEFFERSON CITY, MISSOURI**: History unknown.

MISSOURI GREEN
DIAMOND 5 1940s Gill $400-650
White globe face with yellow outlined verticle green diamond in center. Yellow area shape of Missouri in center of diamond with green "FARM/BUREAU" and yellow "GREEN DIAMOND" below yellow area.

****SOUTHERN STATES CO-OPERATIVES, INC. — RICHMOND, VIRGINIA**: Founded as Virginia Seed Service in 1923, Southern States began marketing petroleum products in the 1930s and continues today.

SOUTHERN STATES
COOPERATIVES 5 1934-1940 15" Metal $900-1500
White outer ring around globe face with orange "MOTOR OIL" arched at top, "GASOLINE" at bottom. Blue center circle area blue/orange Southern States logo in center with white "SOUTHERN/STATES/COOPERATIVE/SCS" on logo, script "Quality" onblue area below logo.

****FARMERS CO-OPERATIVE EXCHANGE – RALEIGH, NORTH CAROLINA**: Founded in the 1920s, they were purchased by Southern States in the 1980s. Only a few retail service stations ever displayed the FCX brand. >>No FCX globes known at this time

****TENNESSEE CO-OPERATIVES — LaVERGNE, TENNESSEE**: Closely affiliated with Farmland Industries, Tennessee Co-Operatives has facilities in almost every county in Tennessee today. Many co-operatives continue to market gasoline and several have been actively involved in establishing 24 hour cardlock service stations.

CO-OP REGULAR 6 1946-1970 13.5" Glass NL
White globe face with large Tennessee Co-op logo (perspective view of Tennessee map in red on green background with progressive green "CO-OP" lettering) offset to upper left of globe face. Green speed lines arched around from top ending at red "REGULAR" offset to lower right of globe face.

CO-OP ETHYL 6 1946-1970 13.5" Glass NL
White globe face with large Tennessee Co-op logo (perspective view of Tennessee map in red on green background with progressive green "CO-OP" lettering) offset to upper left of globe face. Green speed lines arched around from top ending at red "ETHYL" offset to lower right of globe face.

****GROWMARK INDUSTRIES — BLOOMINGTON, ILLINOIS**: Founded as Illinois Farm Supply in the 1920s, gasoline was marketed under the "MAGIC ALADDIN" brand until after World War II. The "FS" brand was introduced about 1960 and continues in use today.

ALADDIN GAS 6 1920s 15" Metal $800-1400
White globe face with black/white pyramid in center. Black "ALADDIN" arched around top, "GAS" horizontal across bottom.

ALADDIN GASOLINE
ILL. FARM
SUPPLY CO. 5 1930s 15" Metal $450-750
Blue outlined white globe face with large "AS" in center. Blue outlined red "ALADDIN" arched around top, "GASOLINE" around bottom.

ALADDIN GASOLINE
ILL. FARM
SUPPLY CO. 5 1930s 15" Metal $750-1250
Blue outlined white globe face with red/white/blue Illinois Farm Supply logo in center. Blue outlined red "ALADDIN" arched around top, "GASOLINE" around bottom.
ANTI-KNOCK 5 1930s 15" Metal $350-500
White globe face with white/red outline ring. Blue outlined red "ANTI-KNOCK" in bowtie lettering across center with red burst lines radiating out from center.
ALADDIN
GASOLINE 5 1940s Banded Glass $500-850
ALADDIN
GASOLINE 5 1940s 13.5" Red Glass $1400-2000
Blue outlined white globe face with red/white/blue Illinois Farm Supply logo in center. Blue outlined red "ALADDIN" arched around top, "GASOLINE" around bottom.
ILLINOIS FARM
SUPPLY
COMPANY 4 1950-1958 Capco $200-300
White globe face with red outline. Blue ring around logo in center with white "ILLINOIS FARM SUPPLY COMPANY" around blue ring. Blue outline white "S" in center with diagonal box across center, red "SERVICE" in box, "PETROLEUM/PRODUCTS" on "S". Three red triangles around circle.
FS 5 1958-1970 Capco $125-225
White globe face with red outlined black "f", red "s" logo in center.

****MISSOURI FARMERS ASSOCIATION—COLUMBIA, MISSOURI:** MFA has marketed petroleum products since 1932, and continues to market through farm co-operatives and some retail locations in Missouri and western Tennessee.

MFA GASOLINE
MOTOR OILS
GREASES 4 1946-1960 13.5" Glass $300-475
White globe face with small "MFA" ahield in center. Red outline around edge, blue "GASOLINE" arched around top, "MOTOR OILS GREASES" around bottom.
MFA 3 1960-1970 Capco $225-325
White globe face with red/white/blue "MFA" shield in center.

****FARMERS GAS and OIL COMPANY of MICHIGAN — ITHACA, MICHIGAN:** History unknown.

FARMERS/CO-OP/
PETROLEUM 4 1930s 15" Metal $600-850
White globe face with blue outline around edge. Red "7-point" co-op shield with white ring around center red circle. Blue "FARMERS" arched around top of white ring, "PETROLEUM" around bottom with red outlined white band across center with blue "CO-OP".
FARMERS/HI OCTANE/
CO-OP 4 1940s Gill $250-450
Blue outlined white outer ring around red center of globe face. Blue outlined red "FARMERS" arched around top, "CO-OP" around bottom with blue lettering "HIGHOCTANE" on white band across center red circle.

****CENTRAL CO-OPERATIVE WHOLESALE — SUPERIOR, WISCONSIN:** Regional distributor of products offered by Consumers Co-Operative.

CO-OP "AFFIL W/
NATL CO-OP INC."
(white on red) 5 1930s 15" Metal $500-800
CO-OP "AFFIL W/
NATL CO-OP INC."
(white on red) 5 1930s Gill $300-450
Red globe face with reverse bowtie "CO-OP" across center. Small white lettering arched around top "AFFILIATED WITH NATIONAL CO-OP INC." and white lettering arched around bottom "DISTRIBUTORS/ CENTRAL CO-OPERATIVE WHOLESALE/SUPERIOR WIS."
CO-OP "AFFIL W/
NATL CO-OP INC.
(white on blue) 5 1930s 15" Metal $550-850
CO-OP "AFFIL W/
NATL CO-OP INC.
(white on blue) 6 1930s Gill $350-500
Blue globe face with reverse bowtie "CO-OP" across center. Small white lettering arched around top "AFFILIATED WITH NATIONAL CO-OP INC." and white lettering arched around bottom "DISTRIBUTORS/ CENTRAL CO-OPERATIVE WHOLESALE/SUPERIOR WIS."

****CO-OPERATIVE GLF (Grange League Farm) FARM SUPPLIES, INC. — ITHACA, NEW YORK:** Founded in the 1920s, GLF served farmers in New York and Pennsylvania as well as parts of New England. In 1964, GLF merged with eastern States Farm Supply to form Agway and continues to market petroleum products under that name throughout the Northeast.

GLF QUALITY w/
painted steel faces 5 1930s 15" Metal LP White
$800-1250
Painted white steel face with red omega logo with line across bottom. Large red "GLF" painted at top, script red "Quality" at bottom.
GLF QUALITY
(red/white -
arched "GLF") 5 1930s 15" Metal $600-850
GLF QUALITY
(red/white -
arched "GLF") 3 1930s 13.5" Glass W $250-400
White globe face with large red omega sign covering most of globe face. Large red "GLF" arched above red script "Quality".
GLF QUALITY (white
lettering on red) 5 1946-1964 13.5" Glass W $300-500
Red globe face with large white omega sign covering most of globe face. Large white "GLF" arched above white script "Quality".
GLF QUALITY (red/
white - straight
"GLF") 5 1930s 15" Metal $600-850
GLF QUALITY (red/
white - straight
"GLF") 3 1930s 13.5" Glass W $250-400
GLF QUALITY (red/
white - straight
"GLF") 4 1930s Capco $200-300
White globe face with large red omega sign covering most of globe face. Large red "GLF" above red script "Quality".

****LANDMARK CO-OPERATIVES—COLUMBUS, OHIO:** Founded as Farm Bureau Co-Operative Association, Inc., by the 1950s the company was known as Landmark Co-Operatives. After merging with Indiana Farm Bureau and changing their name to "Countrymark," they continue to operate various facilities, including retail service stations, throughout Ohio, Indiana and Michigan today.

QUALITY/CO-OP/
PRODUCTS
(ONShield) 4 1930s 15" Metal $600-950
White globe face with red/white/blue CO-OP shield logo on globe face. "QUALITY" arched in top of shield, "PRODUCTS" around bottom with "CO-OP" in diagonal band across shield.
QUALITY/PREMIUM/
PRODUCTS
(ON Shield) 4 1930s 15" Metal $600-950
White globe face with red/white/blue CO-OP shield logo on globe face. "QUALITY" arched in top of shield, "PRODUCTS" around bottom with "PREMIUM" in diagonal band across shield.
QUALITY/CO-OP/
PRODUCTS
(ONShield) 3 1930s 13.5" Glass $250-400
White globe face with red/white/blue CO-OP shield logo on globe face. "QUALITY" arched in top of shield, "PRODUCTS" around bottom with "CO-OP" in diagonal band across shield.
QUALITY/FARM BUREAU/
PRODUCTS
(ON Shield) 4 1940s 13.5" Glass $250-450
QUALITY/FARM BUREAU/
PRODUCTS
(ON Shield) 4 1940s Gill $275-475
White globe face with blue outline and red "8 point" border around

Co-op logo in center. White outlined blue outer ring and center band with white "QUALITY" on top ring, "PRODUCTS" on bottom ring with white "FARM/BUREAU" on diagonal band across center.

LANDMARK FYR-ZON GASOLINE 5 1960s 13.5" Glass $300-500
White globe face with large black and red "L" covering most of globe face. Red sun design just above "L". White "LANDMARK" across lower leg of "L" with red logotype "Fyr-Zon" below.

FYR-ZON GASOLINE 4 1950s Capco $200-300
White globe face with red over blue oval in center. White "Fyr-Zon" on upper red area, "GASOLINE" on lower blue area.

****INDIANA FARM BUREAU CO-OPERATVIE ASSOCIATION, INC. - INDIANAPOLIS, IN:** History unknown.
>>NO Indiana Farm Bureau globes are known at this time.

****FARMERS UNION CO-OPERATIVE ORGANIZATION of IOWA:** History unknown.

FARMOIL W/ PLOW SCENE 5 1920s 15" Metal $2200-3200
White globe face with smaller red outlined white circle on lower globe face. Blue lettering arched around edge of small circle ":FARMERS UNION CO-OPERATIVE ORGANIZATION:" at top, "OF IOWA" at bottom with red/gray farm field scene with plow in center of smaller circle. Red outlined blue "FARMOIL" formed around top of circle.

****IOWA FARM SERVICE COMPANY — DES MOINES, IOWA:** History unknown.

FARM BUREAU SPECIAL 5 1940s 14" Gill $250-450

FARM BUREAU SPECIAL 5 1940s Capco $200-300
Red outer ring with gear tooth edge. Red/white outlines around blue center circle. White "FARM BUREAU" arched around top in red area, "SERVICE" around bottom with white script "Special" diagonal across blue center.

FARM BUREAU HI-COMPRESSION 5 1940s 14" Gill $350-550

FARM BUREAU HI-COMPRESSION 5 1940s Capco $250-350
Red outer ring with gear tooth edge. Red/white outlines around blue center circle. White "FARM BUREAU" arched around top in red area, "SERVICE" around bottom with white script "Hi/Compression" diagonal across blue center.

FARM BUREAU OILOGAS 5 1940s 14" Gill $250-550
Red outer ring with gear tooth edge. Red/white outlines around blue center circle. White "FARM BUREAU" arched around top in red area, "SERVICE" around bottom with black outlined orange "OILOGAS" across blue center.

****MINNESOTA FARM BUREAU — ST. PAUL, MINNESOTA:** History unknown.

FARM BUREAU GASOLINE 6 1930s Gill NL
Black/red outlined white globe face with black/red outlined white map of Minnesota in center. Blue outlined red star in center of map, small blue banner arched across map with white "SAVINGS AND SERVICE" on banner. Black outlined red "FARM BUREAU" arched around top, "GASOLINE" around bottom.

MINNESOTA FARM BUREAU 5 1930s Gill $600-950
Red globe face with white center circle. Blue/red outlined white map of Minnesota in center with red rays out on all sides. Blue outlined red star in center of map, small blue banner arched across map with white "SAVINGS AND SERVICE" on banner. White "MINNESOTA FARM BUREAU" arched around top of globe on red area, "SERVICE COMPANY" around bottom.

****MARYLAND CO-OPERATIVES:** History unknown.

CO-OP REGULAR 7 1940s Capco $175-275
White globe face with large "CO-OP" over smaller "REGULAR" across globe face.

CO-OP ETHYL 7 1940s Capco $175-275
White globe face with large "CO-OP" over smaller "ETHYL" across globe face.

****UFA CENTRAL CO-OPERATIVES ASSOCIATION, LTD. — CALGARY, ALBERTA:** History unknown.

MAPLE LEAF GASOLINE 7 1930s 15" Metal $1800-2500
White globe face with ring around center circle. Large maple leaf in center circle with "MAPLE LEAF" arched around top, "GASOLINE" around bottom.

MAPLE LEAF GASOLINE 5 1930s 15" Metal $1800-2500
White globe face with ring around center circle. Small multi color maple leaf in center circle with green "CO-OP" across leaf. Green"MAPLE LEAF" arched around top, red "GASOLINE" around bottom.

MAPLE LEAF PREMIUM 6 1930s 15" Metal $1800-2500
White globe face with ring around center circle. Small multi color maple leaf in center circle with green "CO-OP" across leaf. Green"MAPLE LEAF" arched around top, red "PREMIUM" around bottom.

MAPLE LEAF GASOLINE White on green leaf sideways 5 1930s 15" Metal $1800-2500
White globe face with large green maple leaf, sideways with stem to left, across globe face. White "MAPLE/LEAF/GASOLINE" across leaf.

MAPLE LEAF ETHYL 5 1930s 15" Metal $600-900
White globe face with black outline around white center circle. Black outlined green "MAPLE LEAF" arched around top, black outlined red "GASOLINE" around bottom with large Ethyl (EGC) logo in center circle.

****SOUTHERN FARMERS ASSOCIATION — LITTLE ROCK, ARKANSAS:** History unknown.

SOUTHERN FARMERS CO-OP 6 1940s Capco NL
Description not available.

****UNITED CO-OPERATIVES, INC. - INDIANAPOLIS, INDIANA:** United Co-operatives was formed in the 1930s as a supply company for the various farm co-operatives in the eastern United States. Lubricants were their primary products, but several farm co-ops offered gasoline under the Unico brand.

QUALITY UNICO PRODUCTS 6 1940s 13.5" Glass $250-450
White globe face with red/white/blue CO-OP shield logo on globe face. "QUALITY" arched in top of shield, "PRODUCTS" around bottom with "UNICO" in diagonal band across shield.

****KANSAS FARMERS UNION:** History unknown.

K.F.U. 70 6 1940s Gill NL
Blue outlined white globe face with large blue outlined orange shield on lower globe face. White outlined blue "KFU" on upper shield above "70". Blue outlined logo at top with orange and blue plow scene in center circle.

CROWN & AFFILIATES

****CROWN CENTRAL PETROLEUM — BALTIMORE, MARYLAND:** Crown Central Petroleum traces its earliest history to the east Texas oil fields where, in 1917, oil exploration got underway. By 1920, a Houston refinery was in operation providing high quality lubricants. In 1930, the Blaustein and Rosenberg families, founders of American Oil, purchased Crown Central Petroleum and Crown's headquarters shifted to Baltimore, Maryland. In 1933, jobber marketing in the East was added to a chain of service stations already operated in Houston. Throughout the 1930s and 1940s, jobbers were actively sought in markets from Pennsylvania to Alabama. In many markets, Crown opted to purchase existing marketers and rebrand their locations to Crown, operating them through leasee dealers. Among the companies purchased were two Long Island, New York, companies, Bang Petroleum (1943) and Para-Marine Service Stations (1957), and a metro New York marketer, Nationwide Petroleum (1963). In Baltimore, company owned outlets were acquired with the 1943 purchase of M. Waller Corporation, and numerous Virginia sites were added with the 1948 purchase of Richmond, Virginia, based James River Oil Company. In 1964, Crown became involved in direct retail marketing with the addition of the "Peoples" stations of Energy Oil Co. So successful was direct retailing that by the early 1970s jobber accounts were abandoned in favor of all company operated stations. Crown also entered the convenience store market in the early 1980s with the purchase of Sunoco's "Fast Fare" chain in the Southeast. Crown products are currently available at nearly 300 locations in the Southeast and Mid-Atlantic areas.

CROWN CCPCO 7 1930-1933 15" Metal NL

Orange globe face with white outlined black band arched across center of globe. White lettering "CROWN" on black band with white "C.C.P.Co." monogram on lower orange field.

CROWNZOL CCPCO 7 1930-1933 15" Metal NL

White globe face with orange outlined dark yellow band arched across center of globe face. Black "CROWNZOL" on yellow band, with black "C.C.P.Co." monogram on lower white field.

CROWN GAS-OIL 4 1933-1956 Gill $300-475

Orange globe face with white outlined black band arched across center of globe. White lettering "CROWN" on black band with white "OIL (star) GAS" on lower orange field.

CROWNZOL 4 1933-1956 Gill $300-475

White globe face with orange outlined dark yellow band arched across center of globe face. Black "CROWNZOL" on yellow band, small orange star on lower white field.

CROWNZOL WITH "HI-OCTANE USA" DECAL 4 1950s Gill $325-500

Same as above with decal over star.

CROWNZOL WITH RIBBON DECAL 4 1950s Gill $325-500

Same as above with gold ribbon decal over star.

CROWN GOLD 2 1956-1973 Gill $325-475

CROWN GOLD 5 1956-1973 Capco $200-325

Mirrored gold globe face with orange/white Crown wing logo above center of globe face with blue outlined white lettering "GOLD" below logo.

CROWN SILVER 2 1956-1973 Gill $300-450

CROWN SILVER 5 1956-1973 Capco $200-325

Mirrored silver globe face with word "SILVER" replacing "GOLD" as described above.

CROWN DIESEL 5 1956-1973 Gill $275-425

White globe face with small orange Crown wing logo at top over blue "DIESEL/FUEL".

****THRIFT**: Thrift Oil was a 1950s Crown secondary brand.

THRIFT REGULAR 5 1950s Gill $300-425

Orange globe face with white outlined black band arched across center of globe. White lettering "THRIFT" on black band with white "REGULAR/GAS" on lower orange field.

THRIFT PREMIUM 5 1950s Gill $300-425

Orange globe face with white outlined black band arched across center of globe. White lettering "THRIFT" on black band with white "PREMIUM/GAS" on lower orange field.

****ENERGY OIL COMPANY — ANNANDALE, VIRGINIA**: Founded as Peoples Self-Service Gas Stations, Inc., in Annandale, Virginia, in the late 1940s, Peoples operated a chain of retail gasoline stations in Virginia, Maryland and Washington, DC. By the mid-1960s, the corporate name had become "Energy Oil Company" and over 20 stations were being operated. In July 1964 Crown, who had supplied Peoples for years, purchased the chain and rebranded the stations Crown. Peoples then became a Crown secondary brand, appearing on a few remaining jobber-operated stations in North Carolina. The name was re-introduced for use on some company operated stations Crown purchased from Ashland in 1985, but these were rebranded Crown in 1987 and the Peoples brand is no longer in use.

>>No Peoples globes are known to exist.

CRYSTAL & AFFILIATES

****CRYSTAL OIL & LAND CO. — SHREVEPORT, LOUISIANA**: Founded in Shreveport in 1926 as a producer and refiner, Crystal was largely a petroleum investment company, although they were operating a small number of retail service stations branded Crystal in the late 1920s and early 1930s. They abandoned retail marketing before World War II, but again began amassing retail marketing entities in 1968 with the purchase of Mercury Oil of Shreveport. The firm became known as simply the Crystal Oil Company in 1969. Following difficult times during the gas shortages of the 1970s, various components of Crystal Oil were sold off, some individually and some intact. Today a small number of Crystal stations remain, although under uncertain ownership.

>>No Crystal brand globes from this company are known at this time.

****MERCURY DISCOUNT STATIONS, INC. — SHREVEPORT, LOUISIANA**: A longtime Louisiana gasoline discounter, the Mercury brand appeared on 26 service stations at the time the firm was purchased by the Crystal Oil and Land Company. The Mercury name was changed to Crystal shortly after the November 1968 purchase.

MERCURYGAS PREMIUM REGULAR 9 1950s Capco NL

MERCURYGAS ETHYL 9 1950s Capco NL

MERCURY SUPER REGULAR 9 1960s Capco NL

MERCURY PREMIUM ETHYL 9 1960s Capco NL

These globes are part of a group that was referenced in a group of lens patterns discovered by a collector. Industry sources give a high level of confirmation that this globe exists. No description is available.

****BERRY ASPHALT CO. — LITTLE ROCK, ARKANSAS**: Founded in Chicago in 1918, the company came under control of Quaker State in the 1930s. It existed under several different owners until being purchased by Crystal Oil Company in 1970.

BERRY HY-GRADE PREMIUM 5 1950s Glass $300-500

Blue globe face with yellow outline ring & yellow outline of Arkansas in center of globe face. Yellow "BERRY" arched across top of map with yellow "HI-GRADE" diagonal across center. Yellow "PREMIUM" across bottom of globe face.

****TULSA OIL CO. — DETROIT, MICHIGAN**: Tulsa was an old Detroit independent brand that sold out to Crystal Oil in July 1971. It was in turn purchased by Rock Island about 1978. The United brand name replaced Tulsa in 1985.

TULSA 6 1940s 15" Metal NL

Identical in design to early "Tydol" globes. Orange ring around outer edge with black line around white center circle. Black "TULSA" (oversize "T") across white center.

TULSA 95 4 1950-1970 Capco $450-700

White globe face with yellow outline ring. Red "TULSA" with black dropshade across top of globe face. Red "95" (in quotes) in center above yellow and black stylized cow head and horns.

TULSA 100 3 1950-1970 Capco $450-700

White globe face with yellow outline ring. Red "TULSA" with black dropshade across top of globe face. Red "100" (in quotes) in center above yellow and black stylized cow head and horns.

TULSA HI-TEST 4 1950-1970 Capco $450-700

White globe face with yellow outline ring. Red "TULSA" with black dropshade across top of globe face. Red "HI-TEST" (in quotes) in center above yellow and black stylized cow head and horns.

TULSA DIESEL 5 1950-1970 Capco $475-750
White globe face with yellow outline ring. Red "TULSA" with black dropshade across top of globe face. Red "DIESEL" (in quotes) in center above yellow and black stylized cow head and horns.

TULSA FUEL OIL 5 1950-1970 Capco $475-750
White globe face with yellow outline ring. Red "TULSA" with black dropshade across top of globe face. Red "FUEL OIL" (in quotes) in center above yellow and black stylized cow head and horns.

****CRYSTAL PETROLEUM CO. — BIRMINGHAM, ALABAMA:** This company was a discounter with 74 stations in Alabama, Mississippi, Florida, and Tennessee that was purchased by Crystal Oil in October 1971. Despite the brand name, prior to this time the companies were unrelated to the best of our knowledge.
>>No Crystal globes known at this time.

****EASTERN PETROLEUM CORPORATION—BILOXI, MISSISSIPPI:** Eastern Petroleum Corporation was a discounter with 23 stations in Alabama, Mississippi and Florida operating under the Eastern brand name. The company was purchased by Crystal Oil in November 1971.
>>No Eastern globes known at this time.

****STONE OIL CO. — FITZGERALD, GEORGIA:** This Georgia jobber operation was branding 56 stations under the "Stone's" name when the company was purchased by Crystal Oil Company in July 1972.

STONES REGULAR 5 1950s Capco $250-350
White globe face with blue outline ring and yellow "STONES" over blue script "Regular".

STONES ETHYL 5 1950s Capco $250-350
White globe face with blue outline ring and yellow "STONES" over blue script "Ethyl."

STONES PREMIUM 5 1950s Capco $250-350
White globe face with blue outline ring and yellow "STONES" over blue script "Premium".

****HIGH OCTANE TERMINAL CO.—PANAMA CITY, FLORIDA:** This was a terminal company founded by Mr. Joe Hutchinson that was operating 52 stations under the "Mo-Jo" brand in Florida and Alabama when Crystal Oil Company purchased the operation in August 1972. Stations were rebranded Crystal shortly afterwards.

MO-JO HI NEIGHBOR REGULAR 7 1950s 13.5" Glass $250-400
White globe face with orange-red "Mo-Jo" in center. Green "HI-NEIGHBOR" arched around top, orange-red "REGULAR" around bottom.

MO JO HI NEIGHBOR X-TANE 7 1950s 13.5" Glass $250-400
White globe face with orange-red "Mo-Jo" in center. Green "HI-NEIGHBOR" arched around top, orange-red "X-TANE" around bottom.

MO-JO HI NEIGHBOR X-TANE PREMIUM6 1950s 13.5" Glass $250-400
White globe face with orange-red "Mo-Jo/X-TANE" in center. Green "HI-NEIGHBOR" arched around top, orange-red "PREMIUM" around bottom.

****PLATOLENE — CARMI, ILLINOIS:** Founded in the 1930s as R. J. Oil Refining Co in Terre Haute, Indiana, this company originally marketed under the "500" brand, with "Platolene" added to the "500" after World War II. In December 1973 the operation was acquired by Crystal Oil Company, but shortly thereafter the retail operations were moved to Carmi, Illinois. Several dozen stations are in operation today, under the Platolene 500 and 500 Express brand names.

"500" R. J. OIL & REFINING CO. 5 1940s 13.5" Glass $550-850
Red globe face with black/white outline rings. Large white "500" at top over banner across lower globe face. "R.J. OIL AND REFINING CO." on banner. Black and white checked flag on left.

PLATOLENE 500 4 1950s 13.5" Glass $400-650
Red globe face with black/white outline rings. Large white "500" at top over banner across lower globe face. "PLATOLENE" on banner.

PLATOLENE 500 4 1960s Oval Capco $350-600
White globe face with red "500" on upper globe face. Blue outline ring and blue band on lower globe face with white "PLATOLENE" on band.

DERBY & AFFILIATES

****DERBY OIL CO. —WICHITA, KANSAS:** Derby Oil Co. was a mid-continent refiner/marketer founded in Wichita, Kansas in 1917. In the late 1930s, Derby was instrumental in the creation of Sovereign Service, a marketing organization for independent dealers supported by a number of mid-continent refiners. After World War II, Derby concentrated on an expanding chain of retail service stations as well as numerous dealer operated units and a large jobber/marketer arrangement with Piester. In 1955 Colorado Interstate Gas Company purchased Derby and, with its large oil production system, allowed for greater expansion of the Derby refinery and marketing network. In the early 1970s, Colorado Interstate Gas became a part of the Coastal Corporation and in 1988 Derby stations, along with others in the Coastal organization, were rebranded Coastal.

DERBY 6 1917 OPE Sphere NL
Spherical globe with early Derby "DOC" logo on globe face.

DERBY RED BACKGROUND WHITE STAR 5 1920-1926 OPE W $1600-2200
White globe with red globe face. Large white star covers most of globe face. Interlocking "D/O/Co." in blue in center of star.

DERBY RED BACKGROUND WHITE STAR 5 1920-1926 OPB W $1400-1800
White globe with red globe face. Large white star covers most of globe face. Interlocking "D/O/Co." in blue in center of star.

DERBY SUPER RATIO MOTOR FUEL 5 1926-1935 15" Metal $500-850
White globe face with blue outlined red circle in center. White star on red center circle with blue interlocking "D/O/Co." in center of star. Blue "SUPER-RATIO" arched around top, "MOTOR FUEL" arched around bottom.

DERBY GASOLINE 3 1935-1946 Gill $275-450
White globe face with red outlined blue circle in center. Small white "textured" star in center blue circle with blue "D/O/Co." interlocking on center of star. Blue "DERBY" arched around top, "GASOLINE" around bottom.

DERBY GASOLINE 3 1935-1946 Gill $275-450

DERBY GASOLINE 5 1935-1946 13.5" Glass $275-450
White globe face with blue outlined red circle in center. Small white "textured" star in center red circle with blue "D/O/Co." interlocking on center of star. Blue "DERBY" arched around top, "GASOLINE" around bottom.

DERBY'S FLEXGAS ETHYL (EGC) 3 1935-1946 Gill $275-450

DERBY'S FLEXGAS ETHYL (EGC) 6 1935-1946 13.5" Glass $275-450
Red globe face with black outlined white circle in center. Small Ethyl (EGC) logo in center with yellow rays. White "DERBY'S" arched around top, "FLEXGAS" around bottom.

DERBY'S FLEXGAS5 1946-1950 Gill $275-450

DERBY'S FLEXGAS6 1946-1950 13.5" Glass $275-450
White globe face with blue outlined red circle in center. Small white "textured" star in center red circle with blue "D/O/Co." interlocking on center of star. Blue "DERBY'S" arched around top, "GASOLINE" around bottom with dot in white ring at center on either side.

DERBY'S FLEXGAS ETHYL large (EC) 3 1946-1950 Gill $275-450

DERBY'S FLEXGAS ETHYL large (EC) 3 1946-1950 13.5" Glass $275-450
Red globe face with black outlined white circle in center. Large Ethyl (EC) logo in center circle. White "DERBY'S" arched around top, "FLEXGAS" around bottom.

DERBY 3 1950-1970 Gill $275-450

DERBY 2 1950-1970 Capco $200-300
Blue/white outline rings around red globe face. Large white "tex-

tured" star on red face with blue band horizontal across star below center. White "DERBY" on blue band.

DERBY 6 1965-1970 Gill Plastic Lens $250-400

Domed plastic lens. Blue/white outline rings around red globe face. Large white "textured" star on red face with blue band horizontal across star below center. White "DERBY" on blue band.

DERBY PREMIUM ETHYL 4 1960-1970 Capco $225-350

Red globe face with black outlined white circle in center. Large Ethyl (EC) logo in center circle. White "DERBY" arched around top, "PREMIUM" around bottom.

DERBY DIESEL 4 1950-1970 Gill $250-450

DERBY DIESEL 4 1950-1970 Capco $200-325

White globe face with blue outline ring. Large red "DIESEL" with blue underline across center of globe face. Small red circle/ white star "DERBY" logo at bottom.

****SOVEREIGN SERVICE**: Sovereign Service was a franchise program allowing independent dealers supplied by the various mid-continent refiners participating in the program to be branded "Sovereign Service." Dealers were held to high standards of operation, supported by cooperative advertising of the program's features (independent dealers only, clean rest rooms, 24 hour service, etc.). Derby was a leading participant in the program, along with Kanotex / Apco, Vickers, Rock Island, Eason, Elreco, Yale, Shallow Water and others. The program was initiated in the late 1930s (about 1937) and survived as late as 1954. Globes from the individual brands are listed with that company's listing in these books.

SOVEREIGN "SS" 4 1937-1954 Yellow Ripple $1100-1500

SOVEREIGN "SS" 3 1937-1954 Gill $300-450

Brown globe face with large white outlined yellow shield covering most of globe face. Brown "SOVEREIGN" diagonal across shield with small brown shield at top of larger shield. Yellow interlocking "SS" on small brown shield.

SOVEREIGN ETHYL 4 1937-1954 White Ripple $900-1300

SOVEREIGN ETHYL 4 1937-1954 Gill $300-450

Red globe face with large white/red outlined white shield covering most of globe face. Red "SOVEREIGN" diagonal across shield with small Ethyl (EC) logo in black outlined circle at top of shield.

SOVEREIGN QUALITY 3 1937-1954 Yellow Ripple $1100-1500

SOVEREIGN QUALITY 3 1937-1954 Gill $300-450

Brown globe face with large white outlined yellow shield covering most of globe face. Brown "SOVEREIGN/QUALITY" diagonal across shield with small brown shield at top of larger shield. Yellow "S" on small brown shield.

KANOTEX SOVEREIGN QUALITY 4 1937-1952 Yellow Ripple $1200-1650

KANOTEX SOVEREIGN QUALITY 3 1937-1952 Gill $400-600

Same as "SOVEREIGN QUALITY" above with small black outlined oval replacing "SS" shield at top of globe face. Multi-color Kanotex logo with red "KANOTEX" in small white oval.

DIXIE DISTRIBUTORS

****DIXIE DISTRIBUTORS, INC. — ANN ARBOR, MICHIGAN**: Dixie Distributors, Inc. was a distributor association put together by a group of Michigan based oil jobbers in Ann Arbor in 1926. Almost immediately independent jobbers from all around the country applied for a Dixie franchise and joined the association in order to be able to market their products under the Dixie name. The organization expanded greatly until World War II when gasoline shortages forced many independents, including many of the Dixie affiliates, out of business. Marketing efforts by the majors after World War II were much greater than an independent organization could hope to compete with and many former Dixie jobbers eventually were branded by the majors. Nevertheless, Dixie stations survived until the 1960s — or later in some areas — and the organization survives today.

DIXIE OILS GASOLINE 4 1926 -1928 OPE N $1500-2200

White globe with dark blue ring around yellow center. White diamond on yellow center with blue "OILS" on diamond. White "DIXIE" arched around top, "GASOLINE" around bottom.

DIXIE HIGH-TEST 5 1926-1928 OPE N $1800-2400

White globe with dark blue ring around yellow center. White diamond on yellow center with blue "OILS" on diamond. White "DIXIE" arched around top, "HIGH TEST" around bottom.

DIXIE POWER TO PASS 5 1928-1931 Blue Oval Keyhole $2200-3000

Blue glass body for oval lenses that has rounded area between lens area and base. White "POWER-TO-PASS" lettering on lower area. Globe face is blue with white oval in center. Blue outlined yellow diamond on center white oval. Blue "OILS" in diamond, white "DIXIE" arched around top, "GASOLINE" around bottom.

DIXIE POWER TO PASS 5 1928-1931 White Oval Keyhole $2200-3000

White glass body for oval lenses that has rounded area between lens area and base. Blue "POWER-TO-PASS" lettering on lower area. Globe face is blue with white oval in center. Blue outlined yellow diamond on center white oval. Blue "OILS" in diamond, white "DIXIE" arched around top, "GASOLINE" around bottom.

DIXIE BLUE STREAK 6 1928-1931 Blue Oval Keyhole $2300-3100

Blue glass body for oval lenses that has rounded area between lens area and base. White "BLUE STREAK" lettering on lower area. Globe face is blue with white oval in center. Blue outlined yellow diamond on center white oval. Blue "OILS" in diamond, white "DIXIE" arched around top, "GASOLINE" around bottom.

DIXIE GASOLINE BENZOL 5 1928-1931 Green Oval Keyhole $2400-3200

Green glass body for oval lenses that has rounded area between lens area and base. White "BENZOL" lettering on lower area. Globe face is blue with white oval in center. Blue outlined yellow diamond on center white oval. Blue "OILS" in diamond, white "DIXIE" arched around top, "GASOLINE" around bottom.

DIXIE KNOCK OUT 5 1928-1931 White Oval Keyhole $2300-3100

White glass body for oval lenses that has rounded area between lens area and base. Blue "KNOCK OUT" lettering on lower area. Globe face is blue with white oval in center. Blue outlined yellow diamond on center white oval. Blue "OILS" in diamond, white "DIXIE" arched around top, "GASOLINE" around bottom.

DIXIE LIBERTY 5 1928-1931 Yellow Oval Keyhole $2400-3200

Yellow glass body for oval lenses that has rounded area between lens area and base. White "LIBERTY" lettering on lower area. Globe face is blue with white oval in center. Blue outlined yellow diamond on center white oval. Blue "OILS" in diamond, white "DIXIE" arched around top, "GASOLINE" around bottom.

DIXIE LIBERTY 6 1928-1931 White Oval Keyhole $2300-3100

White glass body for oval lenses that has rounded area between lens area and base. Blue "LIBERTY" lettering on lower area. Globe face is blue with white oval in center. Blue outlined yellow diamond on center white oval. Blue "OILS" in diamond, white "DIXIE" arched around top, "GASOLINE" around bottom.

DIXIE NEVERNOX 5 1928-1931 Red Oval Keyhole $2400-3200

Red glass body for oval lenses that has rounded area between lens area and base. White "NEVER-NOX" lettering on lower area. Globe face is blue with white oval in center. Blue outlined yellow diamond on center white oval. Blue "OILS" in diamond, white "DIXIE" arched around top, "GASOLINE" around bottom.

DIXIE GASOLINE 4 1920s-1930s 15" Metal $750-1200

White globe face with blue oval covering most of globe face. White oval in center of blue oval with yellow diamond on white oval. Red "OILS" on yellow diamond, white "DIXIE" arched around top on blue oval, "GASOLINE" around bottom.

DIXIE GASOLINE

POWER TO PASS 4 1931-1940 15" Metal $800-1400
DIXIE GASOLINE
POWER TO PASS 5 1931-1940 Jeweled $1400-2200
DIXIE GASOLINE
POWER TO PASS 3 1931-1960 Gill $350-550
DIXIE GASOLINE
POWER TO PASS 4 1931-1960 Yellow Ripple $1000-1500
DIXIE GASOLINE
POWER TO PASS 3 1941-1960 13.5" Glass W $300-500
DIXIE GASOLINE
POWER TO PASS 4 1941-1960 Yellow Capco $250-450
Yellow globe face with blue oval on center of globe face. White oval on center of blue oval with yellow diamond on white oval. Red "OILS" on yellow diamond, white "DIXIE" arched around top on blue oval, "GASOLINE" around bottom. Blue "POWER TO PASS" arched around bottom of globe face under oval.
DIXIE ETHYL (EGC) 5 1931-1940 15" Metal $600-900
DIXIE ETHYL (EGC) 5 1931-1940 Jeweled $1200-2000
DIXIE ETHYL (EGC) 3 1931-1949 Gill $300-500
DIXIE ETHYL (EGC) 4 1931-1949 Yellow Ripple $850-1350
DIXIE ETHYL (EGC) 3 1931-1949 13.5" Glass W $250-450
Yellow outline ring around inner and outer edge of blue outer ring. Black ring around white center circle. White "DIXIE" in arched around top in blue ring, "GASOLINE" around bottom with large Ethyl (EGC) logo in center white circle. Yellow rays out from Ethyl logo.
DIXIE ETHYL (EC) 4 1941-1949 Blue Ripple $1600-2200
DIXIE ETHYL (EC) 4 1941-1949 13.5" Glass W $250-450
DIXIE ETHYL (EC) 4 1941-1949 Yellow Capco $250-450
Yellow outline ring around inner and outer edge of blue outer ring. Black ring around white center circle. White "DIXIE" in arched around top in blue ring, "GASOLINE" around bottom with large Ethyl (EC) logo in center white circle. Yellow rays out from Ethyl logo.
DIXIE PREMIUM 4 1949-1960 Yellow Capco $300-475
Same as "DIXIE GASOLINE POWER TO PASS" except red "PREMIUM" on center yellow diamond and no "POWER TO PASS" on lower globe face.
DIXIE BLUE 5 1930-1932 13.5" Glass W $250-400
White globe face with large blue "DIXIE" arched over "BLUE".
DIXOLINE W/LINES 6 1932-1940 15" Metal $350-500
DIXOLINE W/LINES 4 1932-1940 13.5" Glass W $225-375
White globe face with dark blue "DIXOLINE" diagonal from upper left to lower right on globe face. Light blue dropshading around each letter and light blue line above and below lettering.
DIXOLINE 3 1932-1940 13.5" Glass W $225-375
DIXOLINE 2 1932-1940 Gill $250-400
White globe face with blue outline ring. Blue "DIXOLINE" diagonal from lower left to upper right with "D" and "E" extending through outline ring.

****BURFORD OIL COMPANY — DALLAS, TEXAS**: Later known as East Texas Refining, the Burford Oil Company sold gasoline under the Burford, Crystal, and Silencer brands. Later, it became a Dixie Distributors affiliate.

BURFORD SPECIAL 5 1930s 15" Metal NL
Black yellow outlined white globe face with black BURFORD arched around top, SPECIAL around bottom with yellow center circle with black band and diamond and orange "THE ANTIKNOCK GASOLINE" in three lines in the center diamond.
CRYSTAL 4 1930s OPB NL
White globe with black outlined green in reverse bowtie lettering across center of globe face.
CRYSTAL
UPSIDE DOWN 5 1930s OPB NL
Same as above, design inverted.
CRYSTAL 5 1930s 15" Metal NL
Description not available.
SILENCER
"S-S-SH" 4 1930s 15" Metal $750-1200
White globe face with yellow oval across center of glboe face. Black "SILENCER" in oval with orange "S-S-SH" across top, "Stops Knocks" across bottom.
SILENCER ETHYL
(EGC) 4 1930s 15" Metal $750-1200
White globe face with yellow oval across upper globe face. Black "SILENCER" in oval. Black outlined white circle on lower globe face with small Ethyl (EGC) logo in circle.
DIXIE SILENCER
STOPS KNOCKS 5 1940s Yellow Capco $300-550
White globe face with yellow oval across center. black "SILENCER" in yellow oval. Red "DIXIE" arched across top, "Stops Knocks" across bottom.
SILENCER DIXIE
ETHYL (EGC) 5 1940s Red Capco $275-500
White globe face with black outlined white circle in center. Small Ethyl (EGC) logo in center circle. Blue "SILENCER" arched around top, "DIXIE" around bottom.

DX & AFFILIATES
MID-CONTINENT PETROLEUM — TULSA, OKLAHOMA: Mid-Continent was organized in 1916 from the various holdings of oil pioneer Josh Cosden and others. Gasoline marketing under the Diamond brand name began about 1920. By the mid-1920s, Mid-Continent operated a large number of stations throughout the central United States. A super-premium fuel "Diamond D-X" was introduced about 1933 and by 1940 stations were branded as "Diamond D-X" stations. After World War II, the brand name became simply "D-X". In 1955, Mid-Continent merged with Sunray Oil Co. and the corporate name became Sunray DX Oil Company, revised in 1962 to DX-Sunray Oil Company. In 1968 DX-Sunray merged with Philadelphia based Sun Oil Company and many dual logo Sunoco/DX items were introduced at this time. DX globes had been phased out by the time of the merger. Shortly after the merger the Sunoco brand name replaced DX in states where Sunoco already marketed. The DX brand was retained in other areas until after 1980 when Sunoco began gradually replacing DX. As of this writing, only small areas of DX branding remain, primarily jobber operated stations that just haven't yet been converted.

NEVR NOX
GASOLENE 5 1920s OPE W $1200-1750
White globe with black logotype "NevrNox" across globe face. Black/white outlined red arrow below logotype with white "GASOLENE" on arrow.
NEVR NOX
GASOLENE
DIAMOND
PRODUCT 4 1920s OPE W/N $1700-2600
Same as above with red "DIAMOND" arched around top, "GASOLINE" around bottom.
NEVR NOX
"PEPPY POWER"
(sm letters) 5 1920s OPE N $2000-2800
White globe with red outline around face. Black/white outlined black "NevrNox" logotype across globe face with red arrow below. White "GASOLINE" on red arrow. Small red "PEPPY POWER" arched around top, "IN EVERY DROP" around bottom.
NEVR NOX
"PEPPY POWER" 5 1920s OPE RW $2000-2800
White globe with red outline around face. Black/white outlined black "NevrNox" logotype across globe face with red arrow below. White "GASOLINE" on red arrow. Red "PEPPY POWER" arched around top, "IN EVERY DROP" around bottom.
NEVR NOX
"PEPPY POWER"
(inverted) 6 1920s OPE RW $1200-1800
INVERTED white globe with red outline around face. Black/white outlined black "NevrNox" logotype across globe face with red arrow below. White "GASOLINE" on red arrow. Red "PEPPY POWER" arched around top, "IN EVERY DROP" around bottom.
NEVR NOX
GASOLENE NON-
POISONOUS 5 1920s OPE W $1500-2000
White globe with red diamond outline at center of globe face. Black "NevrNox" logotype inside diamond. Red arrow below logotype with white "GASOLINE" on arrow. Small black "NON-POISONOUS" below arrow. Likely created in response to the national "scare" about "poison" Ethyl gasoline.
DIAMOND NITRO
GASOLINE 6 1920s OPE RW NL
White globe face with outline ring around globe face. Large "NITRO"

across center with "DIAMOND" arched around top and "GASOLINE" around bottom.

DIAMOND 5 1920s OPE RW $1000-1600
White globe with red outline ring around face. Solid red diamond in center with large black "DIAMOND" across diamond.

DIAMOND GAS/OIL 5 1920s 15" Metal $800-1350
White globe face with red outline ring. Solid red diamond in center with large black "DIAMOND" across diamond. Black outlined red "GAS" arched around top, "OIL" around bottom.

DIAMOND 4 1925-1933 OPE RW $750-1250
White globe with red diamond in center of globe face. White and red diamond outlines around center diamond. Black "DIAMOND" across diamond logo.

DIAMOND 5 1930-1933 13.5" Glass W $200-400
White globe face with red outline ring. Red diamond in center outlined by white diamond and red diamond bands. Black "DIAMOND" across white band and red center diamond.

DIAMOND DX LUBRICATING... TAN 4 1933-1936 Banded $300-500
Tan globe face with red outline ring. Red diamond outline with tan center in center of globe face. Large black outlined tan "D-X" with black dropshade superimposed over diamond outline. Black "DIAMOND" arched over top, black "chevron" below diamond with tan "LUBRICATING MOTOR FUEL" on chevron.

DIAMOND DX LUBRICATING MOTOR FUEL 3 1933-1936 13.5" Glass W $250-450

DIAMOND DX LUBRICATING MOTOR FUEL 4 1933-1936 Banded $300-500
White globe face with red outline ring. Red diamond outline with tan center in center of globe face. Large black outlined white "D-X" with black dropshade superimposed over diamond outline. Black "DIAMOND" arched over top, black "chevron" below diamond with tan "LUBRICATING MOTOR FUEL" on chevron.

NEVRNOX ETHYL 4 1933-1936 13.5" Glass W $250-475

NEVRNOX ETHYL 4 1933-1936 Banded $300-525
White globe face with red outline ring. Black outlined red logotype "NevrNox" arched around top. Black and white Ethyl (EGC) logo with red rays below globe in plain black and white, without the yellow lettering and info box.

DX LUBRICATING MOTOR FUEL... WHITE 3 1936-1946 Banded $225-400
White globe face with red outline ring. Red diamond outline with tan center in center of globe face. Large black outlined white "D-X" with black dropshade superimposed over diamond outline. Black "chevron" below diamond with tan "LUBRICATING MOTOR FUEL" on chevron.

DX ETHYL LUBRICATING MOTOR FUEL 4 1936-1946 Banded $300-475

DX ETHYL LUBRICATING MOTOR FUEL 4 1936-1946 13.5" Glass W $300-450
White globe face with red outline ring. Red diamond outline with white center in center of globe face. Large black "D-X" superimposed over diamond. Black outlined white band below "D-X" with red "ETHYL" on band. Black "chevron" below logo with white "LUBRICATING MOTOR FUEL" on chevron.

POWER...tan 4 1933-1936 Banded $250-400
Tan globe face with red diamond outline with tan center in center of globe face. Red outline ring around face. Black "POWER" across diamond logo.

POWER...white 3 1936-1940 Banded $250-375

POWER...white 3 1936-1940 13.5" Glass W $250-350
White globe face with red diamond outline with white center in center of globe face. Red outline ring around face. Black "POWER" across diamond logo.

POWER "G"...white 4 1940-1946 Banded $275-450

POWER "G"...white 3 1940-1946 13.5" Glass W $250-400
White globe face with red diamond outline with white center in center of globe face. Red outline ring around face. Black "POWER" across diamond logo, with large black italic "G" at lower edge of diamond.

DX LUBRICATING GASOLINE 2 1946-1957 Capco $200-325
Tan globe face with red outline ring. Red and black "D-X" logo at top of globe face. Black script "Lubricating" above red "GASOLINE" below logo.

DX ETHYL 3 1946-1955 Capco $225-350
Tan globe face with red outline ring. Red and black "D-X" logo at top of globe face. Black oval with tan "ETHYL" above black "Lubricating Gasoline" below logo.

DX DIESEL 4 1946-1957 Capco $200-300
Tan globe face with red outline ring. Red and black "D-X" logo at top of globe face. Black script "Diesel" below logo.

DX BORON 2 1955-1957 Capco $150-275
White globe face with thin red and thicker black line diagonal down from left side and curved across bottom. Red and black "D-X" logo with tan center at top of globe face above red italic "BORON".

DX BORON with "rocket" frame 5 1955-1957 Capco $850-1350
Same globe as above, enclosed in a cutout "rocket" frame designed to fit around Capcolite globe.

DX LUBRICATING GASOLINE - 1957 LOGO 2 1957-1968 Capco $125-225
White globe face with large red/white/blue 1957 "DX" logo above blue "LUBRICATING/GASOLINE".

DX BORON- 1957 LOGO 2 1957-1968 Capco $125-225
White globe face with large red/white/blue 1957 "DX" logo above blue "BORON".

DX DIESEL - 1957 LOGO 3 1957-1968 Capco $200-325
Red outlined white globe face with large red/white/blue 1957 "DX" logo above italic blue "DIESEL".

DX MARINE - 1957 LOGO 5 1957-1968 Capco $450-850
White globe face with red outline ring. Red/white/blue "DX" logo at top. Two blue wave lines across globe face below logo with two blue boats and "MARINE" above top line and "GASOLINE" between lines.

****SUNRAY OIL:** Sunray, a mid-continent area producer/refiner dating back to the 1920s, was historically very involved with production and unbranded marketing with very little attention paid to branded "Sunray" marketing. Gasoline was sold under the company name through a small number of jobbers and due to the limited branded marketing, the globes listed are rare. Sunray Oil merged with Mid-continent Petroleum in 1955 and the "DX" brand was added to Sunray logos at existing Sunray stations. About 1962 those few stations became simply "DX" stations.

SUNRAY 6 1930-1940 Orange Ripple NL
White edge of globe face with large black/white outlined octagon covering most of globe face. Upper 3/4 of octagon is orange with large yellow sun and rays. Lower 1/4 is green. Large black "SUNRAY" down arched below sun across rays.

SUNRAY GASOLINE 4 1940-1955 Orange Ripple $1500-2300

SUNRAY GASOLINE 3 1950-1955 Orange Capco $750-1100
Black edge of globe face with large black/white outlined octagon covering most of globe face. Uper 3/4 of octagon is orange with large yellow sun and rays. Lower 1/4 is green. Large black "SUNRAY" arched *up* below sun across rays. Black "GASOLINE" in lower green area.

SUNRAY ETHYL (EGC) 6 1932-1946 13.5" Glass $800-1200
Black edge of globe face with large black/white outlined octagon covering most of globe face. Uper 3/4 of octagon is orange with large yellow sun and rays. Lower 1/4 is green. Large black "SUNRAY" arched *up* below sun across rays. Yellow circle on lower green area. Ethyl (EGC) with black rays on circle.

SUNRAY ETHYL (EC) 4 1946-1955 Red Ripple $1300-2100

SUNRAY ETHYL (EC) 3 1946-1955 Red Capco $700-1000
Black edge of globe face with large black/white outlined octagon covering most of globe face. Uper 3/4 of octagon is orange with

large yellow sun and rays. Lower 1/4 is green. Large black "SUNRAY" arched *up* below sun across rays. Yellow circle on lower green area. Ethyl (EC) with black rays on circle.

****PREMIER — LONGVIEW, TEXAS:** Premier was an independent Texas refiner with stations in Texas, Oklahoma, Arkansas and Louisiana. In 1964 the company merged with Sunray DX. Some stations were spun off to Vickers, while DX retained the Premier name. Some unsuccessful DX stations were later converted to Premier and the brand survived at a small number of locations until DX was rebranded Sunoco in the 1980s.

PREMIER
REGULAR 3 1935-1955 13.5" Glass $225-375
White globe face with blue/red outline rings. Red "PREMIER" with blue dropshade across upper globe face with blue script "REGULAR" below Premier".

PREMIER MOTOR 4 1935-1955 13.5" Glass $225-375
White globe face with blue/red outline rings. Red "PREMIER" with blue dropshade across upper globe face with blue script "MOTOR" below Premier".

PREMIER
ETHYL (EGC) 4 1935-1941 13.5" Glass $225-375
White globe face with blue/red outline rings. Red "PREMIER" with blue dropshade across upper globe face with Ethyl (EGC) logo with yellow rays below "PREMIER".

PREMIER
ETHYL (EC) 3 1941-1955 13.5" Glass $225-375
White globe face with blue/red outline rings. Red "PREMIER" with blue dropshade across upper globe face with Ethyl (EC) logo with yellow rays below "PREMIER".

PREMIER
REGULAR 3 1950-1965 Capco $175-250
White globe face with blue/red outline rings. Red "PREMIER" with blue dropshade across upper globe face with blue script "REGULAR" below Premier".

PREMIER
ETHYL (EC) 3 1950-1965 Capco $175-250
White globe face with blue/red outline rings. Red "PREMIER" with blue dropshade across upper globe face with Ethyl (EC) logo with yellow rays below "PREMIER".

PREMIER DIESEL
FUEL 5 1950-1965 Capco $200-275
White globe face with blue/red outline rings. Red "PREMIER" with blue dropshade across upper globe face with blue script "DIESEL/FUEL" below Premier".

PREMIER MARINE
WHITE 5 1950-1965 Capco $400-600
White globe face with blue/red outline rings. Red "PREMIER" with blue dropshade across upper globe face with blue script "MARINE/WHITE" below Premier".

****BARNSDALL REFINING CO. — BARNSDALE, OKLAHOMA:** One of the oldest oil refiners in the business, Barnsdall began retail gasoline marketing in 1921. At one time one of the largest mid-continent marketers, Barnsdall was reorganized during the Depression as Barnsdall Refining. In 1940, the Bareco brand name replaced the Barnsdall brand. In 1952, Bareco closed their last refinery to concentrate on wax production and the gasoline marketing was sold off to Mid-Continent Petroleum. Mid-Continent eventually phased out the Bareco brand after the merger with Sunray.

BARNSDALL SMALL
GLOBE 4" BASE 6 1921-1922 OPE N NL
White globe with red "BARNSDALL REFINING" arched around top, "GASOLINE" around bottom wth red "B" in black square in center. Red "BE" above square and "SQUARE" below square.

BARNSDALL
DOMED GLOBE 5 early 1920s OPE $2500-3500
White "pylon" domed top globe with etched details. Blue "BARNSDALL" across top of globe face, blue "B" in center outlined in square and blue "GASOLINE" across bottom.

BARNSDALL BE
SQUARE GAS 5 1920s OPE N $1800-2500
White globe with black outline ring around face. Large blue "B" outlined in a square in the center of globe face with small "BE" just above square and "SQUARE" just below. Red "BARNSDALL" arched around top, "GASOLINE" around bottom.

BARNSDALL
SUPER GAS 5 1920s OPE N $1800-2600
White globe with large blue "SUPER-GAS" across globe face. Red "BARNSDALL" arched around top, "ANTI-KNOCK" in straight line below "SUPER-GAS".

BLUE MOTOR
FUEL 4 1930-1941 13.5" Glass $300-475
White globe face with blue outline ring. Red square in center with white and blue squares outlining it. Red "BLUE" arched around top, "GASOLINE" around bottom.

BARNSDALL
MOTOR FUEL 4 1930-1941 13.5" Glass $300-475
White globe face with blue outline ring. Red square in center with white and blue squares outlining it. Red "BARNSDALL" arched around top, "GASOLINE" around bottom.

BARNSDALL
MOTOR FUEL
(FLAT FACE) 5 1930-1935 13.5" Glass NL
Flat face for special body. White globe face with blue outline ring. Red square in center with white and blue squares outlining it. Red "BARNSDALL" arched around top, "GASOLINE" around bottom.

BARNSDALE/BE/B/
SQUARE/GASOLINE 3 1930-1941 13.5" Glass
$300-475
White globe face with blue outline ring. Blue "B" in blue square outline in center. Small "BE" above square, small "SQUARE" below. Red "BARNSDALL" arched around top. "GASOLINE" around bottom.

BARNSDALL/
SUPER-GAS/
ANTI-KNOCK 4 1930-1933 Balcrank $300-500
White globe face with blue outline ring. Red "BARNSDALL" arched around top, blue "SUPER-GAS" across center and red "ANTI-KNOCK" across bottom.

BARNSDALL/
SUPER-GAS/
ETHYL (EGC) 4 1933-1940 13.5" Glass $300-475
White globe face with blue outline ring. Red "BARNSDALL" arched around top, blue "SUPER-GAS" across center and Ethyl (EGC) logo below.

****BARECO — BARNSDALL, OKLAHOMA:** Successors to Barnsdall Refining, the Bareco name replaced Barnsdall in 1940. Bareco marketing was sold to Mid-Continent in 1952, although the Bareco name remained in use for several years after that.

BARECO/BE/B/
SQUARE/
GASOLINE 3 1940-1955 13.5" Glass $300-450
White globe face with blue outline ring. Blue "B" in blue square outline in center. Small "BE" above square, small "SQUARE" below. Red "BARECO" arched around top, "GASOLINE" around bottom.

BARECO/
SUPER-GAS/
ETHYL (EGC) 4 1940-1955 13.5" Glass $300-475
White globe face with blue outline ring. Red "BARECO" arched around top, blue "SUPER-GAS" across center and Ethyl (EGC) logo below.

BARECO/
SUPER-GAS/
ETHYL (EC) 4 1940-1955 13.5" Glass $300-450
White globe face with blue outline ring. Red "BARECO" arched around top, blue "SUPER-GAS" across center and Ethyl (EC) logo below.

SUPER GAS/BE/B/
SQUARE/
PREMIUM 4 1940-1955 13.5" Glass $300-450
White globe face with blue outline ring. Blue "B" in blue square outline in center. Small "BE" above square, small "SQUARE" below. Red "SUPER-GAS" arched around top, "PREMIUM" around bottom.

****MONARCH MANUFACTURING CO. — COUNCIL BLUFFS, IOWA:** Owner of the famous "MonaMotor" trademark, Monarch began gasoline marketing about 1920 and merged into Barnsdall around 1929.

MONAMOTOR
GASOLINE 5 1920s OPE N $2000-2600
White globe with red face. Black/white outlined white diamond in center with black logotype "MonaMotor" across diamond and white outlined black "GASOLINE" arched around bottom.
MONAMOTOR
GASOLINE 6 1920s OPB $1800-2400
White globe with red face. Black/white outlined white diamond in center with black logotype "MonaMotor" across diamond and white outlined black "GASOLINE" arched around bottom.
MONAMOTOR
GASOLINE 6 1920s OPE Oval NL
Oval globe with white globe with red face. Black/white outlined white diamond in center with black logotype "MonaMotor" across diamond and white outlined black "GASOLINE" arched around bottom.
MONAMOTOR
BENZOL 6 1920s OPE N NL
White globe with red face. Black/white outlined white diamond in center with black logotype "MonaMotor" across diamond and white outlined black "BENZOL" arched around bottom.
MONAMOTOR
HI-TEST 6 1920s OPE N NL
White globe with red face. Black/white outlined white diamond in center with black logotype "MonaMotor" across diamond and white outlined black "HI TEST" arched around bottom.

****HAWKEYE OIL COMPANY:** Hawkeye was an Iowa marketer operating stations under the Hawkeye and Parco brand names. The company was purchased by Mid-Continent around 1930.

PARCO GAS 5 1910s 15" Metal $600-950
Perforated metal globe face. Solid dark color with white outline ring and white "PARCO/GAS" across globe face.
PARCO RED BALL
GASOLINE 6 1920s 15" Metal NL
Yellow globe face with red circle in center. Black "PARCO" with red dropshade arched around top, "GASOLINE" around bottom with "RED" to left of circle and "BALL" to right.
RED BALL
GASOLINE 5 1920s 15" Metal $600-900
Yellow globe face with black/white outlined red circle in center. White outlined black "RED BALL" arched around top, "GASOLINE" around bottom.
HAWKEYE OILS 6 1920s 15" Metal NL
White globe face with large black triangle/red thunderbird IOMA logo in center. Black "HAWKEYE" arched around top, "OILS" around bottom.

****SWENEY GASOLINE and OIL CO. - PEORIA, ILLINOIS:** Founded in 1909 by P.D. Sweney, they opened their first station in 1914 and operated under the Sweney brand until the late 1950s. In addition Sweney branded DX in 1954 and operated under the DX and later Sunoco brand until 1972 when it appears that Sun Oil purchased the operation outright.
SWENEY SKY
HAWK
GASOLINE 4 1930s 15" Metal $1900-2500
White globe face with red script "Sweney" across upper face. Light green hawk below "SWENEY" above light green outlined dark green "SKY HAWK". Red "GASOLINE" arched around bottom.
SWENEY
KNOCKLESS
GASOLINE 6 1930s 15" Metal $500-850
White globe face with red "Sweney" arched around top of globe face. Dark green "KNOCKLESS" across center with green outlined red "GASOLINE" across bottom.
SWENEY MOTOR
GASOLINE 5 1930s 15" Metal $500-850
White globe face with dark green outlined light green "Sweney" diagonal across center. Red "MOTOR" across top, "GASOLINE" across bottom.
GOLD COIN
MOTOR
GASOLINE 6 1930s 15" Metal $400-600
Details not available. Lettering "GOLD COIN MOTOR GASOLINE" on globe face.
SWENEY SKY
HAWK
GASOLINE 6 1930s 14" Gill NL
White globe face with red script "Sweney" across upper face. Light green hawk below "SWENEY" above light green outlined dark green "SKY HAWK". Red "GASOLINE" arched around bottom.

ERICKSON & AFFILIATES
Founded as a general store in Centuria, Wisconsin, in 1928, the Erickson family—seven Erickson brothers—entered into the gasoline business with a small station in Barron, Wisconsin, in 1931. By the end of World War II, the family had invested in a Twin Cities refinery operation known as Northwestern Refining. Following that investment, the family operation was split into several pieces which are listed individually below. Today, the remaining components, when taken together, make up one of the largest and most innovative gasoline marketers. They are also leaders in the convenience store field.

****ERICKSON OIL CO. — MINNEAPOLIS, MINNESOTA:** The primary Erickson company, Erickson Oil Company, expanded from a single station established in Barron, Wisconsin, in 1931. Following World War II, the expanding company purchased Northwestern Refining and in the 1950s Erickson stations were located in four states. In 1959, Erickson stations began a conversion to Holiday Stationstores, an early conception of today's convenience store. The firm is now known as Holiday Companies and operates nearly 300 convenience stores and supermarkets throughout eleven of the North Central states.

ERICKSON'S 7 1950-1970 Glass Oval NL
Black outlined yellow oval with script "Erickson's" logotype across center of globe face.
ERICKSON'S SUPER
TEST REGULAR 7 1950-1970 Glass Oval NL
Black outlined yellow oval with script "Erickson's" logotype across center of globe face. Small "SUPER TEST/REGULAR" at lower right of logotype.
ERICKSON'S
ETHYL (EC) 7 1950-1970 Glass Oval NL
Black outlined yellow oval with script "Erickson's" logotype across center of globe face. Small white circle with Ethyl (EC) logo at lower right of logotype.
HOLIDAY
(logo shaped) 5 1962-1965 Plastic 2pc NL
Perhaps the most unusual plastic shaped globe ever used in the United States. Globe consists of two formed plastic halves assembled back-to-back, joined by a gold plastic channel. Each non-symmetrical half is in the shape of the Holiday stylized rocket logo. The upper band is red, the lower, extending into the base, is blue, with a white oval in the center. Black "Holiday" is painted across the oval with a red arrow pointing down as the dot of the "I".

****LIFE OIL CO. — MINNEAPOLIS, MINNESOTA:** Life Oil Company was an Erickson affiliate in the early 1950s that consisted primarily of low volume stations converted to this secondary brand.

LIFE SUPER
REGULAR 9 1950-1955 Glass Oval NL
Yellow globe face with large red outlined white burst in center. Large red rectangle in center of burst with white "LIFE" across rectangle. Small "SUPER/REGULAR" at lower right of globe face.
LIFE ETHYL (EC) 9 1950-1955 Glass Oval NL
Yellow globe face with large red outlined white burst in center. Large red rectangle in center of burst with white "LIFE" across rectangle. Small black outlined white circle with Ethyl (EC) logo at lower right of globe face.

****ERICKSON OIL COMPANY — STEVENS POINT, WISCONSIN:** One of the Erickson brothers operated a small chain of stations in eastern and central Wisconsin from a base in Stevens Point. These stations were apparently later absorbed into the Holiday chain.

ERICKSON'S 101
SUPER REGULAR 5 1940-1960 Glass Oval $600-950

White globe face with large red "101" in center. Green "ERICKSON'S" arched around top, "SUPER/ REGULAR" around bottom.

ERICKSON'S 110
HI GRADE ETHYL 5 1940-1960 Glass Oval $600-950
White globe face with large red "110" in center. Green "ERICKSON'S" arched around top, "HI GRADE/ETHYL" around bottom.

ERICKSON'S RANGE
OIL HIGHEST
QUALITY 6 1940-1960 Glass Oval $650-1000
White globe face with large blue "RANGE OIL" across center. Orange "ERICKSON'S" arched around top, "HIGHEST QUALITY" arched around bottom.

****ERICKSON OIL PRODUCTS — HUDSON, WISCONSIN:** One of the Erickson brothers established this separate operation in Hudson, Wisconsin, in 1950. They continue to operate over 50 Erickson and Freedom-Erickson stations in four states today, as well as a successful chain of supermarkets.

HIGH TEST ERICKSON
REGULAR 9 1952-1970 Oval Capco $150-250
Blue globe face with white band across center. Red "ERICKSON" across center white band, white "HIGH TEST" in blue field above, "REGULAR" below.

PREMIUM ERICKSON
ETHYL (EC) 9 1952-1970 Oval Capco $200-300
Blue globe face with white band across center. Red "ERICKSON" across center white band, white "PREMIUM" in blue field above, small Ethyl (EC) logo below.

ERICKSON OIL
PRODUCTS 5 1952-1970 Oval Capco $250-350
Blue globe face with white band across center. Red "ERICKSON" across center white band, white "OIL" in blue field above, "PRODUCTS" below.

****NORTHWESTERN REFINING COMPANY — ST. PAUL PARK, MINNESOTA:** Northwestern Refining Co., the St. Paul Park, Minnesota, refining unit of the Erickson family, began operating the SuperAmerica gasoline/ c-store units about 1962. In 1970 SuperAmerica, a marginal c-store operation, was sold to Ashland Oil. By the middle 1970s, the company began to expand out of its North Central base. Many of the Ashland affiliate brand locations, closed during the gasoline shortages of 1973 and 1978, were rebuilt as SuperAmerica convenience stores and the operation today is considered the model of efficient and customer service oriented gasoline c-store chains. Note that this entire section is duplicated from Ashland.

SA 5 1962-1965 Plastic $150-225
White plastic trapezoid shaped halves assembled back to back with large red trapezoid shield on globe face. White lettering "SA" on red shield above white "SUPERAMERICA".

SA (red logo plastic
pump topper) 5 1965-1985 Plastic $75-125
Unusual shaped multi-piece rectangular globe. Looking down from the top, the globe is hour-glass shaped with a flat waistline. Along each side of the waistline is the white on red shield "SA" logo for SuperAmerica. On each side in the flared area are black numbers indicating the pump position number. A black plastic top covers the internal area, allowing access to light bulb inside. Many are installed non-illuminated.

SA (red/blue logo
plastic pump topper)4 1985-Pres Plastic $75-125
Unusual shaped multi-piece rectangular globe. Looking down from the top, the globe is hour-glass shaped with a flat waistline. Along each side of the waistline is the red "S" / blue "A" logo for SuperAmerica. Three blue lines run across the bottom of the globe face. On each side in the flared area are black numbers indicating the pump position number. A black plastic top covers the internal area, allowing access to light bulb inside. Many are installed non-illuminated.

EXXON & AFFILIATES

STANDARD OIL COMPANY of NEW JERSEY — NEW YORK, NEW YORK: Esso is, of course, Standard Oil of New Jersey, the largest of the Standard companies to emerge from the 1911 breakup of Standard Oil. Standard (NJ). Esso began marketing in several Eastern Seaboard states and in Tennessee, Louisiana, and Arkansas through Standard of Louisiana, a subsidiary assigned to them in the 1911 breakup. The Esso brand was first introduced in 1926 as an Ethyl grade gasoline and in 1933 replaced Standard as the company's brand name. Esso marketing expanded to include stations in 19 states, with marketing after World War II added throughout the country under various brand names detailed below. In 1959, all domestic marketing was grouped under the subsidiary Humble Oil and Refining, formerly an independent company that Standard purchased in 1919. Under Humble, the Esso brand name was retained in the East and South. Enco was introduced to replace the various other regional brands and Humble was used in Ohio when Sohio objected to both Esso and Enco. In 1972, the brand name Exxon was introduced to replace Esso, Enco and Humble, although Enco remained in use in the upper Midwest until that region was abandoned several years later. Exxon continues to market through over 11,000 stations in 40 states.

STANDARD MOTOR
GASOLINE
(CONVEX SIDES) 6 1912-1918 OPE CHIMENY $3000-4000

STANDARD MOTOR
GASOLINE
(RECESSED
SIDES) 5 1912-1918 OPE CHIMENY $3000-4000

STANDARD MOTOR
GASOLINE
(FLAT SIDES) 5 1915-1918 OPE Chimney $3000-4000
White globe with red Standard "S" logo covering globe face. Red "STANDARD" in top of "S", "GASOLINE" below with white "MOTOR" on red center of "S".

STANDARD MOTOR
GASOLINE
(FLAT SIDES) 6 1918-1923 OPE $2000-3000
Globe has flat sides and no chimney top. White globe with red Standard "S" logo covering globe face. Red "STANDARD" in top of "S", "GASOLINE" below with white "MOTOR" on red center of "S".

STANDARD MOTOR
GASOLINE 5 1918-1923 16.5" Metal Red HP $1200-1500

STANDARD MOTOR
GASOLINE 5 1918-1923 15" Metal Red HP $1200-1500
White globe face with red Standard "S" logo covering most of globe face. Red "STANDARD" in top of "S", "GASOLINE" below with white "MOTOR" on red center of "S".

STANDARD
RED AND BLACK 6 1918-1923 15" Metal Red HP $1250-1600
White globe face with red Standard "S" logo covering most of globe face. Black "STANDARD" in top of "S", "GASOLINE" below with white "MOTOR" on red center of "S".

STANDARD
BENZOL 7 1920s 15" Metal Red HP NL
Red globe face with white Standard "S" logo covering most of globe face. White "STANDARD" in top of "S", "GASOLINE" below with red "Benzol" on red center of "S".

"STANDARD" 5 1923-1924 16.5" Metal $400-600
"STANDARD" 4 1923-1924 15" Metal $400-600
White globe face with blue outline ring broken at sides where lettering extends to edge and at bottom for "REG.U.S.PAT.OFF.". Red "STANDARD" (in quotes) across globe face.

"STANDARD" ETHYL
(EGC) GASOLINE 6 1924-1926 15" Metal NL
Wide red outer ring with white and blue outlines. White center circle. White "STANDARD" arched around top, "ETHYL GASOLINE" around bottom with large Ethyl (EGC) logo in white center circle with blue rays.

"STANDARD"
WHITE 5 1924-1930 16.5" Metal $400-650

"STANDARD"
WHITE 4 1924-1930 15" Metal $400-650
White globe face with blue "STANDARD" above "WHITE" across center.

STANDARD Upper/
lower case 5 1924 16.5" Metal $400-650

STANDARD Upper/
lower case 5 1924 15" Metal $400-650
White globe face with red lines forming white band across center

and red half -circles above and below (referred to throughout this text as "bar and circle"). Blue "Standard" on crossbar.

STANDARD BAR & CIRCLE 3 1924-1933 16.5" Metal Red L/HP $400-600
STANDARD BAR & CIRCLE 3 1924-1933 15" Metal Red L/HP $400-600
White globe face with red lines forming white band across center and red half -circles above and below (referred to throughout this text as "bar and circle"). Blue "STANDARD" on crossbar.

STANDARD BAR & CIRCLE REVERSE 4 1924-1933 16.5" Metal Red L/HP $400-650
STANDARD BAR & CIRCLE REVERSE 4 1924-1933 15" Metal Red L/HP $400-650
White globe face with blue lines forming white band across center and blue half -circles above and below (referred to throughout this text as "bar and circle"). Red "STANDARD" on crossbar.

STANDARD BAR & CIRCLE AVIATION 7 1924-1933 15" Metal Red L/HP NL
White globe face with red lines forming white band across center and red half -circles above and below (referred to throughout this text as "bar and circle"). Blue "STANDARD" on crossbar. Red "AVIATION" arched around top, "GASOLINE" around bottom.

ESSO SCRIPT 4 1926-1933 16.5" Metal Red L/HP $700-950
ESSO SCRIPT 4 1926-1933 15" Metal $700-950
ESSO SCRIPT (milk glass lens) 6 1926-1933 15" Metal NL
ESSO SCRIPT 7 1926-1933 13.5" Glass NL
White globe face with thick red and thin blue outline rings. Blue script "Esso" diagonal across globe face.

ACTO 3 1930-1939 16.5" Metal $300-450
ACTO 3 1930-1939 15" Metal $300-450
White globe face with black outlined green "ACTO" across globe face. Green line forming half circles above and below.

ESSO 2 1933-1972 16.5" Metal Red L/HP $300-450
ESSO 2 1933-1972 15" Metal Red L/HP $300-450
ESSO 5 1946-1972 13.5" Glass Red L/HP $300-450
White globe face with "ESSO" logotype in red across center of face.

ESSOLENE 3 1933-1939 16.5" Metal Red L/HP $350-500
ESSOLENE 3 1933-1939 15" Metal Red L/HP $350-500
White globe face with red Standard style bar and circle. Blue "Essolene" in crossbar.

NEW ESSO 5 1938 Decal 16.5" Metal Red L/HP $350-550
NEW ESSO 5 1938 Decal 15" Metal Red L/HP $350-550
Essolene "bar and circle" globe described above with decal over crossbar. Script blue "NEW" to left of red "ESSO" logotype on decal.

NEW ESSO EXTRA 9 1938 Decal 16.5" Metal Red L/HP NL
NEW ESSO EXTRA 9 1938 Decal 15" Metal Red L/HP NL
Description not available.

ESSO EXTRA 6 1930s 18" Metal Neon Red L/HP NL
ESSO EXTRA 3 1939-1952 15" Metal Red L/HP $350-500
ESSO EXTRA 5 1950-1952 13.5" Glass Red L/HP $350-500
White globe face with "ESSO" logotype in blue over blue "EXTRA".

ESSO EXTRA UNDERLINED 3 1952-1972 16.5" Metal Red L/HP $350-500
ESSO EXTRA UNDERLINED 3 1952-1972 15" Metal Red L/HP $350-500
ESSO EXTRA UNDERLINED 4 1952-1972 13.5" Glass $350-450
White globe face with "ESSO" logotype in blue over blue "EXTRA". Red line under "EXTRA".

GOLDEN ESSO EXTRA 4 1956-1962 15" Metal HP Blue NL
White globe face with gold "GOLDEN" over "ESSO" logotype in blue over blue "EXTRA".

ESSO AVIATION 5 1933-1972 16.5" Metal Red L/HP $1000-1500
ESSO AVIATION 5 1933-1972 15" Metal Red L/HP $1000-1500
ESSO AVIATION 5 1933-1972 13.5" Glass $750-1250
White globe face with red "ESSO" logotype over blue "AVIATION". Small blue wings at top, red dot at bottom.

ESSO KEROSENE 4 1940-1972 16.5" Metal Red L/HP $400-650
ESSO KEROSENE 4 1940-1972 15" Metal Red L/HP $400-650
White globe face with red "ESSO" logotype over red "KEROSENE".

ESSO DIESEL 4 1940-1972 15" Metal Red L/HP $400-650
White globe face with blue "ESSO" logotype over blue "DIESEL".

ESSO MARINE 6 1940-1972 15" Metal Red L/HP $800-1250
White globe face with red "ESSO" logotype over red "MARINE".

ESSO PLUS 5 1962-1972 13.5" Glass $300-500
White globe face with red "ESSO" logotype over blue "PLUS".

ESSO (new wide band logo) 6 1965-1972 Gill $400-650
White plastic globe face with 1960s wide light blue band Esso logo. Wide light blue oval outline with red "Esso" logotype in center. Found in Florida, and assumed to be used in the expansion of the Esso brand in Florida in the 1960s. Obviously a branded jobber representing a brand that commonly used Gill body globes, probably Atlantic, switched to the Esso brand and plastic globe faces were made specific for that use. This is a non-standard issue item.

ESSO MILK GLASS TOP FOR TOKHEIM 36B 5 1936-1950 OPB $800-1250
White milk glass top for Tokheim 36B with red "Esso" logotype on front and back faces.

ESSOLENE MILK GLASS TOP FOR TOKHEIM 36B 5 1936-1939 OPB $800-1250
White milk glass top for Tokheim 36B with red "Essolene" logotype on front and back faces.

ESSO EXTRA MILK GLASS TOP FOR TOKHEIM 36B 5 1939-1950 OPB $800-1250
White milk glass top for Tokheim 36B with blue "Esso Extra" logotype on front and back faces.

****EXXON:** The Exxon brand name was introduced in the fall of 1972, replacing Esso in 19 Eastern states, Humble in Ohio, and Enco elsewhere. When Exxon was introduced, the Enco brand name remained in use in a limited number of stations in the Chicago marketing area ... an area that had been scheduled for abandonment since about 1976.

The globes shown below were specifically made for use by a jobber using globes extensively in North Carolina who had converted from the Sinclair brand to the Exxon brand in 1972. They were legitimately used on gas pumps in a multi-county area until about 1980.

EXXON 4 1972-1980 Capco $250-350
White globe face with red "EXXON" logotype across face.

EXXON PLUS 4 1972-1975 Capco $250-350
White globe face with red "EXXON" and blue "PLUS" logotype across face.

EXXON EXTRA 4 1972-1980 Capco $250-350
White globe face with red "EXXON" and blue "EXTRA" logotype across face.

EXXON DIESEL 4 1972-1980 Capco $250-350
White globe face with red "EXXON" and blue "DIESEL" logotype across face.

****STANDARD OIL CO. of LOUISIANA:** This is a marketing affiliate assigned to Standard (NJ) at the 1911 breakup of Standard Oil. Stanocola operated stations in Tennessee, Arkansas and Louisiana under the Stanocola brand until 1923 when the "STANDARD" bar and circle image was introduced throughout the marketing territory.

STANOCOLA 5 1918-1923 OPE W $4500-6000
White globe with green outlined red outer ring around face. White center with large red/green/black/yellow STANOCOLA shield in center. White lettering on red outer ring "STANDARD OIL COMPANY" arched around top, "OF LOUISIANA" around bottom.

STANOCOLA 5 1918-1923 15" Metal $2400-3000
White globe face with green outlined red outer ring around face. White center with large red/green/black/yellow STANOCOLA shield in center. White lettering on red outer ring "STANDARD OIL COMPANY" arched around top, "OF LOUISIANA" around bottom.

STANOCOLA IN THIS PUMP 5 1920-1923 15" Metal $2000-2600
White globe face with black outline ring. Red band across center of globe face with blue/white lettering "STANOCOLA/Gasoline". Full color "STANOCOLA" shield at top of globe face. Black "IN THIS/PUMP" in lower white area below red band.

STANOCOLA POLARINE 5 1920-1923 15" Metal NL
White outlined globe face with white center band and dark upper 6

lower areas. Stanocola shield and outline logo on upper globe face with "STANOCOLA" over script "Polarine" on center band. White "The "STANDARD" Motor Oil" on lower area. Known only in old photos.

****HUMBLE OIL and REFINING — HOUSTON, TEXAS**: Humble Oil and Refining was a Houston based producer/refiner/marketer purchased by Standard (NJ) in 1919. The Humble brand was used in Texas and New Mexico from 1919 until 1961. In 1959, all United States marketing came under the subsidary "Humble Oil and Refining." "Humble" was added to stations of each brand then in use, although each retained its own regional brand name (Esso, Carter, Oklahoma, etc.). In 1961, Enco replaced Humble as the station brand in Texas and Oklahoma. The Humble brand was introduced in Ohio when both Esso and Enco were blocked by Sohio and remained in use in Ohio until the name was converted to Exxon in 1972.

HUMBLE GASOLINE/ HUMBLE OILS GEAR 5 1925-1932 Window 13.5" $500-850
This globe face is actually a window lens used in gas station buildings. they fit a standard 13.5" glass or plastic body and are frequently found assembled into a complete globe. White globe face with blue outline ring and blue "gear" in center. Red "HUMBLE" arched around top, "GASOLINE" around bottom with red "HUMBLE/OILS" across center of gear, white "HUMBLE OIL and REFINING CO" arched around top of gear, "HOUSTON,TEXAS" around bottom.

HUMBLE OIL & REF/ HUMBLE OILS 4 1919-1932 15" Metal $900-1400
Blue outer area of globe face with large white gear covering most of globe face. Red "HUMBLE/OILS" across center of globe face with blue "HUMBLE OIL and REFINING CO." arched around top and "HOUSTON,TEXAS" arched around bottom.

FLASHLIKE ETHYL 4 1929-1932 15" Metal $500-800
White globe face with red "FLASHLIKE" diagonal across upper globe face above large Ethyl (EGC) logo.

HUMBLE BAR AND CIRCLE 7 1932-1935 OPB NL
White globe face with red lines forming white band across center and red half -circles above and below (referred to throughout this text as "bar and circle"). Blue "HUMBLE" on crossbar.

HUMBLE BAR AND CIRCLE 4 1932-1935 15" Metal $600-900
White globe face with red lines forming white band across center and red half -circles above and below (referred to throughout this text as "bar and circle"). Blue "HUMBLE" on crossbar.

HUMBLE 4 1935-1945 15" Metal $500-850
White globe face with red/white/blue outline rings. Red "HUMBLE" across center.

HUMBLE 4 1945-1961 15" Metal $500-850
White globe face with red/white/blue outline rings. Red *"Humble"* across center.

THRIFTANE 4 1939-1947 15" Metal $500-750
White globe face with thick blue ring around outer edge. Red/white/blue rings around center white circle. Red script "Thriftane" diagonal across globe face.

****TEXAS PACIFIC COAL and OIL — FT. WORTH, TEXAS**: Texas Pacific Coal and Oil was an independent Texas marketer that sold out to and was rebranded Humble in 1945.

TP PRODUCTS 4 1930-1945 15" Metal $2500-3500
White globe face with two black outline rings separated by red spacers around globe face. Large red and black teepee in center with black "TP" across and black "PRODUCTS" below.

TP ETHYL 6 1930-1945 15" Metal NL
White globe face with red outline ring. Red and white lines forming teepee in center with red "T" to left and "P" to right and Ethyl (EGC) logo in center. Red "GASOLINE" below teepee.

TP PRODUCTS 6 1930-1945 13.5" Glass NL
White globe face with two black outline rings separated by red spacers around globe face. Large red and black teepee in center with black "TP" across and black "PRODUCTS" below.

****COLONIAL BEACON OIL COMPANY—BOSTON, MASSACHUSETTS**: This company was formed in a late 1920s merger between Boston area based Colonial Oil Company and Beacon Oil. Colonial Beacon, an innovative marketer with stations throughout New England, sold out to Standard (NJ) in 1930. The Colonial name, as a subsidiary, survived until the end of World War II, although products were branded Esso in 1933. Colonial influenced the entire industry with the introduction of service bays at stations and the marketing of tires, batteries and automotive accessories (known as TBA in the trade) through their stations. Colonial was instrumental in the formation of the Atlas Supply Company, providing the various Standard companies that participated in the program with a TBA line.

COLONIAL GAS (MILK GLASS) 6 1920-1930 Octagon $700-1000
Square milk glass lenses in metal frame with cut corners forming octagon shaped globe. White globe face with logotype black "Colonial/Gas" on globe face.

COLONIAL GAS 5 1920-1930 Octagon $650-900
Square lenses in metal frame with cut corners forming octagon shaped globe. White globe face with logotype black "Colonial/Gas" on globe face.

COLONIAL ETHYL 5 1926-1930 Octagon $650-900
Square lenses in metal frame with cut corners forming octagon shaped globe. White globe face with logotype black "Colonial/Ethyl" on globe face.

COLONIAL WINTERGAS 6 1926-1930 Octagon $700-950
Square lenses in metal frame with cut corners forming octagon shaped globe. White globe face with logotype green "Colonial/ Wintergas" on globe face.

COLONIAL MOTOR 7 1926-1930 16.5" Metal NL
White globe face with black logotype "Colonial/Motor" on globe face.

COLONIAL BAR AND CIRCLE 4 1930-1933 15" Metal $450-650
White globe face with red bar and circle design similar to Standard bar and circle above. Black outlined green "COLONIAL" on bar.

****STANAVO**: Formed in 1929 by Standard of Indiana, New Jersey, and California as a joint venture to research aviation petroleum needs and market those products through their individual marketing outlets. Stanavo first marketed aviation gasoline in 1931. With the withdrawl of Standard of New Jersey in 1937, the other two participants decided to discontinue the effort and Stanavo was dissolved in June 1938.

STANAVO 5 1931-1938 16.5" Metal NL
White globe face with outlined bird-like airplane image across center of globe face. Blue outlined red "STANAVO" across top with "AVIATION GASOLINE" to lower right below airplane.

****POWERINE OIL CO.—DENVER, COLORADO**: This was a Denver based refiner/marketer. Purchased by Standard (NJ) in 1938, the Powerine name was retained until the end of World War II when it was replaced by Oval-E.

POWERINE 4 1925-1935 15" Metal $600-850
Red globe face with black outlined white bar and circle band across center. Black outlined red "POWERINE" on band.

POWERINE'S CRYSTAL GAS 5 1920s 15" Metal $700-950
White globe face with black "POWERINE'S" arched around top above large black script "Crystal". White "GAS" on black extension from "L" in "Crystal.

BEARCAT MOTOR FUEL 6 1920s 15" Metal HP NL
White globe face with large black/white/red cartoon-like drawing of cat across center of globe face. Large red "Bearcat" across upper globe face with black outlined red exclamation point at an odd angle. Red "MOTOR FUEL" across lower globe face.

BEARCAT MOTOR FUEL 6 1920s 13.5" Balcrank NL
White globe face with large black/white/red cartoon-like drawing of cat across center of globe face. Large red "Bearcat" across upper globe face with black outlined red exclamation point at an odd angle. Red "MOTOR FUEL" across lower globe face.

****NORTHWEST REFINING CO. — CUT BANK, MONTANA**: Northwest marketed Grizzly gasoline. The firm was purchased by Standard in 1942 and rebranded Carter in 1945.

GRIZZLY DUBBS
CRACKED 5 1930s-1946 15" Metal NL
White globe face with black outline ring. Detailed standing brown bear in center of globe face. Black "GRIZZLY" arched around top, "GASOLINE" around bottom with small red "DUBBS/CRACKED" to left of bear.

GRIZZLY DUBBS
CRACKED 5 1930s-1946 Capco $2000-3000
White globe face with black outline ring. Detailed standing brown bear in center of globe face. Black "GRIZZLY" arched around top, "GASOLINE" around bottom with small red "DUBBS/CRACKED" to *right* of bear.

GRIZZLY ETHYL 7 1930s-1946 Capco $500-750
From an old photo. White globe face with dark outline ring and dark ring around center circle. Lettering "GRIZZLY" arched around top, "ETHYL" around bottom with large Ethyl (EGC) logo in center.

NORDAL 7 1930s-1946 Capco $200-300
From an old photo. White globe face with dark outline ring. Dark color lettering "NORDAL" across center of globe face.

****YALE OIL CO. — BILLINGS, MONTANA:** Yale marketed Litening gasoline, was purchased by Standard (NJ) in 1946, and was rebranded Carter.

LITENING 5 1930-1938 15" Metal $450-700
White globe face with black outlined red "LITENING" in jagged logotype lettering across globe face.

LITENING "B" 5 1920s 15" Metal $550-750
White globe face with black outlined red "LITENING" in jagged logotype lettering across globe face, superimposed over large black "B" in center of globe face.

WILDFIRE 5 1938-1946 15" Metal $600-850
White globe face with large blue reverse bowtie lettering "WILDFIRE" across center of globe face.

LITENING 5 1938-1946 Clear or Yellow Ripple $900-1300
White globe face with black outlined red "LITENING" in jagged logotype lettering across globe face.

LITENING ETHYL 5 1938-1946 Clear Ripple $900-1300
Black outline ring around white globe face with black outline around white center circle. Black outlined red "LITENING" logotype lettering arched around top, "GASOLINE" around bottom with large Ethyl (EGC) logo in center.

WILDFIRE 5 1938-1946 Blue Ripple $1500-1900
White globe face with large blue reverse bowtie lettering "WILDFIRE" across center of globe face.

****OVAL E:** The Oval E name was introduced in 1946 to replace Powerine. The name was changed to Carter in 1950.

Oval-E 5 1946-1950 Red Capco $250-400
White globe face with red logotype "E" in blue oval in center.

Oval-E EXTRA 5 1946-1950 Blue Capco $300-450
White globe face with red logotype "E" in blue oval in center. Red "EXTRA" below oval.

****CARTER OIL CO. — TULSA, OKLAHOMA:** Long a Standard production subsidiary, the Carter brand name was introduced in 1945 to replace Grizzly and Litening. In 1950, Carter replaced Oval-E to bring all of Standard's western marketing under the Carter brand. The Carter brand was replaced by Enco in 1961.

CARTER 6 1946-1950 15" Metal $400-600

CARTER 4 1946-1962 Red Capco $300-400
White globe face with red "Carter" across face.

CARTER EXTRA 4 1946-1962 Blue Capco $350-450
White globe face with blue "Carter" over red "EXTRA" across face.

POWERINE 9 1946-1962 Red Capco NL
White globe face with red outlined white logotype lettering "POWERINE" arched across upper globe face above smaller red "MOTOR FUEL".

CARTER DIESEL 9 1946-1962 Red Capco NL
White globe face with blue "Carter" logotype over blue "DIESEL FUEL".

CARTER BUS
DIESEL 9 1946-1962 Red Capco NL
White globe face with blue "Carter" logotype over blue "BUS DIESEL".

CARTER
KEROSENE 9 1946-1962 Red Capco NL
White globe face with green "Carter" logotype over green "KEROSENE".

****ENCO:** Enco was introduced in 1960, first in the Oklahoma/Pate marketing territory as a replacement for those brand names, and expanded to Carter and Humble territories in 1961. Enco (from the slogan America' s Leading ENergy COmpany) was to be Standard's primary brand name in all areas where Esso could not be used. Enco was replaced by Exxon in 1972, except in select areas where Standard planned to withdraw from marketing. Enco was used as late as 1980 in those areas.
>>No Enco globes known.

****PATE OIL CO.—MILWAUKEE, WISCONSIN:** Founded in Milwaukee in 1933, Pate operated over 50 stations when purchased by Standard in 1956. The Pate brand was replaced by Enco in 1960.

PATE OIL CO.
GASOLINE
w/shield 6 1933-1940 15" Metal $900-1200
White globe face with detailed red/white/blue Pate shield in center with "PATE OIL COMPANY" arched around top and "GASOLINE" around bottom.

PATE OIL CO.
GASOLINE
w/shield 6 1940-1946 13.5" Glass $600-850
White globe face with detailed red/white/blue Pate shield in center with "PATE OIL COMPANY" arched around top and "GASOLINE" around bottom.

(PATE) CHALLENGE
(shield) 7 1946-1956 13.5" Glass $600-800
White globe face with large red/yellow/blue coat of arms in center of globe face. Blue band diagonal across globe face with yellow logotype "Challenge" on band.

(PATE) AIR GLIDE
(shield) 5 1946-1956 13.5" Glass $600-800
White globe face with large red/yellow/blue coat of arms in center of globe face. Blue band diagonal across globe face with yellow logotype "AIR GLIDE" on band.

PATE AIR GLIDE 7 1956-1961 13.5" Glass $350-550
White globe face with blue outline ring. "PATE" in small red letters offset to upper left of globe face. Red "AIR/GLIDE" offset to lower right.

PATE ULTRA
AIR GLIDE 7 1956-1961 13.5" Glass $350-575
White globe face with blue outline ring. "PATE" in small red letters offset to upper left of globe face. Red "ULTRA" with blue underline above red "AIR/GLIDE" offset to lower right.

PATE DIESEL FUEL 6 1956-1961 13.5" Glass $300-475
White globe face with blue outline ring. "PATE" in small red letters offset to upper left of globe face. Red "DIESEL/FUEL" on lower center of globe face.

PATE RANGE OIL 6 1956-1961 13.5" Glass $300-500
White globe face with blue outline ring. "PATE" in small red letters offset to upper left of globe face. Red "RANGE/OIL" on lower center of globe face.

****OKLAHOMA OIL COMPANY—CHICAGO, ILLINOIS:** Oklahoma Oil was a Chicago area discount marketer that Standard (NJ) purchased in 1956. Almost immediately another Chicago discounter, Perfect Power, was purchased and rebranded Oklahoma. The Gaseteria/Hoosier Pete operation, added in 1957, was also rebranded Oklahoma. The Enco name was first introduced to replace Oklahoma in 1960.

OKLAHOMA
REGULAR 5 1930s OPP $400-600
White globe with series of concentric circles. "OKLAHOMA" across center and smaller "REGULAR" below.

OKLAHOMA WHITE 5 1930s OPP $400-600
White globe with orange outline ring. Black "OKLAHOMA" arched around top with orange script "White" across center of globe face.
>>Other Oklahoma globes known to exist - no details available.

****PERFECT POWER OIL COMPANY — CHICAGO, ILLINOIS**: Founded in 1935, Perfect Power was one of the Chicago area's largest gasoline discounters when they were purchased by Oklahoma Oil Company in 1956. >>No Perfect Power globes can be identified at this time, but we have information that leads us to believe that they do wxist.

****GASETERIA OIL CO. — INDIANAPOLIS, INDIANA**: Founded in 1928 in Indianapolis by Russell Williams, Sr., Gaseteria operated the Bonded stations throughout Indiana and Kentucky. A family friend, Paul Deer of Springfield, Ohio, was permitted to open stations in Ohio beginning at the end of World War II using the Bonded brand and images. In 1954 Gaseteria, in expansion mode, purchased Hoosier Petroleum and in 1956 they purchased the remnants of Johnson Oil Refining/Brilliant Bronze. In 1957 Standard purchased Gaseteria and rebranded the operation Oklahoma.

GASETERIA OIL
GASOLINE 7 1930s 15" Metal NL
From an old photo. Dark color outline ring around white globe face. Solid color half circles at top and bottom of globe face forming white band across center of face. Large "Gaseteria" logotype on band with small "OIL" in upper half circle, "GASOLINE" on lower.

****QUALITY GASOLINE COMPANY — PERU, INDIANA**: Founded about 1932, Quality was purchased by Gaseteria in the mid-1930s and the Bonded brand name was adopted by Gaseteria and continued in use until the company was purchased by Standard Oil Company of New Jersey in 1957.

BONDED 68 5 1940-1946 13.5" Glass W/N $750-1000
White globe face with large gold border with blue detail around eight point blue seal. White Old English "Bonded" across upper seal with large red '68' in single quotes covering lower seal and lower globe face.

BONDED 70 5 1940-1946 13.5" Glass W/N $750-1000
White globe face with large gold border with blue detail around eight point blue seal. White Old English "Bonded" across upper seal with large red "70" in single quotes covering lower seal and lower globe face.

BONDED 78 5 1940-1946 13.5" Glass W/N $750-1000
White globe face with large gold border with blue detail around eight point blue seal. White Old English "Bonded" across upper seal with large red "78" in single quotes covering lower seal and lower globe face.

BONDED 98 5 1946-1955 13.5" Glass W/N $750-1000
White globe face with large gold border with blue detail around eight point blue seal. White Old English "Bonded" across upper seal with large red "98" in single quotes covering lower seal and lower globe face.

BONDED ETHYL 5 1946-1955 13.5" Glass W/N $750-1000
White globe face with large gold border with blue detail around eight point blue seal. White Old English "Bonded" across upper seal above white outlined red script "Ethyl" with white circle on lower seal. Ethyl (EC) logo on circle.

BONDED
REGULAR 5 1955-1958 Plastic $300-450
White formed plastic faces assembled back-to-back with yellow seam. Rounded end rectangle with pointed bottom and half circle on top. Probably the most unusually shaped gas pump globe. Blue/gold seal in upper half circle with gold script "Quality" on seal below red band. White "REGULAR" on red band. Blue "BONDED" across rectangular area of globe.

BONDED ETHYL 7 1955-1958 Plastic $300-450
White formed plastic faces assembled back-to-back with yellow seam. Rounded end rectangle with pointed bottom and half circle on top. Probably the most unusually shaped gas pump globe. Blue/gold seal in upper half circle with gold script "Quality" on seal below red band. White "ETHYL" on red band. Blue "BONDED" across rectangular area of globe.

****JOHNSON OIL REFINING CO. — CHICAGO, ILLINOIS**: Dating from the first decade of this century, Johnson Oil was a mid-continent refiner/marketer that marketed gasoline under the Johnson, Jo-Re-Co, and Brilliant Bronze brand names. Gaseteria purchased Johnson in 1956 and sold the refinery operation to Deep Rock. Johnson Oil was rebranded Oklahoma with the 1957 purchase of Gaseteria by Standard (NJ).

JOHNSON TIME TELLS
GASOLENE 4 1930s 15" Metal $2200-2800

JOHNSON TIME TELLS
GASOLENE 5 1930s 13.5" Glass W $1500-2000
Series of black and orange outline rings around black outer ring and orange center circle. Detailed hourglass in center circle with black/white wings to either side. Black "TIME" "TELLS" on wings. White "JOHNSON" arched around top, "GASOLENE" around bottom.

JOHNSON TIME
TELLS ETHYL
GASOLENE 4 1930s 15" Metal $1500-2000

JOHNSON TIME
TELLS ETHYL
GASOLENE 4 1930s 13.5" Glass W $750-1200
Similar to above with white center circle and Ethyl (EGC) logo in circle.

JO-RE-CO TIME TELLS
GASOLENE 4 1940s 13.5" Glass W $1500-2000
Series of black and orange outline rings around black outer ring and orange center circle. Detailed hourglass in center circle with black/white wings to either side. Black "TIME" "TELLS" on wings. White "JO-RE-CO" arched around top, "GASOLENE" around bottom.

JOHNSON WINGED
70 TIME TELLS
GASOLENE 4 1934-1940 13.5" Glass W $800-1250
Series of black and orange outline rings around black outer ring and orange center circle. Black outlined white "Winged" over "70" across center of globe face. Black outlined white wings to either side of "70" with "TIME" "TELLS" on wings. White "JOHNSON" arched around top, "GASOLENE" around bottom.

JOHNSON WINGED
70 GASOLENE 4 1940s 13.5" Glass W $750-1200
White globe face with black/orange outline rings. Black/white outlined orange rectangle above large black shield. Black "JOHNSON" on orange rectangle, White "WINGED" on upper shield above white "70" with orange wings. White "GASOLINE" below.

JOHNSON Black/
WHITE/ORANGE 4 1940s Capco $200-300
Black outline ring around globe face with upper half white and lower half orange. Black/white outlined black band across center with white "JOHNSON" on black band.

****BRILLIANT BRONZE**: Johnson Oil Company secondary discount brand.

BRILLIANT
BRONZE 4 1950-1956 Capco $250-350
White globe face with large red burst at top. Black "brilliant/bronze" in logotype lettering covering most of globe face below burst.

BRILLIANT ETHYL
T/T GASOLINE 4 1950-1956 Capco $600-950
Series of black and orange outline rings around black outer ring and orange center circle. Detailed hourglass positioned low in center circle with black/white wings to either side. Black "TIME" "TELLS" on wings. White "brilliant" arched around top above "ETHYL", "GASOLENE" around bottom.

****HOOSIER PETROLEUM CORP. — INDIANAPOLIS, INDIANA**: An Indiana marketer founded in 1930 as a regional gasoline distributor. Many Indiana independents trace their origins to Hoosier Pete distributorships in the 1930s. Hoosier Petroleum was purchased by Gaseteria in 1956.

HOOSIER PETE 100 5 1930s-1958 13.5" Glass N
$500-750
Mint green globe face with lower 1/3 black. Red "HOOSIER PETE" arched around top above large white "100" positioned over color split. Green "REGULAR" below 100 and small green "GASOLINE" arched around bottom.

HOOSIER PETE
ETHYL 5 1930s-1958 13.5" Glass N $500-750
Mint green globe face with lower 1/3 black. Red "HOOSIER PETE" arched around top above large white "100" positioned over color split. Green "ETHYL" below 100 and above small green non-detailed Ethyl logo and small green "GASOLINE" arched around bottom.

HOOSIER PETE

FUEL OIL 5 1930s-195813.5" Glass N $600-850
Mint green globe face with lower 1/3 black. Red "HOOSIER PETE" arched around top above black "HEATER OIL". Green "FILTERPURE PROCESSED" arched around bottom with red flame above. Green "BTU" on flame.

**VORTEX OIL COMPANY — CHICAGO, ILLINOIS: Vortex Oil was an independent operation that was purchased by Hoosier Petroleum in the 1930s. The Vortex brand name remained in use until the purchase of Hoosier Pete by Gaseteria.
>>No Vortex globes known at this time.

**RYAN'S: A Hoosier Pete affiliate.

RYAN'S JET
REGULAR 5 1950s 13.5" Glass $400-650
White globe face with blue shield outline with blue stars at top. Blue/white stripes forming bottom of shield at bottom. Red "RYAN'S" over blue "JET" over red "REGULAR" across center of shield design.
RYAN'S JET
HI-TEST 5 1950s 13.5" Glass $400-650
White globe face with blue shield outline with blue stars at top. Blue/white stripes forming bottom of shield at bottom. Red "RYAN'S" over blue "JET" over red "HI-TEST" across center of shield design.
RYAN'S JET
KEROSENE 5 1950s 13.5" Glass $400-650
White globe face with blue shield outline with blue stars at top. Blue/white stripes forming bottom of shield at bottom. Red "RYAN'S" over blue "JET" over red "KEROSENE" across center of shield design.

**SENCO, INC. — ATLANTA, GEORGIA: Senco was formed after World War II as Clark's Oil Company with signage similar to the Milwaukee based Clark Oil. The company was rebranded Senco around 1956 and was purchased by Standard (NJ) in 1962 and rebranded Esso.

SUPER SENCO
GASOLINE 7 1950-1962 13.5" Glass NL
Orange globe face with white band across center. "SENCO" on center band with "SUPER" above and "GASOLINE" below.

**SOUTHERN OIL CO. — MIAMI, FLORIDA: Southern Oil was purchased by Standard (NJ) in 1962 and rebranded Esso.

DEEM 5 1955-1962 Capco $200-325
Red globe face with large black outlined white "DEEM" across globe face. Lettering appears in "3-D" with black and yellow dropshading.

**OLENTANGY OIL CO. — COLUMBUS, OHIO: Olentangy was an early Columbus, Ohio, independent marketer that was purchased by Standard (NJ) in 1961 and rebranded Humble.

SUN FLASH
MOTOR 7 1930s 13.5" Glass NL
From an old photo. Dark outer ring area with white ring around dark center circle. Large white lettering "SUN FLASH" across center. White "REGULAR" arched around top, "MOTOR" around bottom.
SUN FLASH
ANTI-NOK 7 1930s 13.5" Glass NL
From an old photo. Dark outer ring area with white ring around dark center circle. Large white lettering "SUN FLASH" across center. White "PREMIUM" arched around top, "ANTI-NOK" around bottom.
SUN-FLASH 7 1950s Capco NL
From an old photo. Appears to have script logotype "Sun Flash" across dark oval on globe face.

**WOOD OIL COMPANY — CAMDEN, OHIO: Founded in 1920, Wood operated stations in Ohio. The company was sold to Standard Oil Company of New Jersey in 1965 and its stations were rebranded Humble.

WOOD OILS
GASOLINE
(w/lion) 5 1920s 15" Metal NL
WOOD OILS
GASOLINE
(w/lion) 5 1930s-1940s 13.5" Glass NL
Blue globe face with white outline ring. Detailed multi-color lion's head in center of globe face. White "WOOD OILS" arched around top, "GASOLINE" around bottom.
WOOD HI TEST 5 1920s 15" Metal NL
WOOD HI TEST 5 1930s-1940s 13.5" Glass NL
Red globe face with detailed multi-color lion's head in center. White "WOOD" arched around top, "HI-TEST" around bottom.

**ALERT OIL COMPANY — WILMINGTON, DELAWARE:
>>No Alert Globes Known.

**IMPERIAL OIL LIMITED — TORONTO, ONTARIO: Founded in 1880, Imperial Oil became the Canadian branch of Rockefeller's Standard Oil. Following the breakup of Standard in 1911, control of Imperial was assigned to the Standard Oil Company of New Jersey, and today Imperial Oil remains an Exxon subsidiary. Credited with many firsts in Canadian gasoline marketing history, Imperial Oil—still operating under the venerable "Esso" brand today—is Canada's largest petroleum marketer.

PREMIER 6 1911-1918 OPE RW $600-1000
White globe with red globe face. Black outlined white script "Premier" above smaller "GASOLINE" across center of globe face.
QUEEN 6 1911-1918 OPP RW $850-1200
Small globe with 4" base. White globe face with large red "Queen" over smaller red "GASOLINE".
IMPERIAL PREMIER 4 1918-1925 OPE RW
$1200-1750
White globe with large black outlined red circle at top of globe face. White outlined black "IMPERIAL" logotype on circle. Black outlined red script "Premier" above smaller "GASOLINE" across center of globe face.
IMPERIAL QUEEN 5 1918-1925 OPE RW $1000-1750
White globe with large black outlined red circle at top of globe face. White outlined black "IMPERIAL" logotype on circle. Black outlined red script "Queen" above smaller "GASOLINE" across center of globe face.
IMPERIAL QUEEN 5 1918-1925 OPE RW $1000-1750
White globe with large black outlined red circle at top of globe face. White outlined black "IMPERIAL" logotype on circle. Red script "Queen" above smaller "GASOLINE" across center of globe face.
IMPERIAL
POLARINE 4 1918-1922 OPB NL
White globe with blue outlined red "IMPERIAL" arched around top above blue script "Polarine". Blue and white drawing of polar bear in center with blue "FOR/MOTOR CAR/AND" to left of bear, "MOTOR/BOAT/LUBRICATION" to right. Other small blue lettering below, above red banner across lower globe face. Blue "THE IMPERIAL OIL COMPANY/LIMITED" on red banner.
PREMIER
GASOLINE 4 1925-1930 OPB $900-1150
White globe with large red circle forming globe face. White "Premier" above smaller white "GASOLINE" across globe face.
IMPERIAL ETHYL 4 1926-1930 OPB $600-900
White globe with black outlined red "IMPERIAL" arched around top, "GASOLINE" around bottom with large Canadian Ethyl (EGC) logo with black rays in center.
IMPERIAL PREMIER
(white) GASOLINE4 1918-1925 15" Metal $600-875
White globe face with large black outlined red circle at top of globe face. White outlined black "IMPERIAL" logotype on circle. Black outlined red script "Premier" above smaller "GASOLINE" across center of globe face.
IMPERIAL PREMIER
(black) GASOLINE 4 1918-1925 15" Metal $600-875
White globe face with large black outlined red circle at top of globe face. Black outlined white "IMPERIAL" logotype on circle. Black outlined red script "Premier" above smaller "GASOLINE" across center of globe face.
PREMIER 3-STAR 6 1930-1938 15" Metal $500-750
White globe face with blue outline ring set in from edge of globe face. Blue "Premier" across upper globe face above three blue outlined red stars on lower globe face.
IMPERIAL 3-STAR 4 1930-1938 15" Metal $500-750
White globe face with blue outline ring set in from edge of globe

face. Blue "IMPERIAL" across upper globe face above three blue outlined red stars on lower globe face.

IMPERIAL ETHYL 4 1930-1938 15" Metal $650-900
White globe face with black outlined red "IMPERIAL" arched around top, "GASOLINE" around bottom with large Canadian Ethyl (EGC) logo with yellow rays in center.

ESSO ETHYL 4 1938-1950 15" Metal $500-750
White globe face with blue outline ring set in from edge of globe face. Large red "Esso" logotype across upper globe face above large Canadian Ethyl (EGC) logo with yellow rays.

ESSO EXTRA ETHYL 4 1938-1950 15" Metal $750-1000
White globe face with blue outline ring set in from edge of globe face. Large red "Esso" logotype across upper globe face above large Canadian Ethyl (EGC) logo with yellow rays. Red "EXTRA" vertical diagonal on each side of Ethyl logo.

ACTO (red letters with blue border) 5 1930-1940 15" Metal $300-500
White globe face with blue outline ring set in from edge of globe face. Large red "ACTO" across center of globe face.

HOME OIL COMPANY, LTD. — VANCOUVER, BRITISH COLUMBIA: Founded in 1925, Home Oil was a marketing firm operating throughout British Columbia. The firm opened its first gas stations in 1928. The operation had close ties to Imperial Oil throughout its history and was purchased by Imperial in the mid-1960s. Stations were rebranded Esso at that time.

SUPER POWER w/flag 5 1940s 15" Metal $900-1300
White globe face with red outline ring. Large green/white outlined red flag diagonal across most of globe face. White script "Super" over white "POWER" on flag.

SUPER POWER w/flag 5 1940s 13.5" Glass $500-800
White globe face with red outline ring. Large green/white outlined red flag diagonal across most of globe face. White script "Super" over white "POWER" on flag.

HOME ETHYL 5 1940s 13.5" Glass NL
From an old photo. Dark color outline ring with large light colored flag in center. Dark colored "HOME/ETHYL" across flag with white circle in center and small Canadian Ethyl (EGC) in center.

CHAMPLAIN OIL COMPANY, LTD. — MONTREAL, QUEBEC: A regional marketer in eastern Canada, Champlain Oil was long affiliated with Imperial Oil. Stations were rebranded to Esso in the 1980s.

CHAMPLAIN +70 GASOLINE 5 1930s 13.5" Glass $300-500
Blue outlined white globe face with blue outlined white center circle. Red plus symbol and blue "70" in center circle. Red "CHAMPLAIN" arched around top, red/blue outlined box around red "GASOLINE" at bottom

CHAMPLAIN ETHYL (EGC) GASOLINE 5 1930s 13.5" Glass $300-500
Blue outlined white globe face with blue outlined white center circle. Large Canadian Ethyl (EGC) logo with yellow rays in center circle. Red "CHAMPLAIN" arched around top, red/blue outlined box around red "GASOLINE" at bottom.

CHAMPLAIN SPECIAL 5 1940s Banded Glass $300-500
White globe face with series of red outline rings around large half circles at top and bottom of globe face. Red/white outline extends around diagonal blue band across center of globe face. White "CHAMPLAIN" on blue band with red "SPECIAL" just below band in white half circle.

CHAMPLAIN ETHYL 5 1940s Banded Glass $300-500
White globe face with series of red outline rings around large half circles at top and bottom of globe face. Red/white outline extends around diagonal blue band across center of globe face. White "CHAMPLAIN" on blue band with red "ETHYL" just below band in white half circle.

FINA & AFFILIATES

AMERICAN PETROFINA — DALLAS, TEXAS: Founded in Belgium in 1920, American Petrofina rebuilt from World War II and entered retail gasoline markets in Ontario and Quebec in 1951. American Petrofina is the Dallas, Texas, based subsidiary of Belgium's PetrofinaSA. American Petrofina has been formed almost entirely by mergers and buyouts, beginning with the purchase of Panhandle Oil Corp. of Wichita Falls, Texas, in October 1956. Subsequent purchases, where an entire company's marketing was purchased and converted, are listed below. With an ever larger focus on petrochemicals, the company became known as Fina Oil and Chemical Company in the mid-1980s. Fina currently operates approximately 2600 stations throughout the mid-continent and southeast.

FINA: The Fina brand first appeared in the United States in October, 1956 and continues in use at stations throughout the Southeast and Midwest today.

FINA 6 1957-1965 Capco NL
White globe face with "Fina red" outline ring. Large blue "FINA" across center of globe face.

PANHANDLE OIL CO. — WICHITA FALLS, TEXAS: Founded in Wichita Falls in 1918, Panhandle was originally a production and refining company. Panhandle opened its first service station in December 1923 in Wichita Falls. Within a year or two, this firm was operating stations from Phoenix, Arizona, east to Jackson, Mississippi. The company continued to expand in the 1930s, but shortages during World War II forced the company to cut back its extensive marketing territory, limiting the marketing area to parts of Texas by the early 1950s. Panhandle was continuing to operate stations in Texas when it sold out to, and was rebranded, Fina in October, 1956.

PANHANDLE PRODUCING REFINING 5 1923-1930s 15" Metal Red $900-1500
White globe face with red triangular Panhandle logo with longhorns covering most of globe face. White "PRODUCING" in left leg of triangle, "REFINING" on right.

PANHANDLE OCTAGNE 5 1930s 15" Metal Red $1000-1750
White globe face with large black outlined red octagon covering most of globe face. White script "Octagne" diagonal across octagon with small black/white Panhandle logo at top.

PANHANDLE SPECIAL 5 1930s 15" Metal Red $1000-1750
White globe face with brown outlined yellow octagon covering most of globe face. Brown "Special" with underline diagonal across octagon with small brown/white Panhandle logo at top.

PANHANDLE NOXLESS 5 1930s 15" Metal Red $1000-1750
White globe face with green rough edged design similar to a circular saw blade covering most of globe face. White/black outlined white "Noxless" diagonal across face with small black/white Panhandle logo at top.

PANHANDLE NOXLESS ETHYL 7 1930s 15" Metal Red NL
White globe face with red rough edged design similar to a circular saw blade covering most of globe face. White/black outlined white "Noxless" diagonal across face with large Ethyl (EGC) logo to lower right.

PANHANDLE 3 1940s 15" Metal Red $800-1400
White globe face with red triangular Panhandle logo with longhorns covering most of globe face.

PANHANDLE 6 1950s 13.5" Glass $650-1000
White globe face with red triangular Panhandle logo with longhorns covering most of globe face.

GUTHRIE REFINING CO. — GUTHRIE, OKLAHOMA: Guthrie Refining operated from 1932-1934 and sold out to Panhandle.

GUTHRIE REFINING CO's G.R.C. GASOLINE 6 1932-1934 13.5" Glass W NL
Handpainted lettering on early glass body. Black "GUTHRIE" arched around top, "GASOLINE" around bottom with black outlined red "REFINING CO'S." above small black outlined box in center. Black outlined red "G.R.C." in box.

AMERICAN LIBERTY — DALLAS, TEXAS: Formed in the 1930s from oil holdings of the Murchison family, American Liberty sold out to Fina in October 1957.

AMERICAN LIBERTY
REGULAR 4 1940s 13.5" Glass $1250-1700
White globe face with red outline ring. Blue "AMERICAN LIBERTY" arched around top with red/white/blue "Statue of Liberty" head in center. Red script "Regular" below Liberty.

AMERICAN LIBERTY
PREMIUM 4 1940s 13.5" Glass $1250-1700
White globe face with red outline ring. Blue "AMERICAN LIBERTY" arched around top with red/white/blue "Statue of Liberty" head in center. Red script "Premium" below Liberty.

AMLICO REGULAR 5 1950s Capco $200-300
Mint green globe face with dark green outline ring. Dark green "AMLICO" with white dropshade diagonal across face with dark green "REGULAR" below, right of center.

AMLICO PREMIUM 5 1950s Capco $200-300
Mint green globe face with dark green outline ring. Dark green "AMLICO" with white dropshade diagonal across face with dark green "PREMIUM" below, right of center.

PETRO-ATLAS — EL DORADO, KANSAS: Petro-Atlas is a consolidation of the old Eldorado Refining and K-T Refining companies. Petro-Atlas sold out to Fina in March 1958.

K-T MOTOR OILS 6 1920s OPE NL
White globe with red "KT" logotype across center of globe face. Red "OILS" below.

THE ELDORADO/
ELRECO/REFINING
COMPANY 4 1930-1946 Gill $300-550
White globe face with black outline ring forming sawtooth border around outer edge of globe face. Red upper and lower half circles split by white diagonal band across center of globe face. Small black "THE EL DORADO" above large "ELRECO" above "REFINING CO." on diagonal band.

SPECIAL/THE EL
DORADO/ELRECO/
REFINING CO./
GASOLINE 4 1930-1946 Gill $325-575
White globe face with black outline ring forming sawtooth border around outer edge of globe face. Red upper and lower half circles split by white diagonal band across center of globe face. Small black "THE EL DORADO" above large "ELRECO" above "REFINING CO." on diagonal band. White "SPECIAL" in upper red half circle, "GASOLINE" in lower half circle.

ELRECO REFORMED
GASOLINE 4 1930-1946 Gill $275-525
Red globe face with white band across center. White "ELRECO" with black dropshade arched around top with black "Re-Formed" on center white band. White "GASOLINE" arched around bottom.

ELRECO REFORMED
ETHYL 4 1930-1946 Gill $275-525
Red globe face with white band across center. White "ELRECO" with black dropshade arched around top with black "Re-Formed" on center white band. Small Ethyl (EGC) logo in white circle at bottom.

ELRECO PREMIUM 4 1960-1946 Gill $250-500
White globe face with red band across center. Red "ELRECO" with black dropshade arched around top, "GASOLINE" around bottom with white "Premium" on center band.

BUY MILES NOT
GALLONS/
ELRECO 3 1946-1955 Gill $300-500
White globe face with round red/black/white "ELRECO" logo in center. Red "Buy miles-Not gallons" arched around top.

BUY MILES NOT
GALLONS/
ELRECO/
REGULAR 3 1946-1955 Gill $300-500
White globe face with round red/black/white "ELRECO" logo in center. Red "Buy miles-Not gallons" arched around top, "REGULAR" arched around bottom.

BUY MILES NOT
GALLONS/
ELRECO/
PREMIUM 3 1946-1955 Gill $300-500
White globe face with round red/black/white "ELRECO" logo in center. Red "Buy miles-Not gallons" arched around top, "PREMIUM" arched around bottom.

BUY MILES NOT
GALLONS/
ELRECO/WHITE 4 1946-1955 Gill $300-500
White globe face with round red/black/white "ELRECO" logo in center. Red "Buy miles-Not gallons" arched around top, "WHITE" arched around bottom.

ELRECO 5 1955-1958 Plastic $175-275
White plastic formed oval halves cemented back to back. Black/white outline around red center circle split by diagonal white band. Black "ELRECO" across globe face on band. Oval version of earlier round logo.

PETROLEUM MARKETING CO. — TULSA, OKLAHOMA: Pemco is a Tulsa based independent that was rebranded Fina around 1960.

REGULAR PEMCO
GASOLINE 2 1950s Capco $150-250
White globe face with red/blue outline rings. Blue/red outlined white rectangular box across center with large blue "PEMCO" in box. Red "REGULAR" arched around top, "GASOLINE" around bottom.

PREMIUM PEMCO
GASOLINE 2 1950s Capco $150-250
White globe face with red/blue outline rings. Blue/red outlined white rectangular box across center with large blue "PEMCO" in box. Red "PREMIUM" arched around top, "GASOLINE" around bottom.

COLONIAL OIL CO. — DES MOINES, IOWA: Affiliated with Petro-Atlas, Colonial Oil sold out to Petrofina in 1959.

COLONIAL 5 1940s 12.5" Glass $250-400
Yellow globe face with red "COLONIAL" in white wing logotype diagonal across center of globe face.

COLONIAL
REGULAR 4 1950s Capco $175-275
White globe face with red "COLONIAL" in white wing logotype at top of globe face. Red outlined white rectangle with red "REGULAR" on lower globe face.

COLONIAL ETHYL 4 1950s Capco $175-275
White globe face with red "COLONIAL" in white wing logotype at top of globe face. Red outlined white rectangle with red "ETHYL" on lower globe face.

COLONIAL
PREMIUM 4 1950s Capco $175-275
White globe face with red "COLONIAL" in white wing logotype at top of globe face. Red outlined white rectangle with red "PREMIUM" on lower globe face.

COSDEN PETROLEUM — BIG SPRING, TEXAS: Cosden Petroleum was founded in about 1927 by Josh Cosden after he lost controlling interest in Mid-Continent Petroleum, which he had founded in 1916. Cosden was operating a Big Spring, Texas, refinery and petrochemical operation, marketing through stations in six states when the company merged with American Petrofina in April 1963. Cosden stations were rebranded Fina shortly after the merger.

COSDEN
GASOLINE OILS 6 1925-1930 14" OPE N $1800-2500
White globe with solid blue etched globe face. White "COSDEN" arched around top, "OILS" around bottom with script "Gasoline" underlined across center of globe face.

COSDEN LIQUID
GAS OILS 5 1930-1938 13.5" Glass $450-750
Orange globe face with green outline and green outlined white center circle and white band across center. White "COSDEN" with black dropshade arched around top, "OILS" at bottom with black outlined green "LIQUID GAS" on center white band.

COSDEN HIGHER
OCTANE 4 1938-1950 13.5" Glass $400-625

Green globe face with red/white outline rings and white/red/white outlined red band diagonal from lower left to upper right across globe face. White "COSDEN" in red band. White "HIGHER" in upper green area, "OCTANE" below.

COSDEN ETHYL (EGC) 4 1938-1941 13.5" Glass $250-450
Red globe face with white center circle. White "COSDEN" arched around top, "GASOLINE" around bottom with Ethyl (EGC) logo in center white circle.

COSDEN ETHYL (EC) 4 1941-1960 13.5" Glass $250-450
Red globe face with white center circle. White "COSDEN" arched around top, "GASOLINE" around bottom with Ethyl(EC) logo in center white circle.

COSDEN (motor grade) 4 1938-1960 13.5" Glass $200-400
Red outlined green globe face with red "COSDEN" diagonal across center with white dropshade. This generic looking globe was used on pumps dispensing motor grade gasolines.

COSDEN 3 1950-1960 Capco $175-275
Green globe face with red/white outline rings and white/red/white outlined red band diagonal from lower left to upper right across globe face. White "COSDEN" in red band.

COSDEN (red & white) 4 1960-1963 Capco $150-250
White globe face with red outline ring. Red/white outlined red band diagonal from lower left to upper right across globe face. White "COSDEN" on red band.

****COL-TEX**: As best as can be determined, Col-Tex was sold off to Cosden after Apco purchased and rebranded Kanotex, greatly expanding the Apco brand from its original Oklahoma marketing territory. When Apco became the primary brand name, Col-Tex stations were either converted to Apco or sold off with the Col-Tex brand to Cosden, who operated the Col-Tex stations as a secondary brand until 1963. See UDS Total listings, Anderson-Prichard section, for earlier Col-Tex listings.

COL-TEX 3 1954-1963 Capco $250-350
White globe face with black outline ring and red/white outlined red diagonal band from lower left to upper right. White "COL-TEX" in red band.

****ONYX**: This Texas independent sold out to Cosden and became a Cosden secondary brand.

ONYX REGULAR GASOLINE 5 1920s 15" Metal $750-1100
Blue outlined gray globe face with red outlined white band across center. Wood grain "ONYX" in center band, with black outlined red "REGULAR" in upper gray area, "GASOLINE" below.

ONYX AEROPLANE GASOLINE 6 1920s 15" Metal NL
Blue outlined gray globe face with red outlined white band across center. Wood grain "ONYX" in center band, with black outlined red "AEROPLANE" in upper gray area, "GASOLINE" below.

ONYX PETROLEUM PRODUCTS 6 1930s 15" Metal NL
White globe face with red outline ring and red lines forming white horizontal band across center. Red "PETROLEUM" arched across top, "PRODUCTS" across bottom with brown "wood grain" "ONYX" in center band.

ONYX 3 1950s Capco $550-800
White globe face with thick black outline ring. Black outline of Texas in center of globe with red "ONYX" across map.

ONYX ETHYL 4 1950s Capco $250-350
White globe face with thick black outline ring. Black lines offsetting white horizontal band above center. Red "ONYX" in white band, small Ethyl(EC) logo in black cirlce below.

****WIDES OIL COMPANY — MURPHYSBORO, ILLINOIS**: Founded prior to 1932, Wides was a Murphysboro, Illinois, based independent. Wides was purchased by Fina about 1972 and converted to the Fina brand after 1977.

WIDES OIL CO. REGULAR (W/TRIANGLE) 6 1940-1955 13.5" Glass $250-450
White globe face with red outline ring and red ring around white center circle. Large red line triangle in center circle with red "WIDES OIL COMPANY" in white area above triangle. Red "REGULAR" arched around top, "MORE POWER GAS" around bottom.

SAVE WITH/WIDES/ GAS FOR LESS 3 1955-1970 13.5" Glass $250-450
White globe face with red outline ring. Red circle in center split lower left to upper right by diagonal white band. Blue "SAVE WITH" arched around top, "WIDES" with speed lines in center band and "GAS FOR LESS" arched around bottom.

FLYING A/GETTY

TIDEWATER OIL COMPANY—NEW YORK, NEW YORK: Founded in New York in 1887, Tidewater is best known for their world famous motor oil brand "Veedol." Tidewater entered the gasoline market about 1915 and by 1920 was marketing all along the eastern seaboard under the Tydol brand. During the 1930s, control of Tidewater fell into the hands of Standard Oil of New Jersey which set up Mission Corporation as a subsidary to operate Tidewater and Skelly. J. Paul Getty purchased control of Mission in 1937 and merged Tidewater with California's Associated Oil Company in 1938. Associated had introduced the Flying A brand in 1932 for its premium gasoline. With the creation of the Tidewater Associated Oil Company, stations in the west became Associated Flying A stations and stations in the east became Tydol Flying A stations. In 1956 the Tydol and Associated designations were dropped and the brand simply became Flying A. In 1966 stations in the former Associated territory, including the secondary Seaside operation, were sold to Phillips Petroleum and rebranded Phillips 66. Also in 1966, the Tidewater name was dropped in favor of Getty Oil Company, Getty having completed the purchase of Mission in 1951 and having operated Tidewater Associated and Skelly as separate subsidiaries since that time. The Getty brand first appeared in 1970, replacing Flying A at stations in the northeast. In 1983 Getty also began replacing the Skelly brand with Getty; however, before the changeover could be completed Texaco purchased Getty Oil. The Getty marketing in the northeast was sold off to New England independent Power-Test. The former Skelly locations were rebranded Texaco or sold off and Power Test, renamed Getty Oil Marketing and based in Jericho, New York, continues to operate Getty brand stations from Maine to Virginia.

VEEDOL 6 1910s OPE NL
White globe with black outline ring around orange outer ring and black ring around white center circle. Black "Veedol" (large "V") across center of globe face.

TYDOL 5 1910s OPC $1100-1800
White globe with black outline ring around orange outer ring and black ring around white center circle. Black "TYDOL" (large "T") across center of globe face.

TYDOL T/W/O CAST FACES 4 1920-1930 15" Metal H/L Black $900-1400
Cast face with black outline ring around orange outer ring and black outline ring around white center circle. Black "TYDOL" (large "T") across center of globe face, with interlocking "T/W/O" on lower globe face.

TYDOL ETHYL CAST FACES (EGC) 4 1927-1930 15" Metal H/L Black $950-1450
Cast face with black outline ring around orange outer ring and black outline ring around white center circle. Black "TYDOL" (large "T") across center of globe face, with cast Ethyl (EGC) logo on lower globe face.

TYDOL ETHYL CAST FACE BAKED ETHYL 5 1927-1930 15" Metal H/L Black $950-1400
Cast face with black outline ring around orange outer ring and black outline ring around white center circle. Black "TYDOL" (large "T") across center of globe face, with baked on Ethyl (EGC) logo on lower globe face.

VEEDOL OILS 6 1920s 15" Metal H/L NL
Black outline ring around orange outer ring and black outline around white center circle. Black "VEEDOL/OILS" (large "V") across center of globe face.

TYDOL 3 1925-1941 16.5" Metal H/L Black $350-550

TYDOL 3 1925-1941 15" Metal H/L Silver $350-550

TYDOL 4 1925-1941 14" Gill $300-500
TYDOL 3 1925-1941 Gill $250-450
TYDOL 2 1935-1941 13.5" Glass W $200-400
Black outline ring around orange outer ring and black outline around white center circle. Black "TYDOL" (large "T") across center of globe face.

TYDOL T/W/O 4 1925-1930 15" Metal H/L Black $400-750
Black outline ring around orange outer ring and black outline around white center circle. Black "TYDOL" (large "T") across center of globe face, with interlocking "T/W/O" on lower globe face.

TYDOL W/ ETHYL 3 1927-1935 15" Metal H/L Black $400-600
Black outline ring around orange outer ring and black outline around white center circle. Black "TYDOL" (large "T") across center of globe face, with small "WITH" over larger "ETHYL" below "TYDOL".

TYDOL ETHYL 6 1927-1935 15" Metal H/L Black NL
Black outline ring around orange outer ring and black outline around white center circle. Black "TYDOL" (large "T") across center of globe face, with orange "ETHYL" below "TYDOL".

TYDOL ETHYL (EGC) 4 1927-1935 15" Metal H/L Silver $400-600
Black outline ring around orange outer ring and black outline around white center circle. Black "TYDOL" (large "T") across center of globe face, with small Ethyl (EGC) logo below "TYDOL".

TYDOL ARCHED OVER ETHYL (EGC) 4 1935-1941 16.5" Metal H/L Black $600-850
TYDOL ARCHED OVER ETHYL (EGC) 4 1935-1941 15" Metal H/L Black $600-850
TYDOL ARCHED OVER ETHYL (EGC) 4 1935-1941 Gill $400-650
TYDOL ARCHED OVER ETHYL (EGC) 4 1935-1941 13.5" Glass W $350-600
Black outline ring around orange outer ring and black outline around white center circle. Black "TYDOL" arched over large Ethyl (EGC) logo with blue rays.

TIDEX 3 1935-1941 16.5" Metal H/L Black $300-550
TIDEX 2 1935-1941 15" Metal H/L Black $300-550
TIDEX 3 1935-1946 Gill $300-450
TIDEX 4 1935-1946 13.5" Glass W $250-425
Navy blue globe face with black outlined white band across center and vertical, forming cross. Black "TIDEX" on white center band.

TYDOL W/ FLYING A DECAL 4 1941-1942 15" Metal H/L Black $500-800
Black outline ring around orange outer ring and black outline around white center circle. Black "TYDOL" (large "T") across center of globe face, with red "FLYING/A" decal on lower globe face.

TYDOL A W/ ORANGE BAND 4 1941-1942 Gill $400-650
TYDOL A W/ ORANGE BAND 4 1941-1942 13.5" Glass W $375-600
White globe face with thick orange outer ring. Black "TYDOL" across upper globe face with orange over black band across lower globe face. Black outlined orange "A" in center with black outlined white wings to either side.

TYDOL ARCHED OVER ETHYL (EC) 4 1941-1942 13.5"Glass W $350-600
Black outline ring around orange outer ring and black outline around white center circle. Black "TYDOL" arched over large Ethyl (EC) logo with blue rays.

TYDOL "A" W/ RED BAND 4 1942-1946 Gill $400-650
White globe face with black outline ring. Black "TYDOL" across upper globe face. Red over black horizontal band across lower globe face with black outlined red "A" in center. Black outlined white wings extending out from "A".

TYDOL ETHYL W/AIRPLANE 4 1942-1946 Gill $1000-1500
White globe face with black outline ring. Large black "TYDOL" over smaller "ETHYL" across globe face. Red airplane diagonal at top of globe with streak extending down through "L" in "TYDOL".

TYDOL FLYING A W/ RED BAND 4 1946-1949 Gill $400-600
TYDOL FLYING A W/ RED BAND 3 1946-1949 Capco $250-450
White globe face with black outline ring. Black "TYDOL" across upper globe face. Red over black horizontal band across lower globe face with black outlined red "A" in center. Black outlined white wings extending out from "A". Black "FLYING" above "A".

TYDOL ETHYL (BLACK/RED) 3 1946-1947 Gill $200-400
White globe face with black outline ring. Black "TYDOL" on upper globe face with smaller red "ETHYL" offset to right below "TYDOL".

TYDOL ETHYL (BLACK) 3 1946-1947 Gill $225-375
TYDOL ETHYL (BLACK) 3 1946-1947 Capco $175-250
White globe face with black outline ring. Black "TYDOL" on upper globe face with smaller black "ETHYL" offset to right below "TYDOL".

TYDOL FLYING A ETHYL red & black 4 1947-1949 Gill $400-600
White globe face with black outline ring. Black "TYDOL" across upper globe face. Red over black horizontal band across lower globe face with black outlined red "A" in center. Black outlined white wings extending out from "A". Black "FLYING" above "A", "ETHYL" below.

TYDOL AVIATION 6 1940s Gill NL
Black outline ring around orange outer ring and black outline around white center circle. Black "TYDOL/AVIATION" across center of globe face.

TIDEX PLAIN 4 1946-1950 Gill $225-325
White globe face with black outline ring. Black "TIDEX" across center of globe face.

TYDOL A W/ GREEN BAND 4 1949-1953 13.5" Glass $400-650
White globe face with black outline ring. Green band across lower globe face. Black "TYDOL" on upper globe face. Black outlined red "A" with black/white wings on center of green band.

TYDOL A DIESEL FUEL 4 1949-1953 13.5" Glass $450-700
White globe face with black outline ring. Green band across lower globe face. Black "TYDOL" on upper globe face. Black outlined red "A" with black/white wings on center of green band. Black "DIESEL FUEL" below green band.

TYDOL FLYING A W/GREEN BAND 4 1949-1953 13.5" Glass $400-650
White globe face with black outline ring. Green band across lower globe face. Black "TYDOL" on upper globe face. Black outlined red "A" with black/white wings on center of green band. Black "FLYING" above green band.

During the 1940s and 1950s, Tidewater Oil used a square white glass globe with attachable rounded corner lenses, about the size of ad glasses in gas pumps in that era. The following listings are descriptions of the various known lenses designed to fit these unusual globes.

TYDOL 9 1935-1941 OP Shoebox NL
White globe face with orange border with fine black outline. Black "TYDOL" logotype across center.

TYDOL ETHYL 6 1935-1941 OP Shoebox NL
White globe face with orange border with fine black outline. Black "TYDOL" logotype across center, above smaller "ETHYL".

TYDOL 5 1935-1941 Shoebox $450-750
White globe face with orange border with fine black outline. Black "TYDOL" logotype across center.

TYDOL ETHYL 6 1935-1941 Shoebox $450-750
White globe face with orange border with fine black outline. Black "TYDOL" logotype across center, above smaller "ETHYL".

TIDEX 4 1935-1949 Shoebox $450-750
White globe face with purple border with fine black outline. Black "TIDEX" logotype across center.

TYDOL "A" 5 1941-1942 Shoebox $450-750
White globe face with black "TYDOL" over orange over black band. Black/white outlined orange "A" with black/white wings at center of color band.

TYDOL ETHYL 5 1942-1947 Shoebox $500-800
White globe face with large black "TYDOL" logotype over smaller black "ETHYL" across center.

TYDOL "A" 5 1942-1946 Shoebox $500-800
White globe face with black "TYDOL" over red over black band. Black/

white outlined red "A" with black/white wings at center of color band.

TYDOL "A" ETHYL 5 1946-1947 SHOEBOX $500-800
White globe face with black "TYDOL" logotype over red band. Red "A" with black and white wings to left on band with black "ETHYL" to right.

TYDOL FLYING A 5 1949-1956 Shoebox $500-800
White globe face with green band across lower globe face. Black "TYDOL" over small black "FLYING" above band. Black outlined red "A" with black & white wings on green band.

FLYING A ETHYL GASOLINE 5 1949-1956 Shoebox $450-750
Red globe face with black outlined white "A" with black outlined white wings in center. White "FLYING" above "A", white "ETHYL" below.

FLYING A GASOLINE 4 1956-1960 Shoebox $400-650
White globe face with black outlined red "A" with red wings in center. Red "FLYING" above "A", "GASOLINE" below.

FLYING A SUPEREXTRA 4 1956-1960 Shoebox $400-650
Red globe face with black outlined white "A" with black outlined white wings in center. White "FLYING" above "A", white "SUPER" to left below and "EXTRA" to right below "A".

ASSOCIATED OIL COMPANY—SAN FRANCISCO, CALIFORNIA: Founded in 1901, Associated was a west coast integrated oil company that was marketing gasoline by 1915. In 1932 they introduced the Flying A brand as a premium grade gasoline and after the 1938 merger with Tidewater, Flying A became the company's primary brand name. The company also operated stations under the Seaside brand as a secondary operation. Flying A stations in the Associated territory, as well as Seaside, were sold to Phillips in 1966.

ASSOCIATED GASOLINE 4 1915-1932 15" Metal HP Red $2500-3500
Red globe face with white/green/white outline rings. White "ASSOCIATED" arched around top, "GASOLINE" around bottom with green outlined white gasoline can in center. Red "MORE/MILES/TO THE/GALLON" on can.

ASSOCIATED GASOLINE blue border 5 1930-1932 15" Metal HP Red $2600-3600
Red globe face with white/blue/white outline rings. White "ASSOCIATED" arched around top, "GASOLINE" around bottom with blue outlined white gasoline can in center. Red "MORE/MILES/TO THE/GALLON" on can.

ASSOCIATED FLYING A 4 1932-1933 15" Metal HP Red $1200-1800
Red globe face with white/green/white outline rings. Large winged "A" with green drop shades in center of globe face. White "FLYING" with green dropshade across top. White "ASSOCIATED" arched around bottom to left, "GASOLINE" to right with small "AERO-/TYPE" between.

ASSOCIATED FLYING A "AERO TYPE" 4 1933-1940 15" Metal HP Red $1200-1800
Red globe face with white/green/white outline rings. Large winged "A" with green drop shades in center of globe face. White "FLYING" with green dropshade across top. White "AERO-TYPE GASOLINE" arched around bottom.

ASSOCIATED GASOLINE W/ETHYL (EGC) 5 1926-1932 15" Metal HP Red $800-1250
Red globe face with *green* outline ring. White circle in center with large Ethyl (EGC) logo with blue rays in circle. White "ASSOCIATED" arched around top, "GASOLINE" around bottom.

ASSOCIATED GASOLINE W/ETHYL (EGC) 5 1926-1932 15" Metal HP Red $800-1250
Red globe face with *white* outline ring. White circle in center with large Ethyl (EGC) logo with blue rays in circle. White "ASSOCIATED" arched around top, "GASOLINE" around bottom.

ASSOCIATED AVIATION ETHYL 4 1932-1940 15" Metal HP Red $1600-2200
Red globe face with soaring eagle at top of globe face. Ethyl logo with black rays in yellow circle at top, to left of eagle wing. Small white "ASSOCIATED" over larger "AVIATION/ETHYL" on lower globe face.

ASSOCIATED GREEN & GOLD 5 1930-1935 15" Metal HP Green $1200-1600
Red outer band with red upper 1/4 background. "ASSOCIATED" in red at top, "GreenGold" in mint green in center, "GASOLINE" in red at bottom. Lower 1/4 face green.

ASSOCIATED WHITE GOLD GASOLINE 4 1920s 15" Metal HP Green $1100-1500
Mint green globe face with large white burst covering most of globe face. Large black "WHITE/GOLD" on center of globe face with smaller black "ASSOCIATED" above and "GASOLINE" below.

ASSOCIATED AROTANE 5 1940s 15" Metal HP Red $1100-1500
Red globe face with large white "AROTANE" with black and white wings extending across from "A" at top of globe face. Black "HIGH OCTANE QUALITY" below "AROTANE" above white "GASOLINE".

FLYING A GASOLINE (R/W LARGE TYPE) 4 1940-1946 15" Metal HP Red $800-1200

FLYING A GASOLINE (R/W LARGE TYPE) 2 1946-1956 Gill $350-550

FLYING A GASOLINE (R/W LARGE TYPE) 2 1946-1956 13.5" Glass $300-500
White globe face with large black outlined red "A" with red wings and black details on center of globe face. Red "FLYING" above "A", "GASOLINE" below.

FLYING A ETHYL (BLK) GASOLINE (WHITE) 4 1940-1946 15" Metal HP Red $900-1400

FLYING A ETHYL (BLK) GASOLINE (WHITE) 3 1946-1956 Gill $400-750

FLYING A ETHYL (BLK) GASOLINE (WHITE) 3 1946-1956 13.5" Glass $350-650
Red globe face with large black outlined white "A" with white wings and black details positioned above center of globe face. White "FLYING" above "A" with white outlined black "ETHYL" below. White "GASOLINE" below "ETHYL".

FLYING A GASOLINE (R/W SMALL TYPE) 4 1946-1956 13.5" Glass $300-500

FLYING A GASOLINE (R/W SMALL TYPE) 4 1946-1956 Capco $275-400
White globe face with black outline ring. Black outlined red "A" with red wings and black details in center of globe face. Small red "FLYING" over "A", "GASOLINE" below.

FLYING A ETHYL GASOLINE (W/R SM TYPE) 3 1946-1956 13.5" Glass $350-600

FLYING A ETHYL GASOLINE (W/R SM TYPE) 3 1946-1956 Capco $300-450
Red globe face with white outline ring. Black outlined white "A" with white wings and black details on center of globe face. Small white "FLYING" above "A", "ETHYL GASOLINE" below "A".

TIDEWATER ASSOCIATED OIL COMPANY - FLYING A MARKETING:

FLYING A GASOLINE 4 1956-1970 Gill $350-600
White globe face with large black outlined "A" with red wings and black details in center. Red "FLYING" horizontal over "A", "GASOLINE" arched around bottom.

FLYING A ETHYL (BLK) GASOLINE (WHITE) 3 1946-1956 Gill $400-650
Red globe face with large black outlined white "A" with white wings and black details positioned above center of globe face. White "FLY-

ING" above "A" with white outlined black "ETHYL" below. White "GASOLINE" below "ETHYL".

FLYING A SUPER EXTRA 2 1956-1970 Gill $300-500
Red globe face with large black outlined white "A" with white wings and black details in center. White "FLYING" horizontal over "A", "SUPER EXTRA" arched around bottom.

FLYING A KEROSENE 5 1956-1970 Gill $350-550
Red globe face with large black outlined white "A" with white wings and black details in center. White "FLYING" horizontal over "A", "KEROSENE" arched around bottom.

FLYING A DIESEL 5 1956-1970 Gill $400-600
Red globe face with large black outlined white "A" with white wings and black details in center. White "FLYING" horizontal over "A", "DIESEL" arched around bottom.

SEASIDE OIL COMPANY — SANTA BARBARA, CALIFORNIA: As a California refiner/marketer, founded in 1899 and purchased by Tidewater Associated in 1946, Seaside remained a Flying A secondary brand until the western marketing was purchased by Phillips in 1966. Phillips used the Seaside brand for some time after the purchase. Seaside was discontinued in 1973 when Phillips began abandonment of their western marketing.

GRADE GASOLINE (3 sided metal) 5 1920s Unusual Metal NL
Early Keystone Logo similar to globe below. Three-sided globe, rectangular panels in metal frame. Straight non-flange type base.

GRADE NON-CARBONIZING 6 1920s 15" Metal $1600-2100
Black outlined white globe face with large orange outlined black keystone on upper center of globe face. Orange banner across lower edge of keystone. Yellow"GRADE' above small script "non-carbonizing" on upper keystone with "GASOLINE" on banner. Irregular shape band across lower globe face below keystone with yellow "GUARANTEED" over orange "1st Quality" on band.

SEASIDE GRADE 4 1930-1950 15" Metal $1700-2100
White globe face with red triangle offset to upper left. Blue triangle offset to lower right superimposed over red triangle. Yellow wave area at top of blue triangle with blue script "Seaside". White outlined red banner with white "GRADE" arched below waves on blue triangle. White "GASOLINE" and small seagull on lower triangle.

SEASIDE GRADE PLUS 5 1935-1940 15" Metal $1800-2200
White globe face with red triangle offset to upper left. Blue triangle offset to lower right superimposed over red triangle. Yellow wave area at top of blue triangle with blue script "Seaside". White outlined red banner with white "GRADE/+PLUS" arched below waves on blue triangle. White "GASOLINE" and small seagull on lower triangle.

SEASIDE SILVER GULL 5 1930-1950 15" Metal $2000-2500
White globe face with red triangle offset to upper left. Blue triangle offset to lower right superimposed over red triangle. Yellow wave area at top of blue triangle with blue script "Seaside". White "SILVER/GULL" below waves on blue triangle. White "GASOLINE" and small seagull on lower triangle.

SEASIDE SILVER GULL TETRAETHYL 5 1930-1935 15" Metal $2100-2600
White globe face with red triangle offset to upper left. Blue triangle offset to lower right superimposed over red triangle. Yellow wave area at top of blue triangle with blue script "Seaside". White "SILVER/GULL" below waves on blue triangle. White "GASOLINE" and small seagull on lower triangle, with "TETRAETHYL" arched around lower globe face.

SEASIDE ETHYL 5 1935-1950 15" Metal $2000-2500
White globe face with red triangle offset to upper left. Blue triangle offset to lower right superimposed over red triangle. Yellow wave area at top of blue triangle with blue script "Seaside". White "GASOLINE" above white circle with Ethyl (EGC) logo and small seagull on lower triangle.

SEASIDE ETHYL (w/decal) 5 1935-1950 15" Metal $2000-2600
Seaside Ethyl globe as described above with decal covering entire image. White lettering "SEASIDE" on lower triangle above white circle with "ETHYL" across center. Small white "GASOLINE" arched below. Line swirls through entire lower triangle area with "KING OF THE HIGHWAYS" where it widens out below Ethyl circle.

PATHFINDER PETROLEUM —SOUTH PASADENA, CALIFORNIA: Sold out to Tidewater Associated in 1960.
No Pathfinder globes known at this time.

MOHAWK OIL CO.: California independent refiner/marketer. History unknown.

MOHAWK GASOLINE 4 1930s 15" Metal HP Red $2500-4000
Red globe face with black outline ring and black outline around white center circle. White "MOHAWK" arched around top, "GASOLINE" around bottom with red and black Indian head in center.

MOHAWK GASOLINE 4 1930s 15" Metal HP Red $2500-4000
Orange globe face with black outline ring and black outline around white center circle. White "MOHAWK" arched around top, "GASOLINE" around bottom with orange and black Indian head in center.

MOHAWK ETHYL 6 1930s 15" Metal HP Red $2800-4500
White globe face with black outline ring and black ring around white center circle. Red "MOHAWK" arched around top, "ETHYL" around bottom with red and black Indian head in center.

TIDAL REFINING CO. — TULSA, OKLAHOMA: Founded in 1916, Tidal Refining Company became a subsidiary of Tidewater Oil Company about 1930. Tidal formed the core of Tidewater's midwestern marketing from that time until the territory was sold to Mid Continent Petroleum in 1958.

TIDAL 6 1920s OPE $900-1300
White globe with red "TIDAL/GAS" on globe face.

TIDAL 5 1920s 15" Metal $300-500
White outlined green globe face with white band across center. Red "TIDAL" in center band.

NORTHWESTERN OIL COMPANY—SUPERIOR, WISCONSIN: Founded in the 1920s, Northwestern distributed products in northern Wisconsin and northeastern Minnesota until the late 1930s. At that point, Northwestern was purchased by Tidewater-Associated and stations were rebranded to Tydol.

NORTHWESTERN NOCO GASOLINE 5 1920s 15" Metal Black $700-1100
White globe face with black outline ring and black lines forming white band across upper globe face. Orange diamond in center of face with black "NOCO" across diamond. Black "NORTHWESTERN" in band, "Gasoline" across lower globe face.

NORTHWESTERN VIMAMITE GASOLINE 5 1920s 15" Metal Black $850-1250
White globe face with black outline ring and black lines forming white band across upper globe face. Orange diamond in center of face with black "VIMAMITE" across diamond. Black "NORTHWESTERN" in band, "Gasoline" across lower globe face.

GULF & AFFILIATES

****GULF OIL CO. — PITTSBURGH, PENNSYLVANIA:** The Gulf Oil Company was founded in 1901 at the Spindletop, Texas, oil fields. Very early in the company's history they established a marketing presence throughout the south and east that was to dominate the company's direction for the next eighty years. Gulf is credited with having opened the first "built for that purpose, off-street gasoline station in the world," having opened a small retail station at the corner of Baum Boulevard and St. Clair Ave in Pittsburgh in December 1913. Other stations pre-date this, but they were converted from other uses. Gulf continued to market in 29 southern and eastern states until the 1950s when several purchases of regional chains made Gulf a coast-to-coast marketer. By 1965 the Gulf brand was known coast-to-coast. The gas shortages of the 1970s forced Gulf to withdraw from many areas and the early 1980s saw Gulf in serious financial trouble. In 1984 Chevron (Standard of California) purchased Gulf Oil, converting stations in several states to Chevron and selling the southeast Gulf branded marketing to BP and the northeast marketing to Cumberland Farms. Cumberland Farms, now Gulf Oil, LLP, continues to operate stations under the Gulf brand today, while the Gulf stations in the southeast were rebranded BP in 1990.

GASOLINE 5 1914-1915 OPE N $450-850
Appears to be a generic globe, but was used exclusively by Gulf. White globe with tall, thin-stroke lettering in blue or black "GASOLINE".

THAT GOOD GULF GASOLINE 5 1914-1918 OPE Chimney $1800-2600
White globe with orange ring around globe face. Orange circle in center with orange lines extending out from circle to form white band across center. Black "THAT" arched around top, "GASOLINE" around bottom and black "GOOD GULF" on center band.

THAT GOOD GULF GASOLINE 3 1918-1930 OPE W/N $600-950
White globe with orange ring around globe face. Orange circle in center with orange lines extending out from circle to form white band across center. Black "THAT" arched around top, "GASOLINE" around bottom and black "GOOD GULF" on center band.

THAT GOOD GULF GASOLINE tall fancy letters 5 1918-1930 OPE N $650-1000
White globe with orange ring around globe face. Orange circle in center with orange lines extending out from circle to form white band across center. Unusual black "THAT" arched around top, "GASOLINE" around bottom and black "GOOD GULF" on center band.

THAT GOOD GULF GASOLINE 5 1925-1930 OPB W $550-850
White globe with orange ring around globe face. Orange circle in center with orange lines extending out from circle to form white band across center. Black "THAT" arched around top, "GASOLINE" around bottom and black "GOOD GULF" on center band.

THAT GOOD GULF GASOLINE 3 1925-1930 OPC $600-950
White globe with orange ring around globe face. Orange circle in center with orange lines extending out from circle to form white band across center. Black "THAT" arched around top, "GASOLINE" around bottom and black "GOOD GULF" on center band.

THAT GOOD GULF GASOLINE 6 1925-1930 OPC/Etched $750-1100
White globe with etched details. Orange ring around globe face. Orange circle in center with orange lines extending out from circle to form white band across center. Black "THAT" arched around top, "GASOLINE" around bottom and black "GOOD GULF" on center band. >>NOTE: All cast globes have copper threaded bases.

THAT GOOD GULF GASOLINE ORANGE EDGER 5 1925-1930 OPC $850-1200
Wide orange area, extending around body, around white globe face. Blue outline ring in orange area. Orange circle in center with lines extending out to each side forming white band across center. Black "THAT" arched around top, "GASOLINE" around bottom with black "GOOD GULF" on center band.

THAT GOOD GULF GASOLINE 5 1920s 15" Metal $700-1250
Wide white border around orange outline ring. White globe face. Orange circle in center with orange lines extending out from circle to form white band across center. Black "THAT" arched around top, "GASOLINE" around bottom with black "GOOD GULF" on center band.

THAT GOOD GULF GASOLINE neon 6 1920s 18" Neon Metal NL
Wide white border around orange outline ring. White globe face. Orange circle in center with orange lines extending out from circle to form white band across center. Black "THAT" arched around top, "GASOLINE" around bottom with black "GOOD GULF" on center band.

GULF NO NOX 4 1925-1930 OPE W/N $1000-1650
White globe with black outline ring around face. Black line around orange disc logo in center. Black "GULF" with black/white striped dropshade on orange disc. Black "NO-NOX" arched around top, "MOTOR FUEL" around bottom.

GULF NO-NOX 5 1925-1930 OPB W $900-1400
White globe with black outline ring around face. Black line around orange disc logo in center. Black "GULF" with black/white striped dropshade on orange disc. Black "NO-NOX" arched around top, "MOTOR FUEL" around bottom.

GULF NO NOX 4 1925-1930 OPC $1000-1650
White globe face with orange outline ring. Orange circle in center with black "GULF" with white dropshade on circle. Black "NO-NOX" arched around top, "MOTOR FUEL" around bottom.

GULF KEROSENE 5 1920-1930 OPE W $1500-2500
White globe with orange outline ring around face. Orange circle in center with black "GULF" arched round top, "KEROSENE" around bottom.

GULF 3 1930-1935 OPC $500-800
White globe with blue outline ring around orange globe face. Large blue "GULF" with white dropshade across globe face.

GULF 3 1930-1935 OPB N $450-750
White globe with blue outline ring around orange globe face. Large black "GULF" with white dropshade across globe face.

GULF 2 1935-1952 13.5" Glass N $250-450
Orange globe face with blue/white outline rings. Large blue "GULF" with white dropshade across globe face.

GULF 4 1935-1952 Gill $275-475
Orange globe face with blue/white outline rings. Large blue "GULF" with white dropshade across globe face.

GULF 1 1935-1952 12.5" Glass W $300-500
Orange globe face with blue/white outline rings. Large blue "GULF" with white dropshade across globe face.

GOOD GULF 7 1930s 12.5" Glass W NL
Orange globe face with blue outline ring. Blue lettering "GOOD/GULF" across globe face.

NO-NOX ETHYL 7 1930s 12.5" Glass W NL
Orange globe face with blue outline ring. Blue "NO NOX" above Ethyl (EGC) logo on globe face.

TRAFFIC 7 1930s 12.5" Glass W NL
Orange globe face with blue outline ring. Blue "TRAFFIC" across globe face.

TRAFFIC BENZOL 7 1930s 13.5" Glass N NL
White globe face with blue outline ring and blue "TRAFFIC/BENZOL" on globe face.

MARINE WHITE-SPECIAL FOR MARINE USE 4 1940-1947 13.5" Glass N $900-1400
White globe face with blue outline ring. Blue ring around small orange Gulf disc in center. Blue "MARINE WHITE" arched around top, small blue "SPECIAL/FOR MARINE/USE below logo.

GULF DIESEL 3 1946-1952 12.5" Glass W $400-650
Blue globe face with orange/white outline rings. Gulf "orange disc" logo at top of globe face. Orange outlined white band across globe face below logo. Orange "DIESEL/FUEL" on white band.

GULF DIESELECT 4 1946-1952 12.5" Glass W $450-700
Same as above with orange "DIESELECT" replacing "DIESEL/FUEL".

GULF MARINE WHITE GASOLINE 4 1946-1952 12.5" Glass W $550-900
White globe face with blue outline ring and wide blue band across center. Blue "GULF" in upper white area, "GASOLINE" on lower with white "MARINE/WHITE" on center blue band.

GULF 6 1930s 18" Neon NL
Orange globe face with blue/white outline rings. Large blue "GULF" with white dropshade across globe face.

****WILSHIRE OIL CO.—LOS ANGELES, CALIFORNIA**: Wilshire was a west coast refiner/marketer founded in 1919 and marketed from the 1920s until the 1950s under the brand name "Polly Gas." Wilshire was purchased by Malco in 1955, and in turn by Gulf in 1957, just as the brand name was changed to Wilshire. Also in 1957, Wilshire purchased much of the former Sunset International (formerly Sunset Oil Company) marketing, converting all of the California Sunset locations to Wilshire. In 1965, the Wilshire brand was replaced by Gulf and the Gulf brand survived on the west coast until the 1970s. The 1970s gas shortages forced Gulf to abandon many unprofitable markets, as well as others where direct distribution was not feasible.

POLLY GAS 4 1920s-1940 15" Metal $5000-7000
Black globe face with white outline ring. White outlined yellow "POLLY" arched around top, "GAS" around bottom. Detailed green/orange parrot in center, perched on orange outlined yellow sign with black lettering "POLLY/GAS". Black "PROPERTY OF WILSHIRE OIL

CO. INC." very small around bottom.

BUY ECONOMY
GASOLINE 5 1920s-1930 15" Metal $500-850
White globe face with green "ECONOMY" in bowtie lettering across center of globe face. Red "BUY" above "ECONOMY", "GASOLINE" below.

ECONOMY ANTI-KNOCK
GASOLINE 4 1930-1940 15" Metal $1500-1800
White globe face. Dark green outline rings around mint green "tombstone" shaped area covering most of globe face. Dark green "ECONOMY" arched around top of tombstone, with interlocking WOCO logo below "ECONOMY". Red script "Anti-Knock" in white band below logo. Dark green "GASOLINE" on mint green area below band. Green "REFINED BY WILSHIRE" below tombstone area. Small green "PROPERTY OF WILSHIRE OIL CO. INC." around bottom.

RADIO GASOLINE 5 1930s-1940 15" Metal $1200-1800
White globe face with black circel in center. Black/white outlined red lightning bolt across center of globe face from left to right, with black outlined jagged edge red lettering "RADIO" arched around top, "GASOLINE" around bottom. Small white "A/WILSHIRE OIL CO., INC." above lightning bolt in center circle. "PRODUCT" below. Black "PROPERTY OF WILSHIRE OIL CO. INC." very small around bottom.

****SUNSET OIL COMPANY—LOS ANGELES, CALIFORNIA**: Founded in 1914, the company survived receivership in 1932 and a complete reorganization in 1937, and again after World War II as Sunset International, Inc. This firm operated as a refiner under the subsidiary name Marine Refining Corporation. In 1957 Wilshire Oil purchased virtually all of the California marketing operations, leaving Sunset as only a small Arizona jobber operation that survived into the 1970s.

SUNSET
GASOLINE 9 1930-1935 15" Metal NL
Yellow/red outlined globe face with green outer ring and detailed red and yellow sunset seen on yellow and green ocean in center circle. Yellow "SUNSET" arched around top, "GASOLINE" around bottom with yellow car to left of center circle, airplane to right. Green outlined yellow banner arched around bottom of center circle with red "SUNSET PACIFIC OIL COMPANY" in banner.

SUNSET GASOLINE
(w/sunset scene) 6 1935-1940s 15" Metal HP Green NL
Yellow/red outlined globe face with green outer ring and detailed red and yellow sunset seen on yellow and green ocean in center circle. Yellow "SUNSET" arched around top, "GASOLINE" around bottom with yellow car to left of center circle, airplane to right. Green outlined yellow banner arched around bottom of center circle with red "SUNRISE MOTOR OILS" in banner.

MARINE (no sunset) 5 1930-1935 15" Metal $1900-2800
White globe face with blue outlined white "MARINE" across upper globe face. Red, white and blue speedboat scene in red outlined oval in center with blue/white anchors to either side and white "GASOLINE" in blue banner across bottom.

MARINE 6 1935-1937 15" Metal NL
White globe face with blue outlined white "MARINE" across upper globe face. Red, yellow and green sunset on ocean scene in oval in center with blue/white anchors to either side and white "GASOLINE" in blue banner across bottom.

ASCOT W/
RACE CAR 5 1930-1937 15" Metal $3500-5000
White globe face with red/green outlined green irregular shield shaped design covering most of globe face. White lettering "ASCOT" across top of shield, "GASOLINE" arched around bottom with white oval in center. Black/white/green detailed 20's race car on oval.

SUNSET DE LUXE 6 1937-1940s 15" Metal NL
Cream globe face with orange/yellow outlined green triangle-point down-on center of face. White "SUNSET" and multicolor sunset scene on upper triangle over green band on lower triangle. White "DELUXE" in band, with tiny white "GASOLINE" below band in point of triangle.

SUNSET ETHYL 6 1937-1940s 15" Metal NL
White globe face with yellow outlined green shield in center of face. White "SUNSET" over multicolor sunset scene on shield, white "ETHYL" below. Small Ethyl (EGC) logo in yellow outlined circle on lower globe face.

SUNSET MARINE
(w/sunset) 6 1937-1940s 15" Metal NL
Black globe face with orange/yellow outlined green triangle-point down-on center of face. White "SUNSET" and multicolor sunset scene on upper triangle over green band on lower triangle. White "MARINE" in band, with tiny white "GASOLINE" below band in point of triangle.

SUNSET THRIFTY
(w/Scotsman) 6 1940s 15" Metal NL
Cream globe face with red/green outlined yellow triangle on center of face. Green "THRIFTY" over multicolor scotsman on upper triangle over gren band across lower triangle. White "GASOLINE" in lower green band.

****CARPENTER OIL CO. — LINCOLN, NEBRASKA:** In the 1950s Carpenter Oil Company purchased the former Coryell operation. In the middle 1960s, Carpenter added the Superior 400 stations to their operation. Carpenter was purchased by Gulf in 1968, with Gulf retaining use of the Superior 400 trademark for several years following the purchase.
>>See Coryell and Superior 400 listings for globes.

****L.L.CORYELL & SONS — LINCOLN, NEBRASKA**: Founded in 1894 as a distribution company, L. L. Coryell ventured into gasoline marketing in the late 1920s and by 1937 had stations scattered from Denver to Cincinnati and from Mankato, Minnesota, south to West Memphis, Arkansas. In the late 1950s Carpenter Oil Company purchased the Coryell brand and operated stations throughout Nebraska. In the early 1960s they merged with Superior Oil of Des Moines, who was bought out by Gulf in 1968.

CORYELL 70
SUPER 6 1930s 15" Metal $700-1000
White globe face with red "CORYELL" arched around top, "SUPER" around bottom with large green "70" and four green dots in center.

CORYELL 70
SUPER neon 6 1930s 18" Neon Metal NL
White globe face with red "CORYELL" arched around top, "SUPER" around bottom with large green "70" and four green dots in center.

CORYELL 70
SUPER 4 1940s 13.5" Glass $300-500

CORYELL 70
SUPER 4 1950s Capco $250-350
Black outline around red outer ring and around white center circle. White "CORYELL" arched around top, "SUPER" around bottom with large black outlined red "70" in center and four square dots in center.

CORYELL ETHYL
SENIOR (EC) 5 1940s 13.5" Glass $300-500

CORYELL ETHYL
SENIOR (EC) 5 1950s Capco $250-350
Black outline around red outer ring and around white center circle. White "CORYELL" arched around top, "SENIOR" around bottom with large Ethyl (EC) logo in center circle.

CORYELL ZENITH
GASOLINE 5 1940s 13.5" Glass $300-500
Red outer ring with light blue circle in center. White band across center of globe face with white "CORYELL" arched around top, "GASOLINE" around bottom and dark blue "ZENITH" across center white band.

****SUPERIOR OIL CO.—DES MOINES, IOWA**: Superior was a Des Moines based marketer, dating from the mid-1930s. They merged with Carpenter Oil Co. in the early 1960s. About 1968 Gulf purchased Superior and rebranded stations Gulf, although Superior continued in use as a Gulf secondary brand into the 1970s.

SUPERIOR 400
GASOLINE 3 1932-1946 13.5" Glass W $225-375

SUPERIOR 400
GASOLINE 3 1946-1968 Capco $225-325
White globe face with red outline ring. Black "SUPERIOR" arched around top, "GASOLINE" around bottom with large red "400" in center.

SUPERIOR ETHYL
(EC) GASOLINE 4 1946-1968 Capco $200-300
White globe face with red outline ring. Black "SUPERIOR" arched

around top, "GASOLINE" around bottom with Ethyl (EC) logo in center.

SUPERIOR 400
FUEL OIL 5 1946-1968 Capco $225-350
White globe face with red outline ring. Black "SUPERIOR" arched around top, "FUEL OIL" around bottom with large red "400" in center.

****BULK PETROLEUM CO. — CHICAGO, ILLINOIS:** Bulk Petroleum was another old Chicago private brand that sold out to a major oil company during the market expansion of the 1950s-60s. Gulf purchased the operation, branded "BULKO" in 1968. Most of the Chicago area stations were converted to Gulf, but some unsuccessful Gulf locations around the country were operated as Bulko stations in the 1970s.
>>No Bulko globes are known at this time.

****SMITH OIL CO. — ROCKFORD, ILLINOIS:** Founded in Rockford in 1911, Smith Oil was an independent marketer in the Illinois region until the World War II era, when they became a Gulf distributor. In the expansion years of the 1960s, control of Smith passed to Gulf. It is believed that when Gulf withdrew from the upper Midwest in the 1970s, Smith Oil was spun off as an independent, after which they became a branded Sunoco jobber.

SMITH SPECIAL
GASOLINE 6 1930s 15" Metal NL
White globe face with black outline ring. Black "SMITH" with red design at top of globe face over red "SPECIAL/GASOLINE".

SMITH PARAMOUNT
GASOLINE 5 1930s 14" Glass $350-550
White globe face with black outline ring. Black "SMITH" with red design at top of globe face over red "PARAMOUNT/GASOLINE".

SMITH GASOLINE 5 1930s Two Piece Glass NL
White globe face with black outline ring. Black "SMITH" with red design at top of globe face over red "GASOLINE".

****BRITISH AMERICAN OIL CO, LTD.—TORONTO, ONTARIO:** Founded in 1906, British American Oil Company was an earlier pioneer in the exploration and production of Canadian oil reserves. They introduced branded gasoline marketing in 1920 and expanded coast-to-coast in Canada before 1930. In 1952 Gulf Oil began purchasing interest in British American, which they completed in 1968. Stations were converted from the British American "BA" brand to Gulf in 1970. After Chevron purchased Gulf in 1984, the former Gulf Canada operation was sold off to the Canadian national oil company and stations were rebranded PetroCanada.

BRITISH MOTOR
GASOLENE 5 1925-1928 OPE N $900-1400
White globe face with blue "BRITISH" arched around top, "GASOLENE" around bottom and large red "MOTOR" across center.

BRITISH AMERICAN
GASOLENE 5 1920-1928 OPE N $900-1400
White globe face with blue "BRITISH" arched around top, "AMERICAN" around bottom and small red "GASOLENE" across center.

BRITISH AMERICAN
PEERLESS 5 1922-1926 OPE N $900-1400
White globe face with blue "BRITISH" arched around top, "AMERICAN" around bottom with red logotype lettering "Peerless" across center of globe face.

BRITISH AMERICAN
PEERLESS 5 1926-1928 OPE N $900-1400
White globe face with red "BRITISH" arched around top, "AMERICAN" around bottom and red block "PEERLESS" across center.

BRITISH AMERICAN
MOTOR 5 1928-1932 OPB $750-1250
White globe face with blue "BRITISH" arched around top, "AMERICAN" around bottom and large red "MOTOR" across center.

BRITISH AMERICAN
GASOLENE 5 1928-1932 OPB $750-1250
White globe face with blue "BRITISH" arched around top, "AMERICAN" around bottom and large red "GASOLENE" across center.

BRITISH AMERICAN
PEERLESS 5 1928-1932 OPB $750-1250
White globe face with blue "BRITISH" arched around top, "AMERICAN" around bottom and red script "PEERLESS" across center.

BRITISH AMERICAN
ETHYL 5 1928-1932 OPB $750-1250
White globe face with blue "BRITISH" arched around top, "AMERICAN" around bottom and large red "ETHYL" across center.

BRITISH AMERICAN
SUPER POWER 5 1928-1932 OPB $750-1250
White globe face with blue "BRITISH" arched around top, "AMERICAN" around bottom and large red script "Super/Power" across center.

BRITISH AMERICAN
MOTOR 6 1920s 15" Metal $650-900
White globe face with red outline ring. Red "BRITISH" arched around top, "AMERICAN" around bottom with blue script "Motor" through center.

BRITISH AMERICAN
"BRITISH MOTOR" 6 1932-1935 13.5" Glass W
$450-750
White globe face with blue outline ring. Blue "BRITISH" arched around top, "AMERICAN" around bottom with red script "British/Motor" diagonal across center.

BRITISH AMERICAN
MOTOR 4 1935-1940 13.5" Glass W $300-500
White globe face with red outline ring. Red "BRITISH" arched around top, "AMERICAN" around bottom with blue script "Motor" diagonal across center.

BRITISH AMERICAN
PEERLESS 4 1932-1935 13.5" Glass W $300-500
White globe face with blue outline ring. Blue "BRITISH" arched around top, "AMERICAN" around bottom with red script "Peerless" diagonal across center.

BRITISH AMERICAN
ETHYL (EGC) 4 1932-1935 13.5" Glass W $300-500
White globe face with red outline ring and red outline around white center circle. Blue "BRITISH" arched around top, "AMERICAN" around bottom, with large Ethyl (EGC) logo in center.

BRITISH AMERICAN
SUPER POWER 6 1932-1940 Balcrank $400-650
White globe face with blue outline ring. Blue "BRITISH" arched around top, "AMERICAN" around bottom with red script "Super/Power" diagonal across center.

BRITISH AMERICAN
NEVR NOX 5 1932-1940 13.5" Glass W $400-650
White globe face with green outline ring. Red "BRITISH" arched around top, "AMERICAN" around bottom with green script "Nevr Nox" diagonal across center.

BRITISH AMERICAN
PEERLESS ETHYL
(EGC) 5 1935-1940 13.5" Glass W $300-500
White globe face with blue outline ring and blue ring around white center circle. Red "BRITISH AMERICAN" arched around top, red script "Peerless Ethyl" around bottom with Ethyl (EGC) logo in center circle.

BA MOTOR 4 1940-1954 13.5" Glass W $300-500
White globe face with green outline ring. Small red, white and green "B-A" logo at top, with green "Motor" across center of globe face below.

BA NEVR-NOX 4 1940-1954 13.5" Glass W
$350-550
White globe face with green outline ring. Small red, white and green "B-A" logo at top, with green "Nevr-nox" across center of globe face below.

BA ETHYL 4 1940-1946 13.5" Glass W $300-500
White globe face with green outline ring set well in from edge of lens. Small red, white and green "B-A" logo in center, with green "Ethyl" across center of globe face below.

BA PEERLESS
ETHYL 4 1946-1954 13.5" Glass W $300-500
White globe face with green outline ring set well in from edge of lens. Small red, white and green "B-A" logo in center, with green outlined red "Peerless" arched around top and "Ethyl" around the bottom.

BA PEERLESS
ETHYL 4 1946-1954 13.5" Glass W $300-500
White globe face with green outline ring near edge of lens. Small red, white and green "B-A" logo in center, with green outlined red "Peerless" arched around top and "Ethyl" around the bottom.

****ROYALITE OIL COMPANY, LTD. —CALGARY, ALBERTA:** Royalite Oil was formed in 1920 when Imperial Oil purchased control of Calgary Petroleum Products and renamed the operation, establishing it as a natural gas production and pipeline operation. In 1949 Imperial sold Royalite. Under new ownership the operation took steps to become a fully integrated oil company. After purchasing several refineries, the first Royalite service stations opened in 1953. In 1954, Royalite expanded its marketing operations with the purchase of service stations formerly operated by Hi-Way Refineries of Regina, Saskatchewan. In 1962 British American purchased control of Royalite and the companies were merged to form Gulf Oil Canada in late 1968.

ROYALITE
(new logo) 5 1964-1968 Capco $250-400
Blue outlined white globe face with blue and red compass rose logo in center. Blue "ROYALITE" across center of logo.

****HI-WAY REFINERIES, LTD. — REGINA, SASKATCHEWAN:** Founded in 1931, Hi-Way was established primarily to take advantage of the agri-business trade in the prairie provinces. In 1954 Hi Way was purchased by Royalite and stations were rebranded accordingly.

HI WAY GASOLINE 5 1930s 15" Metal NL
Yellow outer area with black outline around white center circle. Black outlined green star in center circle. Black "HI-WAY" arched around top, "GASOLINE" around bottom.

HI-WAY 99
ETHYL 5 1930s 15" Metal NL
Yellow globe face with black outlined white highway shield in center. Black "99" in shield. Diagonal black line lower to upper right across globe face. Black "HI-WAY" at top, "ETHYL" at bottom.

****PURITAN OIL COMPANY. LTD. — REGINA, SASKATCHEWAN:** History unknown, purchased by British American in the 1960s.

STELLARENE
NON KNOCK 5 1940s 13.5" Glass $400-650
Red outlined white globe face with black "STELLARENE" arched around top of globe face, over red "NON-/KNOCK". Large red comet with star in head across upper globe face.

STELLARENE
ETHYL 5 1940s 13.5" Glass $400-650
Green outlined white globe face with dark green "STELLARENE" arched through globe face, superimposed over red comet. Black outlined red "ETHYL" across lower globe face.

STELLARENE
AMBER 5 1950s Capco $300-500
Green outlined white globe face with dark green "STELLARENE" arched through globe face, small red comet with red star in head above. Black "AMBER" across lower globe face.

STELLARENE
ETHYL 5 1950s Capco $300-500
Green outlined white globe face with dark green "STELLARENE" arched through globe face, small red comet with red star in head above. Black "ETHYL" across lower globe face.

****GAS AND OIL PRODUCTS, LTD.—CALGARY, ALBERTA:** Gas & Oil Products Limited (G.O.P.) was organized in 1934 by Albert H. Mayland to operate an absorption plant in Turner Valley, Alberta, to refine naphtha. When the natural gas product could not be sold to a larger refiner, a topping plant was added in 1938 to produce third grade gasoline and distillates. To market these products, a sales organization was formed, with its head office in Calgary, to mainly sell their products to other marketers and to truck and farm dealers as well as to operate a small number of filling stations. A growing increase in demand for higher grades of motor fuels—which previously were purchased from other companies—resulted in the addition in 1939 of a cracking unit to enlarge the refinery's capacity and produce their own high grade gasoline. By the 1940s G.O.P. was the largest independent producer, refiner and marketer in Alberta and its Patron Oil Co. Ltd. subsidiary was the largest in Saskatchewan. Its gasoline was sold under the trade names of Purity 99, Mercury 99 Ethyl, and Miracle 99. The firm also retailed Veedol motor oil products.

Anglo-American Exploration Ltd., owned by Sam Nickle of Calgary, purchased the company's assets in 1953 for 5 million dollars after A.H. Mayland's death. Through the fifties, Purity 99 expanded its base of retailers from western Ontario through Manitoba, Saskatchewan, Alberta and into eastern British Columbia, growing to 857 outlets distributing Purity products. In 1961, Anglo American closed its refinery due to economic conditions and contracted the British American Oil Company Ltd. to fill its petroleum requirements. BA purchased control of Anglo American in 1962 and changed the name of the parent company to Scenic Oils Ltd. It however continued to market Purity products under the Purity 99 name until 1965 when the assets were sold to B.A.'s subsidiary Royalite Oil Company and the name was dropped. In 1968, BA and Royalite amalgamated under the name Gulf Oil Canada Limited after Gulf's completion of its takeover of BA.

GOP WHITE
GASOLINE 5 1934-1940s 15" Metal $500-750
White globe face with black "G.O.P." across top, "GASOLINE" across bottom and red "WHITE" in center.

PURITY 99
GASOLINE 5 1934-1940s 15" Metal $550-800
White globe face with black "PURITY" arched around top, "GASOLINE" around bottom with large red "99" in center. Black dot to either side of "99".

GOP MIRACLE 99 5 1950s 15" Metal $600-850
White globe face with small red "G.O.P." across top and black outlined red band across lower globe face with white "99" on band. Black outlined yellow "MIRACLE" across center, with red and black blocking dropshade below lettering.

MERCURY 99
ETHYL 5 1950s 15" Metal $900-1300
White globe face with large black outlined yellow "MERCURY" across top of globe face. Black outlined red "99" with black outlined mint green wings at lower left of globe face, large Canadian Ethyl (EGC) logo to right.

PURITY 99
GASOLINE 5 1940s 13.5" Glass $275-450
White globe face with black "PURITY" arched around top, "GASOLINE" around bottom with large red "99" in center. Black dot to either side of "99".

GOP MIRACLE 99 5 1950s 13.5" Glass $300-500
White globe face with small red "G.O.P." across top and black outlined red band across lower globe face with white "99" on band. Black outlined yellow "MIRACLE" across center, with red and black blocking dropshade below lettering.

MIRACLE 99 5 1950s 13.5" Glass $300-500
White globe face with black outlined red band across lower globe face with white "99" on band. Black outlined yellow "MIRACLE" across center, with red and black blocking dropshade below lettering.

****ANGLO-AMERICAN OIL COMPANY — CALGARY, ALBERTA:** Anglo-American Oil was a great plains marketer that merged with Gas and Oil Products, Ltd in 1953. While the corporate name, Anglo-American, survived the merger, the newly formed company continued use of the Purity 99 trademarks.

ANGLO REGULAR 5 1930s 13.5" Glass $250-450
White globe face with blue outline ring. Red lines forming white band across upper center with red "ANGLO" in band. Blue script "Regular" below band.

ANGLO ETHYL 5 1930s 13.5" Glass $250-450
White globe face with blue outline ring. Red lines forming white band across upper center with red "ANGLO" in band. Blue script "Ethyl" below band.

****ARCTIC OIL CO.:** History unknown, merged with Anglo American.

ARCTIC 6 1930s 15" Metal NL
Description not available.

ARCTIC GOLD 6 1930s 15" Metal NL
Description not available.

HESS & AFFILIATES

****AMERADA HESS — PERTH AMBOY, NEW JERSEY:** Hess is the only oil company that is better known for its annual promotion than for its products. Chances are, if you collect gasoline memorabilia, the only thing you know about Hess is that they've produced and sold a toy truck, a replica of one of their tankers or other company vehicles, every Christmas since 1964. The company's belief in absolute silence to the public and to the press has prevented much more than information about the trucks to ever be made public. The extensive research for this book has brought up fragments of information that we will attempt to consolidate here.

Hess began operations as a coal yard/service station in Asbury Park, New Jersey, around 1925. They evolved from coal into fuel oil during the Depression and from fuel oil into heavy residual oils for boiler fuels. Terminal facilities at Perth Amboy, New Jersey, were added after World War II and the company opened its first Hess branded retail service stations in about 1959. In 1963, Hess purchased independent refiner Delhi-Taylor, which owned a majority of Billups Eastern as well as Southern Oil out of Ft. Lauderdale, Florida. Southern stations were converted to Hess, and in 1965 Hess purchased the remaining portion of Billups Eastern and rebranded the stations Hess. Along the way Olixir, an Albany, New York, area chain was purchased and rebranded Hess. In 1964, the first of the Christmas trucks were offered, branded either Billups or Hess. Since that time there has been a promotional truck sold most every year at Christmas and these have become very desirable collectibles. In 1968, Hess Oil and Chemical merged with Amerada Petroleum, a production/exploration company with operations around the world. Hess currently markets through over 600 branded outlets in 15 states as well as supplying several strong southern private brands.

****HESS OIL AND CHEMICAL CO. — PERTH AMBOY, NEW JERSEY:** The first Hess branded stations appeared about 1959. The operation expanded all along the eastern seaboard during the 1960s. Today Hess operates over 600 stations from New England to Florida.
>>No Hess globes known.

****OLIXIR — GREEN ISLAND, NEW YORK:** Olixir was a New York independent that sold out to Hess in the early 1960s.

OLIXIR 4 1930s 15" Metal Black $400-600
Yellow globe face with black outline. Black reverse bowtie "OLIXIR" across globe face.
OLIXIR LUBRICATED GAS 4 1930s 15" Metal Black $550-800
OLIXIR LUBRICATED GAS 5 1930s 16.5" Metal Black $550-800
Yellow globe face with black outline ring and black band across center. Black elongated "OLIXIR" in upper half, "GAS" in lower half with yellow "LUBRICATED" across center black band.

****SOUTHERN OIL CO.—FT. LAUDERDALE, FLORIDA:** Southern Oil was a Florida private brand that had been purchased by Delhi-Taylor in the late 1950s. When Hess purchased Delhi-Taylor in 1963, the Southern stations were rebranded Hess.

SOUTHERN 7 1950s Capco NL
White globe face with blue bands across above and below center. White stars on blue bands. Red "SOUTHERN" on white between blue bands.

****BILLUPS EASTERN PETROLEUM — JACKSONVILLE, FLORIDA:** Founded in Greenwood, Mississippi, in 1935, Billups Eastern could trace its roots to a single station opened in 1926. Billups survived as a family owned operation until 1959. In 1959 the company was split into two operating units, Billups Eastern and Billups Western. Billups Western was sold out to Signal Oil and operated by them until 1970 when Signal got out of retail gasoline marketing and sold the operation to Charter Oil. A majority of Billups Eastern was sold to Independent refiner Delhi-Taylor, who was bought out by Hess Oil and Chemical in 1963. The remaining portion of Billups Eastern was bought by Hess in 1965 and Billups stations were rebranded Hess. Billups stations from Delaware to Florida were then rebranded Hess. See also the Billups listings in the Signal section. All of the following globes also appear in that listing.

BILLUPS SUPREME 4 1940s Gill $900-1400
BILLUPS SUPREME 4 1940s Capco $850-1200
White globe face with blue outline ring and red "BILLUPS" arched around top and "SUPREME" around bottom with blue outline of outstretched hand in center,
BILLUPS ETHYL (EC) PREMIUM 9 1950-1970 Gill $1600-2400
BILLUPS ETHYL (EC) PREMIUM 5 1950-1970 Capco $1500-2200
Red globe face with white circle in center. Ethyl (EC) logo in white center circle. White script "Billups" arched around top of globe face and extending around to small airplane on each side as if "Billups" lettering was skywriting. Good detail on airplanes. White "PREMIUM" with blue shading arched around bottom.

****SPAR — ORLANDO, FLORIDA:** Originally Orlando Ice and Fuel, this firm was bought out by Billups Eastern soon after the Billups split in 1959. After the Billups merger with Hess, Spar was operated by another independent for several years.

SPAR 7 1950s 13.5" Glass NL
White globe face with red lettering "SPAR" across center.

****MEADVILLE CORP.—HAVERFORD, PENNSYLVANIA:** Meadville operated large multi-pump discount stations in mid-Atlantic metropolitan areas under the Save-Way and Safe-Way brands. In the 1950s, Meadville acquired Reading, Pennsylvania, based Merit and continues to operate stations under the Merit brand today. Control of Meadville was purchased by Hess in 1964.

MERIT 5 1950s Gill $300-500
Red globe face with white outlined green elongated octagon in center of face. Black outlined white "MERIT" on octagon.
MERIT HI TEST 5 1950s Gill $350-550
Red globe face with white outlined green elongated octagon in center of face. Black outlined white "MERIT" on octagon.
Yellow circle on lower globe face, below "MERIT". Black "HI TEST" in circle.
MERIT ETHYL (EGC) 5 1950s Gill $350-550
Green globe face with white outlined red elongated octagon in center of face. Black outlined white "MERIT" on octagon. White circle with Ethyl (EGC) logo on lower globe face.

HUDSON & AFFILIATES

****HUDSON OIL COMPANY—KANSAS CITY, KANSAS:**
Hudson Oil Company's storied history began in 1930 when Mary Hudson married Wayne Driver and moved to Kansas City. Mr. Driver went to work with Mary's brothers at a Kansas City oil marketer operating under the Silver King name. The family was involved in two gas stations in Kansas City when, in 1933, Mr. Driver was killed in an accident with a fuel oil truck. Mary quickly sold off the two stations; but, within a year she had purchased another station at 25th and Broadway in Kansas City. At age 21, Mary Hudson was selling gasoline under the Hudson name. By 1937 the chain consisted of five stations, the original Kansas City location and stations at Leavanworth, Kansas, Norfolk and Omaha, Nebraska, and Moberly, Missouri. In the late 1930s Mary formed a partnership with her brothers and Hudson Oil was formed. The chain expanded coast to coast with hundreds of locations by the time family members went their separate ways in 1968, with the Hudson brothers operating competing service stations under the Highway (Workingman's Friend) and FISCA (First In Service Company of America) brands. The 1977 purchase of a Cushing, Oklahoma, refinery led to a long downward spiral for Hudson. In 1985 the Mary Hudson operation was bankrupt. More than 300 Hudson brand stations closed, leaving only a dozen or so in operation today. Hudson Affiliates Highway and Fisca continue to market across the country.

****SILVER KING — KANSAS CITY, KANSAS:** Silver King was a Kansas City marketer from which the various members of the Hudson family launched their careers. It is believed that Silver King was purchased when the family consolidated their holdings in 1938.

SILVER KING 6 1930s 15" Metal $600-850
White globe face with large "SILVER" arched around top, "KING" arched around bottom and large "SK" with arrow passing through it in center.
SILVER KING 5 1930s 15" Metal $400-650
White globe face with white/green outline rings. Red "Silver-" diagonal across upper globe face above "-King".

****HUDSON OIL CO.—KANSAS CITY, KANSAS:** A Hudson brand introduced in 1933.**

HUDSON
GASOLINE 4 1933-1946 Gill $900-1300
Globe face has black outline and white outer ring. Orange-red center circle split by black band across center with graduated black horizontal lines in upper half of circle. Black and white 30s transport truck "HI-OCTANE GAS" on center black band. Red outlined black "HUDSON" arched at top, "GASOLINE" at bottom. "GASOLINE" at bottom has red rays through lettering as if center circle is sun.

HUDSON TRANSPORT
GASOLINE 5 1933-1946 Gill $1000-1500
Same basic design as Hudson Gasoline above except "HUDSON OIL CO" replaces "HUDSON" in upper ring, "TRANSPORT/GASOLINE" below black band with truck.

HUDSON ETHYL
GASOLINE (EGC) 5 1933-1946 Gill $900-1300
Same as Hudson Gasoline, black outlined red "HUDSON" arched around top, "GASOLINE" around bottom with large ETHYL (EGC) logo superimposed over truck in center circle. Yellow rays around Ethyl logo.

AAA-1 ETHYL 4 1940-1946 Gill $400-650
Yellow globe face with black outline and black 1/3 lower area. Large black/white outlined red "AAA-1" across most of yellow area with script white "New" just above "AAA-1". Small Ethyl (EGC) logo in black outlined white circle offset to right at color line. White script "Lubricated" over yellow "GAS" on lower black area.

HUDSON
REGULAR 4 1946-1962 13.5" Glass $800-1300

HUDSON
REGULAR 3 1955-1962 Orange Capco $700-1200
Globe face has black outline and white outer ring. Orange-red center circle split by black band across center with graduated black horizontal lines in upper half of circle. Black and white 30s transport truck "HI-OCTANE GAS" on center black band. Red outlined black "HUDSON" arched at top, "REGULAR" at bottom. "REGULAR" at bottom has red rays through lettering as if center circle is sun.

AAA-1 ETHYL 4 1946-1950 13.5" Glass $350-600
Yellow globe face with black outline and black 1/3 lower area. Large black/white outlined red "AAA-1" across most of yellow area with script white "New" just above "AAA-1". Small Ethyl (EC) logo in black outlined white circle offset to right at color line. White script "Lubricated" over yellow "GAS" on lower black area.

HUDSON'S AAA-1
ETHYL 4 1950-1962 13.5" Glass $400-700

HUDSON'S AAA-1
ETHYL 4 1955-1962 Capco $275-450
Yellow globe face with black outline and black 1/3 lower area. Large black/white outlined red "AAA-1" across most of yellow area with small black *"HUDSON'S"* arched around top and script white "New" just above "AAA-1". Small Ethyl (EC) logo in black outlined white circle offset to right at color line. White script "Lubricated" over yellow "GAS" on lower black area.

HUDSON ETHYL
GASOLINE (EGC) 5 1946-1955 13.5" Glass $750-1250
Same as Hudson Gasoline, black outlined red "HUDSON" arched around top, "GASOLINE" around bottom with large ETHYL (EGC) logo superimposed over truck in center circle. Yellow rays around Ethyl logo.

HUDSON ETHYL
GASOLINE (EC) 3 1955-1962 Red Capco $800-1100

HUDSON ETHYL
GASOLINE (EC) 5 1955-1962 Capco $750-1000
Same as Hudson Gasoline, black outlined red "HUDSON" arched around top, "GASOLINE" around bottom with large ETHYL (EC) logo superimposed over truck in center circle. Yellow rays around Ethyl logo.

HUDSON ETHYL
(EC) 5 1950-1962 13.5" Glass $750-1250

HUDSON ETHYL
(EC) 3 1955-1962 Red Capco $4800-1100

HUDSON ETHYL
(EC) 5 1955-1962 Capco $750-1000
Same as Hudson Gasoline, black outlined red "HUDSON" arched around top, "ETHYL" around bottom with large ETHYL (EC) logo superimposed over truck in center circle. Yellow rays around Ethyl logo.

HUDSON
KEROSENE 4 1950s 13.5" Glass $450-700

HUDSON
KEROSENE 4 1950s Capco $400-600
White globe face with light blue outline and blue outline around red flame in center. Red "HUDSON" arched around top, blue "KEROSENE" around bottom.

HORNET OIL COMPANY — TULSA, OKLAHOMA: Hornet is a Tulsa based affiliate that rebranded Hudson in 1958.

HORNET
GASOLINE 3 1950s Capco $250-350
Yellow globe face with black outline ring and black interlocking squares placed randomly around globe face. Large black interlocking squares in center with white center area. Black "HORNET" on white area, black "GASOLINE" on yellow below large squares.

MAJOR'S: The Major's globe is identical to the Hudson globes — the company's relation is unknown but it is assumed that Major's was operated by one of the Hudson family members. Note that the truck pictured is labeled "T. MAJORS".

MAJORS
REGULAR 5 1950s Capco $800-1150
Globe face has black outline and white outer ring. Orange-red center circle split by black band across center with graduated black horizontal lines in upper half of circle. Black and white 30s transport truck "HI-OCTANE GAS" on center black band. Red outlined black "MAJOR'S" arched at top, "REGULAR" at bottom. "REGULAR" at bottom has red rays through lettering as if center circle is sun.

MAJORS ETHYL
(EC) 5 1950s Capco $750-1100
Same as Hudson Gasoline, black outlined red "MAJOR'S" arched around top, "ETHYL" around bottom with large ETHYL (EC) logo superimposed over truck in center circle. Yellow rays around Ethyl logo.

HIGHWAY OIL COMPANY: A Hudson affiliate marketing under the Highway, Hi-Lo, and Workingman's Friend brands.
>>No Highway globes known.

FISCA OIL — KANSAS CITY, KANSAS: A Hudson affiliate with stations coast-to-coast today. Fisca is an acronym for "First In Service Company of America."
>>No Fisca globes known.

HUSKY & AFFILIATES

HUSKY OIL COMPANY — CODY, WYOMING: Husky was originally a brand name of Western Oil and Fuel, Minneapolis, Minnesota. In 1938 Husky Refining Company was established as a separate operation in Cody, Wyoming. Western Oil and Fuel, however, continued to use the Husky brand name until at least 1954. Some dealers are known to have used the name as late as 1960, when Conoco purchased Western and was in the process of reorganizing the company. A Canadian division of Husky Refining (by now known as Husky Oil Company) was established at Lloydminister, Saskatchewan, in 1947. In December 1954 Husky Oil Company acquired H. Earl Clack, an independent marketer in Harve, Montana, greatly expanding their station count and territorial reach. All stations then became identified as Husky Hi-Power stations. In 1968, Husky Oil of Cody, Wyoming, merged with Frontier Oil (a company that had earlier—about 1960—bought Western States Petroleum of Salt Lake City, Utah). The Husky name replaced Frontier and Beeline about 1970 and Husky expanded into Canada. In 1984 Husky Oil sold off the U.S. marketing to concentrate on Canadian marketing, which remains in operation today. In 1990 a Denver based marketer revived the Frontier brand name and "Rarin' to Go" trademark. This marketer currently operates a chain of stations under the Frontier trademark.

WESTERN OIL AND FUEL — MINNEAPOLIS and HUSKY OIL COMPANY — CODY, WYOMING: See above introduction and also the Western listings in the Conoco section for other Western brand globe listings.

HUSKY "QUALITY

PETROLEUM PRODUCTS" 5 1935-1940 13.5" Glass $1700-2300

Globe face has blue outline ring around upper white area. Lower 1/3 is dark blue, separated from upper area by red line. Black/white/red husky dog head with extreme detail with orange/yellow "northern lights" design on upper globe face. Yellow "HUSKY" in lower blue area. Black "QUALITY PETROLEUM PRODUCTS" arched above dog.

HUSKY w/ northern lights 6 1940-1960 12.5" Glass NL

HUSKY w/ northern lights 4 1940-1960 13.5" Glass $1800-2400

HUSKY w/ northern lights 4 1940-1960 Gill $1700-2500

Globe face has blue outline ring around upper white area. Lower 1/3 is dark blue. Blue/gray/white/red husky dog head with good detail with orange/yellow "northern lights" design on upper globe face. Yellow "HUSKY" in lower blue area.

HUSKY ETHYL (EGC) 6 1935-1940 Gill $1800-2500

Red globe face with black outline around white center circle. Inside circle is white with lower 1/3 blue. Blue/gray/white/red husky dog head with good detail and orange/yellow "northern lights" design in upper 2/3 of center circle. Small black outlined white circle superimposed over lower blue area, extending into red outer area. Ethyl (EGC) logo with yellow rays in small circle.

****H. EARL CLACK, INC. — HARVE, MONTANA:** H. Earl Clack operated stations and truckstops under the HECCO brand until 1954 when Husky Oil purchased Clack and rebranded the stations Husky Hi-Power.

HECCO 5 1930s 15" Metal LP $550-800

Good details not available. White globe face with orange "HECCO" arched around top, "GASOLINE" around bottom with diamond in center.

HECCO ECONOMY GASOLINE 5 1930s 15" Metal LP $600-900

White globe face with orange and green outline rings. Orange "HECCO" arched around top, "GASOLINE" around bottom with green "ECONOMY" across center. Small green diamonds above and below "ECONOMY".

HI-POWER GASOLINE-HECCOLENE OIL 5 1930-1954 15" Metal LP $650-1000

White globe face with dark green/orange outline rings. Dark green "HI-POWER" arched around top, "GASOLINE" around bottom with orange/white circle in center. White diamond logo with "HECCOLENE/OIL" around border and orange center with white "HECCO".

HI POWER ETHYL 5 1930-1954 15" Metal LP $550-850

Orange globe face with dark green outline ring. Dark green "HI-POWER" diagonal across top over smaller "Ethyl". Small Ethyl (EC) logo in black outlined white circle with blue rays near bottom.

DIESEL FUEL 6 1930-1954 15" Metal LP $350-550

Orange globe face with green outline ring. White outlined green "DIESEL/FUEL" across globe face.

****HUSKY - HI POWER, INC. — CODY, WYOMING:** In 1954 when Husky purchased H. Earl Clack Oil Company, all gasoline marketing was consolidated under the subsidiary "H. Earl Clack, Inc." using the "Husky Hi-Power" brand name. Marketing was reorganized in 1960 when the U.S. and Canadian divisions were merged and the brand again became simply "Husky."

HUSKY HI-POWER 4 1954-1960 15" Metal $1800-2400

Blue globe face with red outline ring. Upper white area has blue and white standing husky dog. White "HUSKY" over smaller "HI-POWER" in lower blue area.

HUSKY HI-POWER 5 1954-1960 Capco $1200-1600

Blue globe face with red outline ring. Upper white area has blue and white standing husky dog. White "HUSKY" over smaller "HI-POWER" in lower blue area.

HI POWER ETHYL 4 1954-1960 15" Metal $500-800

Red-orange globe face with black outline ring. White outlined black "HI-POWER" diagonal across to over small black "Ethyl". Small Ethyl (EC) logo in black outlined white circle with blue rays near bottom.

****FRONTIER OIL — DENVER, COLORADO:** A Denver based refiner/marketer founded in the mid-1930s, Frontier operated a chain of stations in the Rocky Mountain region. About 1960, Frontier purchased Western States Petroleum and operated Beeline stations on the western slope of the Rockies. In 1968, Husky Oil purchased Frontier Oil and replaced both the Frontier and Beeline brands with Husky around 1970. In 1990, a Denver marketer reintroduced the Frontier brand name on a chain of retail stations in Colorado.

FRONTIER RARIN' TO GO 5 1935-1946 Gill $1400-2000

White globe face with large red "FRONTIER" arched around top. Black rearing horse with detailed red/black/white rider with *tan* details. Small red "RARIN' TO" below horse above larger black "GO" with red speed lines. Red Ripple add $1000.

FRONTIER RARIN' TO GO 6 1935-1946 Gill $1400-2000

White globe face with large red "FRONTIER" arched around top. Black rearing horse with detailed red/black/white rider with *gold* details. Small red "RARIN' TO" below horse above larger black "GO" with red speed lines. Red Ripple add $1000.

FRONTIER GAS 3 1946-1956 13.5" Glass $900-1500

Black outline around red outer ring and around center white circle. Large black outlined white "FRONTIER" arched around top, "GAS" around bottom with 2 stars to either side. Black and white raring horse with rider in center circle with red "RARIN' TO GO!" offset below horse.

FRONTIER OIL CO "90" 6 1946-1950 13.5" Glass $500-750

White globe face with red outline ring. Red "FRONTIER OIL COMPANY" arched around bottom, with large black outlined red "90" at top center with two red stars arched to either side. Black and white line drawing of buffalo in center.

FRONTIER ETHYL 4 1950-1956 13.5" Glass $500-750

Black outline around red outer ring and around center white circle. Large black outlined white "FRONTIER" arched around top, "GAS" around bottom with 2 stars to either side. Large Ethyl (EC) logo in center circle.

FRONTIER DIESEL 5 1946-1956 13.5" Glass $1000-1600

Black outline around red outer ring and around center white circle. Large black outlined white "FRONTIER" arched around top, "DIESEL" around bottom with 2 stars to either side. Black and white raring horse with rider in center circle with red "RARIN' TO GO!" offset below horse.

FRONTIER RARIN' TO GO 3 1956-1970 Capco $600-900

Globe face with red lower half and white upper half. Black outlined white "FRONTIER" across lower red area just below center with white "RARIN'-TO-GO" arched underneath. Black silhouette of raring hose and rider in upper white area.

FRONTIER FLEET FUEL DIESEL 4 1956-1970 Capco $300-500

Black outline around mint green globe face. Red "FRONTIER" arched around top, "DIESEL" arched around bottom with red "FLEET FUEL" with speed lines and black dropshade across center.

****WESTERN STATES PETROLEUM — SALT LAKE CITY, UTAH:** Dating from after World War II, Western States operated the Beeline stations on the western slope of the Rockies. They merged with Frontier Oil around 1960 but the Beeline name remained in use until replaced by Husky about 1970.

BEELINE GASOLINE 4 1946-1965 Capco $1200-1750

Bright yellow globe face with black outline ring. Black logotype script "Beeline" across center of globe with small black "GASOLINE" below. Red dot at top with detailed Black/yellow/white bee flying across dot.

BEELINE DIESEL 4 1946-1965 Capco $1200-1750

Bright yellow globe face with black outline ring. Black logotype script "Beeline" across center of globe with small black "DIESEL FUEL" below. Red dot at top with detailed Black/yellow/white bee flying across dot.

IOMA

INDEPENDENT OIL MEN of AMERICA (IOMA) - CHICAGO, ILLINOIS: The

"Independent Oil Men of America" was formed in 1922 in Chicago, Illinois, to offer independent jobbers and gas/oil marketers a marketing organization and nationwide trademark to assure recognizable quality. The trademark selected was "Red Hat," registered in June 1925. The logo consisted of a large red hat with a black band and white stars. Standard Oil of Indiana, having long used the "Red Crown" trademark, sued for trademark infringement. In 1929, IOMA lost the right to use this valuable trademark. After that time, their trademark consisted of the organization logo, the red "thunderbird" on a black triangle—the corporate logo that had been in use since 1922 as a secondary trademark. The organization survived into the 1970s, but little is known of it in later years.

GENERIC RED HAT GLOBES: The following globes do not display a distributing jobber's brandname.

RED / HAT
GASOLINE 4 1925-1932 15" Metal HP Red $1800-2300
White globe face with large red top hat with black hatband in center. Three white stars on black hatband. Small black triangle/red thunderbird IOMA logo below hat. Black "GASOLINE" arched around bottom with "RED" to left of hat, "HAT" to right.

MOTOR OIL/
GASOLINE
RED HAT 4 1925-1932 15" Metal HP Red $1900-2400
White globe face with wide black ring around outer edge. Large red top hat with black hatband in center. Three white stars on black hat band. Black "RED" to left of hat, "HAT" to right with small IOMA logo below. White "MOTOR OIL" arched around top in black ring, "GASOLINE" around bottom.

REGULAR
GASOLINE
RED HAT 5 1930-1932 Capco $1600-2000
White globe face with large red top hat with black hatband in center. Three white stars on black hatband. Small black triangle/red thunderbird IOMA logo below hat. Black "REGULAR" arched around top and "GASOLINE" arched around bottom with "RED" to left of hat, "HAT" to right.

HIGH HAT: IOMA trademark offered as an alternative to the RED HAT brand.

HIGH HAT
GASOLINE
W/HAT & LOGO 6 1920s 15" Metal NL
White globe face with large red top hat with black hat band and verticle red and black stripes above band. Black "HIGH" to left of hat, "HAT" to right with black "GASOLINE" arched below.

BRANDED RED HAT and INDEPENDENT OILS GLOBES: Listed below are the currently known Red Hat and Independent Oil globes that display the name of the local jobber that used them. While most of the member jobbers elected to use the Red Hat trademark, not all did. Some of the brands listed below identified their products with the IOMA black triangle/red thunderbird logo. This does not indicate that a globe with the IOMA logo (as opposed to the Red Hat brand) is necessarily from after the 1932 discontinuance of the Red Hat.

A GAS CO.

"A GAS CO."
GASOLINE 6 1920s OPE N NL
Small globe with 4" base. White globe with black outline ring around face. Large black triangle with red thunderbird IOMA logo in center of globe face. White "INDEPENDENT" above bird, "OILS" below. Black "A GAS CO." arched around top, "GASOLINE" around bottom.

AMERICAN OIL COMPANY

AMERICAN OIL CO
RED HAT
GASOLINE 6 1925-1932 15" Metal HP Red $1800-2500
White globe face with large red top hat with black hatband in center. Three white stars on black hatband. Small black triangle/red thunderbird IOMA logo below hat. Black "AMERICAN OIL CO." arched around top, "GASOLINE" around bottom with "RED" to left of hat, "HAT" to right.

BLAKE OIL COMPANY —KIRKLAND, ILLINOIS

BLAKE OIL CO.
GASOLINE 5 1920s OPE RW $2000-2600
White globe with black outline ring around face. Large black triangle with red thunderbird IOMA logo in center of globe face. White "INDEPENDENT" above bird, "OILS" below. Black "BLAKE OIL CO" arched around top, "GASOLINE" around bottom.

BRANDT OIL COMPANY —ATLANTA, ILLINOIS

BRANDT RED HAT
GASOLINE 5 1925-1932 OPE RW $3500-5000
White globe with red or black outline ring around globe face. Large red top hat with black hatband in center of globe. Three white stars on black hatband. Small black triangle/red thunderbird IOMA logo below hat. Black "BRANDT" arched around top, "GASOLINE" around bottom with "RED" to left of hat and "HAT" to right.

CONSUMERS OIL CO. —WINCHESTER, TENNESSEE

CONSUMERS OIL CO.
WINCHESTER,
TENN 6 1920s OPE N $2000-2600
White globe with large black triangle with red thunderbird IOMA logo in center of globe face. White "INDEPENDENT" above bird, "OILS" below. Black "CONSUMERS OIL CO" arched around top, "WINCHESTER, TENN" around bottom.

CSC OIL COMPANY —LOUISVILLE, KENTUCKY

CSC RED HAT
GASOLINE 6 1925-1932 OPE RW $3500-5000
White globe with red or black outline ring around globe face. Large red top hat with black hatband in center of globe. Three white stars on black hatband. Small black triangle/red thunderbird IOMA logo below hat. Black script "C.S.C." arched around top, "GASOLINE" around bottom with "RED" to left of hat and "HAT" to right.

DOBSON BROTHERS OIL COMPANY

DOBSON BROS. OIL
CO. GASOLINE 5 1925-1935 15" Metal HP Red $1000-1500
White globe face with large black triangle with red thunderbird IOMA logo in center of globe face. White "INDEPENDENT" above bird, "OILS" below. Black "DOBSON BROS. OIL CO" arched around top, "GASOLINE" around bottom.

FREI - ULRICH OIL CO. —BLACK RIVER FALLS, WISCONSIN

FREI-ULRICH 6 1925-1935 15" Metal HP Red $1000-1500
White globe face with large black triangle with red thunderbird IOMA logo in center of globe face. White "INDEPENDENT" above bird, "OILS" below. Black "FREI-ULRICH" arched around top.

GERKINS OIL COMPANY

GERKINS QUALITY
GASOLINE 6 1925-1935 15" Metal HP Red $1000-1500
White globe face with large black triangle with red thunderbird IOMA logo in center of globe face. White "INDEPENDENT" above bird, "OILS" below. Red "GERKINS" arched around top, "QUALITY" around bottom with black "GASOLINE" just above "QUALITY".

GIBSON BROTHERS OIL CO. —LYNDON, ILLINOIS: Founded in 1924, later a Pure Oil jobber.

GIBSON BROS
OIL CO 5 1925-1935 15" Metal HP Red $1000-1500
White globe face with large black triangle with red thunderbird IOMA logo in center of globe face. White "INDEPENDENT" above bird, "OILS" below. Black "GIBSON BROS. OIL CO" arched around top, "GASOLINE" around bottom.

H.S. GOODELL and SONS —HANCOCK, MICHIGAN: History unknown.

H.S. GOODELL 6 1925-1935 15" Metal HP Red $1000-1500
White globe with black outline ring around face. Large black triangle with red thunderbird IOMA logo in center of globe face. White "INDEPENDENT" above bird, "OILS" below. Black "H.S.GOODELL" arched around top.

H.S. GOODELL
RED HAT 5 1925-1935 15" Metal HP Red $1900-2500
White globe face with large red top hat with black hatband in center.

Three white stars on black hatband. Small black triangle/red thunderbird IOMA logo below hat. Black "H. S. GOODELL" arched around top, "GASOLINE" around bottom with "RED" to left of hat, "HAT" to right.

HAWKEYE OIL COMPANY
HAWKEYE
GASOLINE 5 1925-1935 15" Metal HP Red $1000-1500
White globe face with large black triangle with red thunderbird IOMA logo in center of globe face. White "INDEPENDENT" above bird, "OILS" below. Black "HAWKEYE" arched around top, "GASOLINE" around bottom.

HOME OIL COMPANY
HOME OIL CO.
RED HAT
GASOLINE 5 1925-1932 15" Metal HP Red $2000-2500
White globe face with large red top hat with black hatband in center. Three white stars on black hatband. Small black triangle/red thunderbird IOMA logo below hat. Black "HOME OIL COMPANY" arched around top, "GASOLINE" around bottom with "RED" to left of hat, "HAT" to right.

HOOSIER OIL COMPANY — INDIANAPOLIS, INDIANA
HOOSIER
GASOLINE 6 1925-1935 15" Metal HP Red $1000-1500
White globe face with large black triangle with red thunderbird IOMA logo in center of globe face. White "INDEPENDENT," above bird, "OILS" below. Black "HOOSIER" arched around top, "GASOLINE" around bottom.

BARKHAUSEN OIL COMPANY —GREEN BAY, WISCONSIN
IMPERIAL
GASOLINE 7 1920s 15" Metal HP Red $1100-1600
Metal concave globe face painted white with black triangle IOMA logo in center. Red thunderbird on black triangle. White "INDEPENDENT" over bird, "OILS" below. Black "IMPERIAL" arched around top, "GASOLINE" ached around bottom.
LIBERTY
GASOLINE 7 1920s 15" Metal HP Red $1100-1600
Metal concave globe face painted white with black triangle IOMA logo in center. Red thunderbird on black triangle. White "INDEPENDENT" over bird, "OILS" below. Black "LIBERTY" arched around top, "GASOLINE" ached around bottom.

INDEPENENT OIL COMPANY —SEYMOUR, MISSOURI
INDEPENDENT OIL CO.
SEYMOUR, MO 5 1925-1940 15" Metal HP Red $1000-1500
White globe face with large black triangle with red thunderbird IOMA logo in center of globe face. White "INDEPENDENT" above bird, "OILS" below. Black "INDEPENENT. OIL CO" arched around top, "SEYMOUR, MO" around bottom.

JONES BROTHERS OIL COMPANY —SCOTTSBURG, INDIANA
JONES BROS.
HIGH HAT 6 1920s 15" Metal HP Red NL
White globe face with large red top hat with black hat band and vertical red and black stripes above band. Black "HIGH" to left of hat, "HAT" to right with black "GASOLINE" arched below. Black "JONES BROS" arched around top.
JONES BROS.
RED HAT 5 1920s 15" Metal HP Red $2000-2500
White globe face with large red top hat with black hatband in center. Three white stars on black hatband. Small black triangle/red thunderbird IOMA logo below hat. Black "JONES BROS." arched around top, "GASOLINE" around bottom with "RED" to left of hat, "HAT" to right.

KIMMEL OIL COMPANY —HIAWATHA, KANSAS
KIMMEL RED
HAT GASOLINE 5 1925-1932 OPE RW $3500-5000
White globe with red or black outline ring around globe face. Large red top hat with black hatband in center of globe. Three white stars on black hatband. Small black triangle/red thunderbird IOMA logo below hat. Black "KIMMEL" arched around top, "GASOLINE" around bottom with "RED" to left of hat and "HAT" to right.
KIMMEL RED
HAT GASOLINE 5 1925-1932 15" Metal HP Red $2000-2500
White globe face with large red top hat with black hatband in center. Three white stars on black hatband. Small black triangle/red thunderbird IOMA logo below hat. Black "KIMMEL" arched around top, "GASOLINE" around bottom with "RED" to left of hat, "HAT" to right.

MATHESON OIL COMPANY —BENNETTSVILLE, SOUTH CAROLINA
MATHESON
GASOLINE 5 1920s 15" Metal HP Red $1200-1700
Green globe face with large black triangle with red thunderbird IOMA logo in center of globe face. White "INDEPENDENT" above bird, "OIL" below. White "MATHESON" arched around top, "GASOLINE" around bottom.
MATHESON RED
HAT GASOLINE 5 1925-1932 OPE RW $3500-5000
White globe with black outline ring around globe face. Large red top hat with black hatband in center of globe. Three white stars on black hatband. Small black triangle/red thunderbird IOMA logo below hat. Black "MATHESON" arched around top, "GASOLINE" around bottom with "RED" to left of hat and "HAT" to right.

McCORMICK OIL CO. —McCORMICK, SOUTH CAROLINA
Mc CORMICK RED
HAT OVAL 7 1925-1932 OPE Oval NL
Oval white globe with red or black outline ring around globe face. Large red top hat with black hatband in center of globe. Three white stars on black hatband. Small black triangle/red thunderbird IOMA logo below hat. Black "Mc CORMICK" arched around top, "GASOLINE" around bottom with "RED" to left of hat and "HAT" to right.
Mc CORMICK RED
HAT OVAL 7 1925-1932 OPE Oval NL
Oval white globe with red or black outline ring around globe face. Large black triangle with red thunderbird IOMA logo in center of globe face. Black "Mc CORMICK" arched around top, "GASOLINE" around bottom with "RED" to left of hat and "HAT" to right.

McQUARTER OIL COMPANY
MC QUARTER RED
HAT GASOLINE 6 1925-1932 15" Metal HP Red $1900-2500
White globe face with large red top hat with black hatband in center. Three white stars on black hatband. Small black triangle/red thunderbird IOMA logo below hat. Black "Mc QUARTER" arched around top, "GASOLINE" around bottom with "RED" to left of hat, "HAT" to right.

MUSCO OIL COMPANY
MUSCO RED
HAT GASOLINE 6 1925-1932 15" Metal HP Red $2000-2500
White globe face with large red top hat with black hatband in center. Three white stars on black hatband. Small black triangle/red thunderbird IOMA logo below hat. Black "MUSCO" arched around top, "GASOLINE" around bottom with "RED" to left of hat, "HAT" to right.

NORTHLAND GASOLINE —WATERLOO, IOWA: Primarily a lubricants manufacturer, Northland franchised their lubricants to a number of regional marketers, including Mid West Oil Company of Muskegon, MI. They were also involved with the Erickson refinery in St Paul, Minnesota and other marketing firms in Wilmar, MN. Northland operated a small number of gasoline stations in the 1920s and 1930s. The company is still in business and manufactures specialty lubricants today.
NORTHLAND GAS
(w/IOMA logo) 5 1920s 15" Metal HP Red $1000-1500
White globe face with black triangle/red thunderbird IOMA logo in center. Black "NORTHLAND" arched around top, "GAS" around bottom.
NORTHLAND SUPER
REGULAR
LEADED 85 6 1920s 15" Metal HP Red $2100-2700
White globe face with black outline around white center circle. Black lines form white band across center circle. Black "REGULAR" in center band with green "SUPER" above and "LEADED/"85"" below. Red

"NORTHLAND" arched around top, "GASOLINE" around bottom with pine tree at center to either side.

NORTHLAND
GASOLINE 5 1930s Gill $400-650
White globe face with red/white/black "NORTHLAND" shield in center. Black lettering on shield "THE BADGE OF CERTAINTY". Black outlined red "NORTHLAND" arched around top, "GASOLINE" around bottom.

NORTHLAND
GOLDEN FLASH 5 1930s 13.5" Glass W $350-550
Black/white outlined red globe face with white band diagonal across center. Black outlined red "Northland" diagonal across band. White "GOLDEN" in upper red area, "FLASH" in lower area.

NORTHLAND
GOLDEN FLASH 5 1930s Gill $400-600
Black/red/white outlined black globe face with white band diagonal across center. Black outlined red "Northland" script logotype diagonal across band. White "GOLDEN" in upper black area, "FLASH" in lower area.

NORTHLAND
ETHYL 5 1930s 13.5" Glass W $350-550
Red/white outlined black globe face with white band diagonal across center of globe face. Black outlined red script "Northland" on center band. White circles on top and bottom black areas with small Ethyl (EGC) logo in each circle.

O'NEILS PAINT & HARDWARE CO.—MILWAUKEE, WISCONSIN: History unknown.

O'NEILS EVERY
DROP TESTED 6 1920s 15" Metal HP Red $1500-2000
Orange globe face with large black outlined red "O'NEIL'S" across upper globe face. Still scene over "O" in "O'NEILS". Black "Every drop tested" just below "O'NEILS". Black triangle with red thunderbird logo at bottom of globe face with white "INDEPENDENT" across top of triangle, "OILS" below.

PENDERGAST OIL COMPANY—LAKE GENEVA, WISCONSIN

PENDERS RED
HAT GASOLINE 5 1925-1932 15" Metal HP Red $2000-2500
White globe face with large red top hat with black hatband in center. Three white stars on black hatband. Small black triangle/red thunderbird IOMA logo below hat. Black "PENDERS" arched around top, "GASOLINE" around bottom with "RED" to left of hat, "HAT" to right.

PENDERS
GASOLINE 5 1925-1932 15" Metal HP Red $1000-1500
White globe face with large black triangle with red thunderbird IOMA logo in center of globe face. White "INDEPENDENT" above bird, "OILS" below. Black "PENDER'S" arched around top, "GASOLINE" around bottom.

PENN VALLEY OIL COMPANY

PENN VALLEY
GASOLINE 5 1925-1935 15" Metal HP Red $1100-1600
White globe face with large black triangle with red thunderbird IOMA logo in center of globe face. White "INDEPENDENT" above bird, "OILS" below. Black "PENN VALLEY" arched around top, "GASOLINE" around bottom.

PETERSONS OIL COMPANY—MILWAUKEE, WISCONSIN

PETERSONS OIL
CO. GASOLINE 5 1920s OPE RW $2000-2600
White globe with black outline ring around face. Large black triangle with red thunderbird IOMA logo in center of globe face. White "INDEPENDENT" above bird, "OILS" below. Black "PETERSONS OIL CO" arched around top, "GASOLINE" around bottom.

PETERSONS OIL
CO. GASOLINE 1925-1935 15" Metal HP Red $1100-1600
White globe face with large black triangle with red thunderbird IOMA logo in center of globe face. White "INDEPENDENT" above bird, "OILS" below. Black "PETERSONS OIL CO" arched around top, "GASOLINE" around bottom.

PETTICO OIL COMPANY

PETTICO
GASOLINE 5 1920s 15" Metal HP Red $1000-1500
White globe face with large black triangle with red thunderbird IOMA logo in center of globe face. White "INDEPENDENT" above bird, "OILS" below. Black "PETTICO" arched around top, "GASOLINE" around bottom.

PURE PEP: Company and history unknown but it is not believed to be affiliated in any way with the Pure Oil Company.

PUREPEP
INDEPENDENT
OILS GASOLINE 6 1920s 12.5" Metal HP Red NL
White globe face with with red thunderbird IOMA logo in center of globe face. White "INDEPENDENT" above bird, "OILS" below. Black "PUREPEP" arched around top, "GASOLINE" around bottom.

QUINCY—BOSTON, MASSACHUSETTS: Independent Boston area marketer founded in the 19th century. Continued to operate stations into the late 1980s. The last marketer to use the old IOMA logo.

QUINCY RED
HAT GASOLINE 5 1920s 15" Metal HP Red $2000-2500
White globe face with large red top hat with black hatband in center. Three white stars on black hatband. Small black triangle/red thunderbird IOMA logo below hat. Black "QUINCY" arched around top, "GASOLINE" around bottom with "RED" to left of hat, "HAT" to right.

QUINCY POW-R-
PLUS (w/eagle) 5 1930s 15" Metal HP Red $1200-1750
Green/white outline rings around green outer ring. White center circle with green triangle/red thunderbird IOMA logo. White "QUINCY" arched in upper ring, "POW-R-PLUS" around lower ring.

QUINCY POW-R-
PLUS ETHYL 5 1930s 15" Metal HP Red $600-850
Black outline around green globe face with black outline around white center circle. Large Ethyl (EGC) logo with yellow rays in center circle. White "POW-R-PLUS" arched around top, "EXTRA-SPECIAL" arched around bottom.

QUINCY POWERFUL
GASOLINE 6 1930s 15" Metal HP Red $600-850
White globe face with large red "Q" covering most of globe face. Green diagonal band across center of "Q" with white "QUINCY/POWERFUL/GASOLINE" on band.

ROYAL and ROYAL 400—FT. DODGE, IOWA

ROYAL
GASOLINE 5 1920s OPE RW $2100-2700
White globe with black outline ring around face. Black circle around IOMA black triangle/red thunderbird logo in center. White "GASOLINE" arched around top of circle, "MOTOR OILS" around bottom with red "ROYAL" arched above logo and "GASOLINE" below.

ROYAL 400
GASOLINE
(EMB OF DEP) 5 1920s OPE RW $2200-2800
White globe with black outline ring around face. Black circle around IOMA thunderbird logo in center. White "THE EMBLEM OF" arched around top of black circle, "DEPENDABILITY" arched below with red "ROYAL 400" arched above logo and "GASOLINE" below.

ROYAL 400
GASOLINE
(BIRD ON BLACK) 5 1920s OPE RW $2200-2800
White globe with solid black circle in center of globe face. Red thunderbird from IOMA logo on circle. Red "ROYAL 400" arched around top, "GASOLINE" around bottom.

ROYAL 400 RED
HAT GASOLINE 5 1930s 13.5"Glass W $1600-2200
White globe face with black outline ring. Large red top hat with black hatband in center. Three white stars on hatband. Black triangle/red thunderbird IOMA logo below hat. Company or brand, as listed, arched around top, "GASOLINE" around bottom.

S. M. OIL COMPANY

S.M. OIL CO. RED
HAT GASOLINE 6 1925-1932 OPE NL
White globe with large non-detailed red hat in center. Small black triangle/red thunderbird IOMA logo below hat. Black "S. M. OIL CO." arched around top, "GASOLINE" around bottom with "RED" to left

of hat and "HAT" to right.

BEEBEE OIL CO.: Company & history unknown.

SNAP GAS 6 1920s 15" Metal HP Red NL

White globe face with small black triangle/red eagle logo at bottom of globe face. Large red logotype "Snap Gas" across upper globe face with thick underline. White "BEEBEE OIL CO." on underline.

SWANSON OIL CORPORATION —OMAHA, NEBRASKA

SWANSON OIL CORP. GASOLINE (METAL FACES) 5 1920s 15" Metal HP Red $1200-1700

Metal globe face painted green with black triangle IOMA logo in center. Red thunderbird on black triangle. White "INDEPENDENT" over bird, "OILS" below. White reflective "SWANSON" arched around top above "OIL/CORPORATION" above triangle logo, "GASOLINE" arched around bottom.

TRAFFIC OIL COMPANY —UTICA, NEW YORK

TRAFFIC RED HAT GASOLINE 6 1925-1932 OPE RW $3500-5000

White globe with red or black outline ring around globe face. Large red top hat with black hatband in center of globe. Three white stars on black hatband. Small black triangle/red thunderbird IOMA logo below hat. Black "TRAFFIC" arched around top, "GASOLINE" around bottom with "RED" to left of hat and "HAT" to right.

TRI STAR

TRI-STAR RED HAT GASOLINE 5 1925-1932 15" Metal HP Red $2000-2500

White globe face with large red top hat with black hatband in center. Three white stars on black hatband. Small black triangle/red thunderbird IOMA logo below hat. Black "TRI-STAR" arched around top, "GASOLINE" around bottom with "RED" to left of hat, "HAT" to right.

VAPOR GASOLINE

VAPOR GASOLINE 6 1920s 15" Metal HP Red $1200-1750

Green globe face with large black triangle with red thunderbird IOMA logo in center of globe face. White "INDEPENDENT" above bird, "OIL" below. White "VAPOR" arched around top, "GASOLINE" around bottom.

WEEKS OIL COMPANY —AUBURN, NEW YORK

WEEKS OIL CO., INC. RED HAT GASOLINE 6 1925-1932 15" Metal HP Red $2000-2500

White globe face with large red top hat with black hatband in center. Three white stars on black hatband. Small black triangle/red thunderbird IOMA logo below hat. Black "WEEKS OIL CO. INC." arched around top, "GASOLINE" around bottom with "RED" to left of hat, "HAT" to right.

WESTLAND OIL CO. —MINOT, NORTH DAKOTA: In 1913 Dick Coughlin founded a repair garage in Scobey, Montana and with it came the distribution of various petroleum products. In 1917 Coughlin sold the garage to concentrate on the distribution of petroleum products, and in 1919 merged his Scobey Oil Company with a nearby competitor, Gray-Lindemann Oil Company to form Westland Oil Company. The first service station to display the Westland brandname opened in Scobey, Montana in 1922. They were an early member of the Independent Oil Men of America, and in 1928 the operation was moved to Minot, North Dakota where it was to remain throughout most of this century. In 1970 Westland became a subsidiary of Consolidated Hydrocarbons, and by 1972 the company was part of Thunderbird Petroleum. In 1982 Thunderbird moved to Denver and withdrew from gasoline marketing, and the former Westland refinery was sold to Flying J (formerly Big West Petroleum). Although it appears that the Buffalo brand ceased to exist in 1982, there have been examples of jobbers and single stations that continue to operate under the Buffalo name even today.

WESTLAND GASOLINE (w/red hat) 5 1920s OPE RW $3500-5000

White globe with red or black outline ring around globe face. Large red top hat with black hatband in center of globe. Three white stars on black hatband. Small black triangle/red thunderbird IOMA logo below hat. Black "WESTLAND." arched around top, "GASOLINE" around bottom with "RED" to left of hat and "HAT" to right.

WESTLAND GASOLINE (w/buffalo) 5 1920s OPE RW $3000-4000

White globe with red or black outline ring around globe face. Large black and white Buffalo in center of globe face. Black "WESTLAND." arched around top, "GASOLINE" around bottom.

BUFFALO GASOLINE (w/buffalo) 5 1920s OPE W $2500-3500

White globe with red or black outline ring around globe face. Large black and white Buffalo in center of globe face. Red "BUFFALO." arched around top, "GASOLINE" around bottom.

WESTLAND GASOLINE (w/buffalo) 5 1930s 13.5" Glass WN $1800-2400

White globe face with black outline ring. Black and white detailed buffalo in center of globe face. Red "WESTLAND" arched around top, "GASOLINE" around bottom.

BUFFALO GASOLINE 4 1930s-1950s 13.5" Glass WN $1600-2200

White globe face with black outline ring. Black and white detailed buffalo in center of globe face. Red "BUFFALO" arched around top, "GASOLINE" around bottom.

ETHYL GASOLINE (w/buffalo) 5 1930s-1950s 13.5" Glass WN $1400-2000

White globe face with black outline ring. Black and white detailed buffalo in center of globe face. Red "ETHYL" arched around top, "GASOLINE" around bottom.

BUFFALO GASOLINE 4 1955-1970 13.5" Glass WN $1700-2300

Green globe face with red outline ring and red ring around white center circle. White "BUFFALO" arched around top, "GASOLINE" around bottom with detailed red/black buffalo in center.

BUFFALO PREMIUM GASOLINE 4 1955-1970 13.5" Glass WN $350-550

White globe face with green outline ring and green outline around white center circle. Green band across center. White "PREMIUM" in green band. Red "BUFFALO" arched around top, "GASOLINE" around bottom.

WESTLAND DIESEL 5 1955-1970 13.5" Glass WN $350-550

Red outlined white globe face with dark green "DIESEL" across center of globe face. Light green "WESTLAND" arched around top, "FUEL" around bottom.

WEYAUWEGA OIL CO. —WEYAUWEGA, WISCONSIN: Local oil company and farmers cooperative operation

WEYCO GASOLINE 5 1925 OPE N NL

Small one piece globe with 4" base showing the typical Red Hat IOMA logo.

KENDALL & AFFILIATES

****KENDALL OIL — BRADFORD, PENNSYLVANIA**: Kendall Refining was founded as a lubricants manufacturer and refiner in Bradford, Pennsylvania, in 1881. The company had been well established in motor oils when the first venture into gasoline came in 1922. The company supplied stations in Pennsylvania and New York until after the first gas shortage in 1973 when the decision was made to concentrate on lubricants. The last Kendall branded station closed in 1976.

KENDALL SPHERE SHAPE 5 1922-1925 OPE $750-1100

White sphere shaped globe with red "KENDALL" across globe.

KENDALL 5 1922-1928 OPE $800-1250

White globe with red outline around white globe face and red "KENDALL" lettering across face.

KENDALL GASOLINE W/AIRPLANE 5 1926-1935 15" Metal HP Red $4500-6000

White globe face with red and black "Spirit of St.Louis" type airplane in center. Red "KENDALL" arched around top, "GASOLINE" around bottom.

KENDALL ETHYL (EGC) 4 1926-1935 15" Metal HP Red $650-900
White globe face with black outline ring and black outline around white center circle. Black outlined red "KENDALL" arched around top, "GASOLINE" around bottom with large Ethyl (EGC) logo with blue rays in center.

KENDALL 3 1935-1946 Gill $225-400
KENDALL 3 1935-1946 13.5" Glass $250-350
White globe face with black outlined red "KENDALL" in tall thin lettering.

KENDALL ETHYL 4 1935-1936 Gill $225-425
KENDALL ETHYL 4 1935-1936 13.5" Glass $250-375
White globe face with red outline ring and red outline around white center circle. Black outlined red "KENDALL" arched around top, "GASOLINE" around bottom with large Ethyl (EGC) logo with yellow rays in center.

KENDALL DELUXE (EARLY THIN TYPE) 4 1936-1946 14" Gill $300-500
White globe face with blue outlined red tall thin lettering "KENDALL" over small blue script "DeLuxe".

KENDALL DELUXE (THICKER TYPE) 2 1946-1970 Gill $300-475
KENDALL DELUXE (THICKER TYPE) 2 1946-1970 13.5" Glass $250-375
KENDALL DELUXE (THICKER TYPE) 3 1946-1970 Capco $225-325
White globe face with blue outlined red thicker lettering "KENDALL" over larger blue script "DeLuxe".

KENDALL POLLY POWER 2 1946-1970 Gill $300-475
KENDALL POLLY POWER 2 1946-1970 13.5" Glass $250-375
KENDALL POLLY POWER 3 1946-1970 Capco $225-325
White globe face with large black outlined red "KENDALL" over black script diagonal "Polly Power".

****WAVERLY OIL WORKS — PITTSBURGH, PENNSYLVANIA**: Founded in 1880 as a lubricants manufacturer, Waverly ventured briefly into gasoline marketing from the 1910s until the 1930s. Known afterwards for many years as a quality lubricants manufacturer, Waverly was purchased by Witco Chemical, owners of the Kendall and Amalie brands, in 1975.

WAVERLY GASOLINE "SPHERE" 6 1910s OPE NL
White spherical globe with red "WAVERLY/GASOLINE" arched across globe face.

WAVERLY 6 1920s OPE RW $850-1400
White globe with red "Waverly" script logotype across globe face.

WOW ANTI-KNOCK GASOLINE 6 1920s OPE N NL
White globe face with blue/yellow checkerboard design arched around top and bottom. Large yellow "WOW" with blue shading in center of globe face. Blue "ANTI-KNOCK" above "WOW", "GAS" below.

WAVERLY MOTOR GAS 6 1920s OPE RW $850-1400
White globe with red script "Waverly" over "MOTOR/GAS".

WAVERLY ETHYL (EGC) GASOLINE 6 1920s OPE RW $1200-1800
White globe with black outline around white globe face and black outline around white center circle. Red "WAVERLY" arched around top, "GASOLINE" around bottom with large Ethyl (EGC) logo in center circle.

WAVERLY BLUE MOON 5 1920s 2pc Glass N $850-1250
White globe with blue/white outlined blue globe face. White script "Waverly" over white "BLUE/MOON" in center of globe face.

WOW ANTI-KNOCK GASOLINE 5 1920s 15" Metal $850-1300
White globe face with blue/yellow checkerboard design arched around top and bottom. Large yellow "WOW" with blue shading in center of globe face. Blue "ANTI-KNOCK" above "WOW", "GAS" below.

WAVERLY GAS 6 1920s 15" Metal $600-900
White globe face with black outlined red script "Waverly" diagonal across much of globe face. Thick underline extending from "y" under entire word. Black outlined red "GAS" across bottom below underline.

WAVERLY ETHYL 6 1920s 15" Metal $750-1250
White globe face with black outline around white globe face and black outline around white center circle. Red "WAVERLY" arched around top, "GASOLINE" around bottom with large Ethyl (EGC) logo in center circle.

"W" 5 1920s 15" Metal $300-450
White globe face with large red "W" on globe face.

GOLDEN QUALITY / ECONOMY 5 1930s 13.5" Glass N $250-350
White globe face with large blue keystone covering most of globe face. Large yellow-gold "GOLDEN" across center of keystone in elongated area with white "QUALITY" above and "ECONOMY" below.

KERR McGEE & AFFILIATES

****KERR McGEE OIL INDUSTRIES — OKLAHOMA CITY, OKLAHOMA:** Kerr McGee was founded by Robert S. Kerr and others in the Oklahoma City oil fields as Anderson-Kerr Drilling Company in 1929. Through the early years they operated a single Oklahoma City service station, branded Phillips 66, which closed during the gas rationing days of World War II. Before the war, Phillip's geologist Dean McGee joined the company and in 1946 the corporate name was changed to Kerr-McGee. It was also in that year that the company entered retail gasoline marketing having purchased Oklahoma City based Harrell Bros. The Harrell Bros. "Yellow Cab" gasoline trademark was replaced by the "Kermac" brand. This was the extent of KM's gas station involvement until 1955 when they purchased Deep Rock Oil Co. Deep Rock had been founded in the Cushing, Oklahoma, oil fields in 1913 and had been through a stormy existence. Deep Rock had sold out to Schaffer Oil and Refining of Chicago in the late 1920s and had again been in receivership by the late 1930s. At the time of the KM merger, the Deep Rock brand appeared on over 1000 stations in 16 mid-continental states. In 1957 Kerr-McGee acquired Triangle Refineries, a primarily wholesale operation that had been founded in Abbottsford, Wisconsin, in 1937. Knox Industries of Texas also sold out to KM about this time and the Knox brand survived into the 1960s. Triangle was active in purchasing Southern private brand operations which Kerr-McGee operated well into the eighties, including Peoples Oil of Nashville, 1959; Illinois based Mileage Mart in 1959; Trackside Gasoline Stations of Gulfport, Mississippi, 1962; Wallace Service Stations of Atlanta, Georgia, 1962; along with Power Oil Co. of Macon, Georgia, Coast Oil Co. of Shreveport, Louisiana, and Tan-Kar Stations of Birmingham, Alabama.

****DEEP ROCK OIL CO. — CHICAGO, ILLINOIS:** Founded in the Cushing field in 1913, Deep Rock had a series of financial problems and existed under several owners, including Schaffer Oil Company, before being purchased by Kerr McGee in 1955. The Kerr McGee corporate name was added to Deep Rock stations about 1960 and the "brand" Kerr McGee was introduced with the construction of 17 new stations in Oklahoma City in 1965. Deep Rock stations were gradually converted over to the KM name, nearing completion by 1985. The Deep Rock brand survives in some areas, but only through jobbers who simply haven't updated signage at smaller locations.

>>Note: Arrows (><) in the text are indicators of lettering that appears in a bowtie design.

GASOLINE/ DEEP ROCK/ OILS-GREASES 5 1913-1920 15" Metal Inserts $800-1200
White metal globe face with red "DEEP><ROCK" across center. Reverse bowtie blue "GASOLINE" at top, "OILS GREASE" at bottom.

GASOLINE/ DEEP ROCK/ MOTOR OIL 5 1913-1920 15" Metal $750-1000
White globe face with red "DEEP><ROCK" across center. Reverse bowtie blue "GASOLINE" at top, "MOTOR OILS" at bottom.

GASOLINE/

DEEP ROCK/
MOTOR OIL 4 1920-1928 OPE N $1200-1500

White globe with red "DEEP><ROCK" across center. Reverse bowtie blue "GASOLINE" at top, "MOTOROILS" at bottom.

GASOLINE/
DEEP ROCK/
OILS-GREASES 6 1920-1928 OPE N $1350-1700

White globe with red "DEEP><ROCK" across center. Reverse bowtie blue "GASOLINE" at top, "OILS - GREASES" at bottom.

EASY STARTING/
KANT NOCK/
GASOLINE 5 1925-1928 OPE N $1600-2100

White globe with blue "KANT><NOCK" across center. Reverse bowtie red "EASY STARTING" at top, "GASOLINE" at bottom.

GASOLINE/
DEEP ROCK/
MOTOR OIL 6 1926-1928 OPB R/W $900-1250

White globe with red "DEEP><ROCK" across center. Reverse bowtie blue "GASOLINE" at top, "MOTOROILS" at bottom.

GASOLINE/
DEEP ROCK/
OILS - GREASES 5 1928-1932 15" Metal $900-1100

White globe face with red "DEEP><ROCK" across center. Reverse bowtie blue "GASOLINE" at top, "OILS - GREASES" at bottom.

KANT NOCK WITH
ETYHL (EGC) 5 1928-1932 15" Metal $900-1200

White globe face with red "KANT><NOCK" at top and red "GASOLINE" at bottom with large Ethyl (EGC) logo in center with thick yellow rays out from logo.

GASOLINE/
DEEP ROCK/
MOTOR OIL 3 1928-1932 13.5" Glass $300-550

White globe face with red outline ring and red "DEEP><ROCK" across center. Reverse bowtie blue "GASOLINE" at top, "MOTOR OILS" at bottom.

KANT NOCK WITH
ETHYL (EGC) 4 1928-1932 13.5" Glass $350-550

Black outlined white globe face with red "KANT><NOCK" at top and red "GASOLINE" at bottom with large irregular Ethyl (EGC) logo in center with thick yellow rays out from logo.

DEEP ROCK W/
RINGS 4 1932-1934 Capco $200-325

White globe face with thick blue outline ring and three concentric thinner blue rings. Large blue outlined red DEEP><ROCK across globe face.

REGULAR/
DEEP ROCK/
GASOLINE 3 1934-1936 13.5" Glass $300-450

White globe face with red "DEEP><ROCK" across center. Blue "REGULAR" at top, "GASOLINE" at bottom.

SUPER KANT
NOCK GASOLINE 4 1932-1936 13.5" Glass $350-600

Yellow outlined white globe face with red "KANT><NOCK" across center. Narrow black band above and below Kantnock with white "SUPER" in top band and "GASOLINE" in bottom band.

DEEP ROCK SUPER
GASOLINE 4 1936-1950 13.5" Glass $300-450

Yellow outlined white globe face with red "DEEP><ROCK" lettering at top and "GASOLINE" at bottom. Yellow circle in center with white band across center with black "SUPER" on band.

DEEP ROCK ETHYL
GASOLINE 4 1936-1942 13.5" Glass $300-450

Red outlined white globe face with red "DEEP><ROCK" lettering at top and "GASOLINE" at bottom. Black outlined white circle in center with small Ethyl (EGC) logo in circle.

DEEP ROCK ETHYL
GASOLINE 4 1936-1942 13.5" Glass $300-450

Red outlined white globe face with *black outlined* red "DEEP><ROCK" lettering at top and "GASOLINE" at bottom. Black outlined white circle in center with small Ethyl (EGC) logo in circle.

DEEP ROCK ETHYL
GASOLINE 4 1942-1950 13.5" Glass $300-450

Red outlined white globe face with red "DEEP><ROCK" lettering at top and "GASOLINE" at bottom. Black outlined white circle in center with small Ethyl (EC) logo in circle.

DEEP ROCK GREEN
GASOLINE 4 1936-1950 13.5" Glass $350-550

Green outlined white globe face with red "DEEP><ROCK" lettering at top and "GASOLINE" at bottom. Light green circle in center with white band across center with dark green "GREEN" on band.

DEEP ROCK 1 1950-1972 Capco $175-250

White globe face with large yellow/blue Deep Rock rectangle logo.

DEEP ROCK
WITH ARROWS 5 1972-1980 Capco $250-350

White globe face with large yellow and blue Deep Rock logo set at angle on globe face with blue dropshading on bottom and left of logo. Red arrow tail at upper left of sign, point at lower right.

****KERMAC — OKLAHOMA CITY, OKLAHOMA:** In 1945 when Kerr McGee purchased the former "Yellow Cab" gas stations of Harrell Bros. Oil Company, the several stations were rebranded Kermac. Kerr McGee was more involved in production and refining than in marketing and the Kermac brand never appeared on more than a few stations.

KERMAC
REGULAR 5 1947-1955 Capco $200-300

Light blue globe face with white "KERMAC" arched around top. Script "Regular" across lower globe face.
Note:>>Premium globe probably exists.<<

****KERR MCGEE:** The Kerr McGee name was added to Deep Rock stations about 1960 and the Kerr McGee brand, standing alone, first appeared on a group of Oklahoma City stations in 1965. The KM brand and logo eventually replaced Deep Rock and continues in use today.

KERR MCGEE 5 1965-1970 Capco $200-325

White globe face with red/blue Kerr-McGee "KM" shield at center.

****TRIANGLE REFINERIES — HOUSTON, TEXAS:** Founded in Abbottsford, Wisconsin, in 1937, Triangle Refineries operated stations under the Triangle brand in the 1930s and 1940s. As best as can be determined, following World War II Triangle purchased Houston based Southport Oil Company, which operated under the "HIOTANE" brand. In the 1950s and 1960s, the "CLOVERLEAF" brand was seen on most Triangle affiliated stations. In April 1957 Kerr McGee purchased Triangle Refineries and all of the private brand KM affiliates operated through Triangle into the 1990s.

TRIANGLE NEW
NAVY GASOLINE 6 1937-1940 15" Metal NL

White globe face with blue "TRIANGLE/GASOLINE" arched around top and bottom with "NEW/NAVY" in center.

TRIANGLE GASOLINE
W/CLOVER 4 1940s 13.5" Glass $500-750

White globe face with black "TRIANGLE" arched at top, "REFINERIES" at bottom. Green outline triangle with green 4-leaf clover in center.

CLOVERLEAF
GASOLINE 4 1960s 13.5" Red Capco $350-500

CLOVERLEAF
GASOLINE 4 1960s 13.5" Green Capco $400-650

White globe face with red outline ring. Large green outlined white triangle covering most of globe face with smaller green outlined triangle inside. Green four-leaf clover in smallest triangle with red "CLOVERLEAF" above smaller triangle inside large one.

****SOUTHPORT OIL CORPORATION — HOUSTON, TEXAS:** Southport Oil was a Texas based refiner-marketer that operated under the Hiotane brand. Southport merged with Triangle Refineries after World War II. Stations carried the Hiotane brand into the 1960s but most were eventually converted to Cloverleaf.

SOUTHPORT REGULAR
GASOLINE
HIOTANE 5 1940s 13.5" Glass $400-700

Black outlined yellow globe face with black center circle. Series of black outlined yellow triangles in center circle with black outlined white "HIOTANE" with oversize "T" on center triangle. Black outlined white "SOUTHPORT" arched around top, "REGULAR GASOLINE" around bottom.

SOUTHPORT SUPER

GASOLINE
HIOTANE 5 1940s 13.5" Glass $400-700

Black outlined red globe face with black center circle. Series of black outlined red triangles in center circle with black outlined white "HIOTANE" with oversize "T" on center triangle. Black outlined white "SOUTHPORT" arched around top, "SUPER GASOLINE" around bottom.

HIOTANE 3 1950s 13.5" Glass $750-1000
HIOTANE 3 1950s Capco $500-800

Orange globe face with black/white interlocking border ring. Sunrise scene with white pointed tower in center of globe face. Black lettering vertical on tower "HIOTANE". Black/white script arched around top "As new as Tomorrow".

****MILEAGE MART:** An Illinois based, early gasoline c-store operation bought out in 1959 and rebranded Deep Rock.

SAVE GO
MILEAGE MART 5 1955-1962 13.5" Capco $950-1400

White globe face with red ring outline and red lower 1/3. Black/white/red owl on upper globe face with white "SAVE" lettering. White "BE SMART" on upper red ring, white script "Go MILEAGE/MART on lower 1/3.

****TRACKSIDE SERVICE STATIONS — GULFPORT, MISSISSIPPI:** Trackside was just that, an early trackside discount operation founded in Gulfport in the early 1930s. Kerr McGee, through Triangle, purchased Trackside in 1962 and continued to operate the chain under the Trackside name until the late 1980s.

TRACKSIDE/
ECONOMY/
MOTOR 6 1930s 15" Metal $600-850

White globe face with blue "TRACK-SIDE" arched at top of globe face. Blue outlined yellow diamond at center with blue "ECONOMY" across it, over curved blue "MOTOR"

TRACKSIDE ROCKET
W/TANK CAR 5 1940s 13.5" Glass $1750-2500

White globe face with green/yellow octagon shaped border around black rail tank car with yellow "TRACKSIDE" on it over red "ROCKET/ANTI/KNOCK".

TRACKSIDE
REGULAR 4 1955-1965 13.5" Glass $300-500

White globe face with blue "TRACKSIDE" arched at top of globe face over red script "Regular", curved blue "GASOLINE" around bottom of globe face.

TRACKSIDE
SUPER ETHYL 4 1955-1965 13.5" Glass $300-500

White globe face with blue "TRACKSIDE" arched at top of globe face over red "SUPER/ETHYL", curved blue "GASOLINE" around bottom of globe face.

****PEOPLES OIL CO. — NASHVILLE, TENNESSEE:** Another mid-south discount operation, Peoples dates from the early 1930s and was purchased by Triangle in 1959. Many of the stations were rebranded Kerr McGee by the 1980s.

PEOPLES GENUINE
POWER PLUS 5 1950s 13.5" Glass W $400-650

Blue outlined white globe face with red "PEOPLES" arched around top, above small script "Genuine" over larger "POWER-PLUS" across center. Small red scrollwork design at bottom of globe face.

PEOPLES ETHYL 7 1950s 13.5" Glass W $250-400

White globe face with red "PEOPLES" arched around top over Ethyl (EC) logo.

****POWER OIL CO. —MACON, GEORGIA:** Originally a Pure Oil jobber, Power Oil was founded around 1930 in Macon. Following World War II Power Oil developed a successful chain of private brand stations carrying the Power name. The firm was purchased by Triangle Refineries in 1962, although the Power brand name remained in use until the 1980s.

GOLDEN FLASH
GASOLINE 6 1950s 13.5" Glass $400-650

Yellow globe face with large blue "GOLDEN/FLASH" across center of globe face. Unusual lettering of "FLASH". Blue "POWER OIL CO." arched around top, "GASOLINE" across bottom.

****WALLACE and WALCO — ATLANTA, GEORGIA:** A private brand Atlanta area marketer. Little is known about this company. It was purchased by Kerr McGee.

WALCO 6 1950s Capco $150-225

Solid red globe face with white band diagonal across globe face from upper left to lower right. Red "WALCO" on diagonal band.1962, along with Power Oil Co. of Macon, GA, Coast Oil Co. of Shreveport, Louisiana, and Tan-Kar Stations of Birmingham, Alabama.

****IMPERIAL OIL — SOMERSET, KENTUCKY:** History unknown, marketed in the 1960s and 1970s in Kentucky.

IMPERIAL 5 1950s Capco $150-250

Red globe face with large white "IMPERIAL" across center. White shield with vertical stripes and large white wings to either side across bottom of globe face.

****TAN-KAR STATIONS — BIRMINGHAM, ALABAMA:** Tan-Kar was a private brand marketer purchased by Triangle in 1962.
>>No globes are known from this brand.

****COAST OIL COMPANY — SHREVEPORT, LOUISIANA:** Coast Oil was a private brand marketer purchased by Triangle in 1962.
>>No globes are known from this brand.

****MUTUAL OIL CO. — BIRMINGHAM, ALABAMA:** Mutual, another private brand from the 1930s, was purchased by Kerr McGee in 1985 and stations were rebranded Kerr McGee. Mutual was one of the last companies to regularly use metal band globes, purchasing them as late as 1957 when they were discontinued by Cincinnati Advertising Products.

MUTUAL 4 1931-1957 15" Metal $1250-1750

White globe face with arched red lettering "MUTUAL" at top of globe face. Large blue jumping rabbit at center of globe face over red and blue ribbon with "GASOLINE", near bottom of globe face.

MUTUAL ETHYL 7 1931-1942 15" Metal $600-850

White globe face with arched red lettering "MUTUAL" at top of globe face. Large outlined Ethyl (EGC) logo at center of globe face over curved red "GASOLINE" near bottom of globe face.

MUTUAL ETHYL 4 1942-1957 15" Metal $600-850

White globe face with arched red lettering "MUTUAL" at top of globe face. Large outlined Ethyl(EC) logo at center of globe face over curved red "GASOLINE" near bottom of globe face.

MUTUAL 3 1957-1970 13.5" Glass $750-1000
MUTUAL 3 1957-1970 Capco $600-850

White globe face with arched red lettering "MUTUAL" at top of globe face. Large blue jumping rabbit at center of globe face over red and blue ribbon with "GASOLINE", near bottom of globe face.

MUTUAL ETHYL 4 1957-1970 13.5" Glass $300-450
MUTUAL ETHYL 4 1957-1970 Capco $200-325

White globe face with arched red lettering "MUTUAL" at top of globe face. Large outlined Ethyl(EC) logo at center of globe face over curved red "GASOLINE" near bottom of globe face.

****UNITED OIL COMPANY — CAPE GIRADEAU, MISSOURI:** United was a Missouri based independent that distributed Triangle Refineries products.

UNITED SUPER
CHARGED 4 1940s Capco $450-700

White globe face with blue outline ring. Red circle with orange and green bands in center. White "Super/Charged" in circle. Red "UNITED" arched around top, blue "GASOLINE" around bottom with small blue "REGULAR" below red center circle.

****GOLDEN RULE — ST. LOUIS, MISSOURI:** Golden Rule was a St Louis, Missouri, based independent that distributed Triangle Refineries products.

GOLDEN RULE 5 1940s 13.5" Glass W $300-500

White globe face with red outline ring. large yellow circle/bar design covers most of globe face with black outlined red "GOLDEN/RULE" superimposed over design.

GOLDEN RULE
ETHYL 5 1940s 13.5" Glass W $300-450
Black outlined white globe face with black outlined white center circle. Black script "Golden Rule" arched around top with large Ethyl (EGC) logo in center circle. Black "GASOLINE" arched around bottom.

**SPARKY: Sparky was a Triangle supplied brand that operated in Alabama in the 1950s.

SPARKY DYNAMO
ETHYL 5 1950s Capco $300-450
White globe face with small black "SPARKY" at top of globe face over black "DYNAMO" with yellow inlines over outlined Ethyl (EC) logo.

****VAUGHT OIL COMPANY — FLORA, ILLINOIS:** Vaught was a small Illinois private brand operation that sold out to Kerr McGee in the early 1960s.

VAUGHT'S SUPER
REGULAR 6 1940s 13.5" Glass $400-600
White globe face with red outline ring. Large red band with pointed ends across center of globe face superimposed over large blue "V". White "SUPER" on band with blue "VAUGHT'S" above and "REGULAR" below.
VAUGHT'S BLUE
BLAZE
KEROSENE 5 1940s 13.5" Glass $400-600
White globe face with blue outline ring. Large blue "BLUE BLAZE" across center with much smaller motto "Hotter Than Blue Blazes" just below. Red "VAUGHT'S" across top of globe face, "KEROSENE" below.

LION
LION OIL CO. —EL DORADO, ARKANSAS: Lion Oil Company was founded in 1921 in the south Arkansas oil fields. Primarily a exploration/drilling company and an early innovator in petrochemicals, Lion was operating a few service stations by the early 1920s. From the 1920s until the early 1950s, gasoline marketing was limited to Arkansas, Mississippi, western Tennessee, and northern Alabama and Louisiana, although Lion Naturalube Motor Oils were distributed throughout the U.S. In 1954 the Monsanto Chemical Corporation purchased Lion and began to expand the marketing territory throughout the south central U.S., expanding as far west as central Kansas and as far north and east as Bristol, Virginia. In 1959 Lion introduced the jumping lion logo, replacing the "lion and mound" logo that had been in use since the 1920s. During the 1960s the Monsanto name appeared on gas pump signage along with the Lion brand and some St.Louis, Missouri, area stations added Monsanto signage alongside the Lion signs, creating speculation that the Lion name would be replaced. The experiment was short lived as Monsanto decided to get out of the retail gasoline business. In 1971 Monsanto sold the petroleum divisions of Lion to "The Oil Shale Co." (TOSCO), retaining the petrochemical divisions. Lion branded marketing was discontinued in 1981, although TOSCO/ Lion is still involved in unbranded marketing. Since purchasing Lion as its first venture into petroleum marketing, Tosco enjoyed steady growth through the 1970s and 1980s, and explosive growth in the 1990s. In the mid-1990s Tosco began experimentation with franchising the operation of major gasoline brands in limited areas, beginning with BP stations that had formerly been Mobil outlets in the Pacific Northwest. Also added to the fold in 1996 was the Circle K convenience store operations nationwide, and in 1997 the Unocal 76 branded marketing in the far west, including sole ownership of the famed "76" trademark, and more BP branded franchise operations in New England. See also the Unocal section for more information on Tosco.
LION 4 1924-1930 OPE N $600-1100
Black "LION" on white globe.
LION ONE PIECE
W/LION 5 1926-1930 OPE N $3500-6000
White globe with black "LION" arched around top of globe face, "GASOLINE" around bottom with detailed brown standing lion, facing to right, in center of globe face.
LION PETROLEUM
PRODUCTS 6 1930-1946 12" Metal $1800-2500
Orange globe face with large black mound with black and white lion in center. White "LION" and "PETROLEUM PRODUCTS" on base of mound.
LION GASOLINE 4 1930-1946 13.5" Glass W $1250-1700
Black outlined white outer ring with black "LION" arched at top, "GASOLINE" at bottom with orange center circle. Small black mound with black and white lion on center circle. White "LION" and orange "PETROLEUM PRODUCTS" on base of mound.
LION ETHYL (EGC) 4 1930-1942 13.5" Glass W $500-850
Black outlined white outer ring with orange "LION" at top, "GASOLINE" at bottom with large Ethyl(EGC) logo with yellow rays in white center.
LION ETHYL (EC) 4 1942-1946 13.5" Glass W $500-850
Black outlined white outer ring with orange "LION" at top, "GASOLINE" at bottom with large Ethyl(EC) logo with yellow rays in white center.
LION 3 1946-1960 13.5" Glass W $750-1000
Orange globe face with large black mound with black and white lion on top. Large white "LION" lettering on mound.
LION KNIX KNOX 3 1946-1960 13.5" Glass W $1200-1600
Black outlined white outer ring with black "LION" at top, "KNIX KNOX" at bottom. Orange center of globe with small black mound in center with black and white lion on top. White "LION" lettering on mound, orange "PETROLEUM PRODUCTS" on base of mound.

MAPCO & AFFILIATES
****MAPCO — TULSA, OKLAHOMA:** Mapco is an independent refiner that purchased Nashville, Tennessee, based Red Ace Petroleum and Western Gasoline Stations as well as Memphis, Tennessee, based Delta Oil in the 1970s. By 1985 most locations had been rebranded Delta. In the late 1980s, the Delta stations were converted to convenience stores selling branded Mapco Express or Delta Express.
>>No Mapco brand globes known to exist.

****RED ACE PETROLEUM — NASHVILLE, TENNESSEE:** Founded in the early 1930s, Red Ace developed a chain of discount service stations all across Tennessee. They were purchased by Gasoline Marketers, Inc., a predecessor to the Mapco operation, in about 1973. Stations were rebranded Delta in the early 1980s.

RED ACE
ANTI-KNOCK
GASOLINE 9 1930s 15" Metal NL
Red/white outlined blue globe face with large red heart in center of globe face. Blue outlined red diamond across center of heart. White "ANTI-KNOCK" in diamond with white "Red Ace" arched around top inside heart, "GASOLINE" around bottom.
RED ACE 4 1946-1965 13.5" Glass $300-550
Blue globe face with red/white outline rings. Large white outlined red "X" covers most of globe face with white "RED/ACE" in center of "X".

****DELTA OIL CO. — MEMPHIS, TENNESSEE:** Delta Oil was another regional marketer purchased by Gasoline Marketers, Inc. in the early 1970s. The Delta brand was chosen to replace Red Ace and Western in the early 1980s and remains in use as Delta Express convenience stores today.

DELTA 5 1950s 13.5" Glass $250-450
White globe face with black outlined red triangle, point up, in center of globe face. Black/white outlined black box over upper triangle with white "DELTA" in box.

****SOUTHERN INDEPENDENT OIL & REFINING CO.—EVANSVILLE, INDIANA:** As best as we can determine, Southern Independent Oil Company was founded as Southern Marketer, Inc. in Huntington, Indiana, in the 1930s. Sometime around 1940, the company merged with another local operation, Capitol Oil Company, and adopted their Red Bird brand name. The Red Bird name remained in use until the 1970s on a small number of stations in the Evansville area. The firm was purchased by Monsanto in the early 1970s and in turn by Mapco in about 1980.

RED BIRD 9 1950s 13.5" Glass NL
White globe face with detailed red bird in center. Lettering "RED BIRD" arched around top, "SOUTHERN INDEPENDENT OIL COMPANY" arched around bottom. No other details available.
RED BIRD 6 1950s Oval Capco NL
Details are unknown. It is known that this globe features a Cardinal

and the "RED BIRD" brand name.

****SOUTHERN MARKETER, INC. — EVANSVILLE, INDIANA:** This company is thought to be the predecessor to Southern Independent Oil Company, marketers of Red Bird gasoline.

SOUTHERN
MARKETER INC. 5 1940s 13.5" Glass $1400-2000
White globe face with black "SOUTHERN" arched around top, "MARKETER, Inc." around bottom. Black outline map of Indiana in center with red road details. Brown rock and green tree image superimposed across map.

****CAPITOL OIL COMPANY — INDIANAPOLIS (CAMBY), INDIANA:** Capitol Oil was a marketing firm that originated the "Red Bird" brand name.

RED BIRD 5 1950s Balcrank $900-1250
RED BIRD 5 1950s 13.5" Glass $800-1100
White globe with red/black cardinal on green tree limb at top of globe face. Red "RED BIRD" with heavy black upshading across center of globe face with green "CAPITOL OIL CO." across lower globe face.

MARATHON & AFFILIATES

MARATHON OIL CORPORATION — FINDLAY, OHIO: The Ohio Oil Company was founded in 1887 when several northern Ohio oil production firms joined together to market their crude oil, primarily to Standard Oil. In 1889 Standard purchased Ohio Oil and operated the company as a production subsidary until the breakup of Standard in 1911.

After the breakup, Ohio Oil continued as before with oil exploration and production, satisfied to sell their product as crude oil and leave the refining and marketing to others. Ohio Oil first ventured into marketing with the June 1924 purchase of Lincoln Oil Refining Company of Robinson, Illinois. Lincoln Oil operated a refinery and about 20 retail gasoline stations in the Robinson-Terre Haute area at the time of the purchase by Ohio Oil. With an assured source of crude oil supply for the small refinery, the "Linco" brand began to expand. By 1930 Ohio Oil had expanded the Linco brand throughout most of the current Marathon territory—Ohio, Indiana, Illinois, Michigan and Kentucky.

In the meantime, Ohio Oil had been involved in numerous successful oil production ventures and found they needed even more outlets for their products. In 1930, Ohio Oil purchased Transcontinental Oil, a refiner/marketer that had marketed gasoline under the trademark "MARATHON" since about 1920. They had acquired the brand from a Pittsburgh operation, Riverside Oil Co., that we believe to be a forerunner of Republic Oil Company.

Transcontinental sold gasoline under the Marathon trademark across the Midwest and South, from North Carolina to New Mexico. Trasanscontinental can best be remembered for a significant first when in 1929 they opened several Marathon stations in Dallas, Texas, in conjunction with Southland Ice Company's "Tote'm" stores (later 7-Eleven), creating the first Gasoline-Convenience Store tie-in. The Marathon brand proved so popular that by World War II the name had replaced Linco at stations in the original five-state territory.

During the war, the widespread Marathon territory couldn't be properly supplied and marketing operations outside the Great Lakes area were sold, primarily to Tydol. Concentrating their marketing in a smaller territory allowed Marathon to expand within the territory. The late 1950s brought the first of many acquisitions—the Tower Oil Company of Cincinnati, Ohio—that would expand the company's marketing presence. Numerous other purchases are detailed in the listings below, with a group of marketing companies brought together by the 1980s under a subsidiary, Emro Marketing.

Ohio Oil, renamed Marathon Oil in 1962, was purchased by U.S. Steel (now USX) in 1981 and continues to market through dealer and jobber outlets under the Marathon name in the north central states and through Emro Marketing under several brand names, primarily Speedway and Starvin Marvin, throughout the south and east.

****LINCOLN OIL REFINING CO. — ROBINSON, ILLINOIS:** Lincoln Oil was urchased by Ohio Oil in 1924. Linco served as the company's primary brand until about 1939.

LINCO GASOLINE/
OILS 5 1924-1925 OPE N $1700-2100
White globe with red/white/blue Linco "crossed Linco" logo in center of globe face. Blue "GASOLINE" arched around top, "OILS" around bottom.

LINCO GASOLINE 5 1924-1926 OPE N NL
Description not available.

GASOLINE LINCO
OILS 5 1924-1934 15" Metal HP Blue $900-1300
White globe face with blue/white Linco logo in center. Red "LINCO" crossed vertical and horizontal in center of logo. Blue "GASOLINE" arched around top, "OILS" around bottom.

LINCO 4 1934-1939 13.5" Glass W $250-450
White globe face with red outline ring. Blue "LINCO" across center of globe face.

LINCO ETHYL
(EGC) 5 1934-1939 13.5" Glass W $300-500
White globe face with red outline ring. Large Ethyl (EGC) logo in center. Blue "LINCO" arched around top, "GASOLINE" around bottom.

LINCO BENZOL 5 1934-1935 13.5" Glass W $450-700
White globe face with red outline ring and blue "LINCO/BENZOL" across center of globe face.

LINCO BLUE 6 1934-1938 13.5" Glass W $400-600
White globe face with blue "BLUE" across globe face.

LINCO AVIATION 6 1934-1939 13.5" Glass W NL
White globe face with red outline ring and blue "LINCO/AVIATION" across center of globe face.

****RED FOX OIL COMPANY — FT. WAYNE, INDIANA:** Red Fox was a regional marketer purchased by Linco in 1928 and rebranded Linco at that time.

RED FOX
GASOLINE 6 1920-1925 15" Metal LP Cream NL
White globe face with red outline ring around white center circle. Detailed red and black fox head scene in center circle with red "RED FOX" arched around top, "GASOLINE" around bottom.

RED FOX
GASOLINE 7 1920s OPE NL
White globe with outline around center circle on globe face. "RED FOX" arched around top, "GASOLINE" around bottom with detailed fox head in center circle. Colors unknown.

(RED FOX) BENZOL
GASOLINE 7 1920s OPE NL
White globe with outline around center circle on globe face. "BENZOL" arched around top, "GASOLINE" around bottom with detailed fox head in center circle. Colors unknown.

****TRANSCONTINENTAL OIL CO. — TULSA, OKLAHOMA:** Transcontinental was purchased by Ohio Oil in 1930. The Transcontinental/Marathon marketing was kept separate from Linco marketing until about 1937 and Marathon had completely replaced Linco by World War II.

MARATHON
cast face 6 1920-1925 15" Metal NL
White globe face with orange silhouette of runner covering center of globe face. Black outlined green "MARATHON" in logotype lettering across runner.

MARATHON 3 1920-1938 15" Metal LP GREEN$1200-1500
White globe face with orange silhouette of runner covering center of globe face. Black outlined green "MARATHON" in logotype lettering across runner.

MARATHON
ETHYL (EGC) 5 1926-1938 15" Metal LP White $2000-2500
White globe face with small flesh color runner at top of globe. Green "MARATHON" arched around top, split between "A" and "T" by runner. Orange outlined black "ETHYL" below runner above Ethyl (EGC) logo.

MULTIPOWER 4 1928-1935 15" Metal LP Black NL
White globe face with orange outlined black "MULTIPOWER" across globe face. Orange/black line below and extended "T" crossbar form band across center.

MARATHON
MULTIPOWER 5 1935-1938 15" Metal H/L Black $2800-3500

White globe face with orange outlined black "MULTI" "POWER" across center of globe face, split by detailed runner. Green "MARATHON" arched around top, "GASOLINE" around bottom.

****THE OHIO OIL COMPANY — FINDLAY, OHIO:** Ohio Oil marketing was consolidated under the Marathon brand name in the late 1930s. It had virtually replaced the Linco brand by 1939.

MARATHON
(green letters) 4 1930s 13.5" Glass N $300-500
White globe face. Green "MARATHON" across center of globe face.

MARATHON BEST IN THE LONG RUN (deco logo) 4 1935-1938 13.5" Glass N $900-1350
White globe face with red circle in center. Large black and white drawing of running man superimposed over and extending beyond center circle. Red/white outlined black deco lettering "MARATHON" across lower part of center circle, with black banner furled below. White "BEST IN THE LONG RUN" on banner.

MARATHON ETHYL (EGC) (deco letters) 5 1935-1938 13.5" Glass N $400-600
White globe face with black outlined red deco lettering "MARATHON" across center of globe face. Black outlined white circle at bottom of globe face with Ethyl (EGC) logo in circle.

MARATHON w/cat 7 1938-1940 13.5" Glass N NL
White globe face with red outline ring. Red/white/black Marathon runner "Best in the Long Run" logo top center of globe face. Black "MARATHON" below logo. Black "CAT" design, possibly a decal, below "MARATHON", symbolizing Marathon's catalyst refining technique.

MARATHON w/ runner logo 4 1938-1940 13.5" Glass N $500-750
White globe face with red outline ring. Red/white/dark blue Marathon runner "Best in the Long Run" logo top center of globe face. Dark blue "MARATHON" below logo.

(MARATHON) ETHYL w/runner logo 4 1938-1940 13.5" Glass N $550-800
White globe face with red outline ring. Red/white/dark blue Marathon runner "Best in the Long Run" logo top center of globe face. Dark blue "ETHYL" below logo.

MARATHON w/ runner logo 4 1940-1954 Capco N $400-650
White globe face with red outline ring. Red/white/bright blue Marathon runner "Best in the Long Run" logo top center of globe face. Bright blue "MARATHON" below logo.

(MARATHON) ETHYL w/runner logo 4 1940-1946 Capco $400-650
White globe face with red outline ring. Red/white/bright blue Marathon runner "Best in the Long Run" logo top center of globe face. Bright blue "ETHYL" below logo.

MULTIPOWER (RED ON WHITE) 5 1938-1954 13.5" Glass N $250-350
White globe face with black outlined red "MULTIPOWER" across globe face.

MARATHON MILE MAKER 3 1946-1962 13.5" Glass N $500-800

MARATHON MILE MAKER 3 1946-1962 Capco $400-650
White globe face with red/white/dk blue "Best in the Long Run" logo in center. Dark blue "MARATHON" arched around top, red "MILE-maker" arched around bottom.

MARATHON SUPER M 3 1954-1962 13.5" Glass N $450-750

MARATHON SUPER M 3 1954-1962 Capco $350-600
Blue globe face with white outlined red/white/blue "Best in the Long Run" logo in center. White "MARATHON" arched around top, "SUPER-M" around bottom.

MULTIPOWER (BLUE ON WHITE) 4 1954-1962 13.5" Glass N $200-350
White globe face with blue "MULTIPOWER" across face.

MARATHON DIESEL #1 4 1946-1962 13.5" Glass N $300-500
White globe face with green dot in center. Black "MARATHON" arched around top, green "Diesel Fuel" around bottom with white "1" in green dot in center.

MARATHON DIESEL #2 3 1946-1962 13.5" Glass N $300-500
White globe face with green dot in center. Black "MARATHON" arched around top, green "Diesel Fuel" around bottom with white "2" in green dot in center.

MARATHON REGULAR MILE MAKER 1 1962-1970 Capco $150-250
White globe face with small red/white/blue Marathon "Big M" logo at top, above small blue "REGULAR". Larger red "MILE-maker" below "REGULAR".

MARATHON PREMIUM SUPER-M 1 1962-1970 Capco $150-250
White globe face with small red/white/blue Marathon "Big M" logo at top, above small blue "PREMIUM". Larger blue "Super-M" below "PREMIUM".

MARATHON DIESEL FUEL 4 1962-1970 Capco $175-275
White globe face with small red/white/blue Marathon "Big M" logo at top. Blue "DIESEL" over smaller red "FUEL" on lower globe face.

MARAFUEL 5 1962-1970 Capco $200-300
White globe face with dark blue "MARAFUEL" across center of globe face.

(MARATHON) MILE MAKER 6 1962-EXP Capco NL
Experimental logo. White globe face with large blue "boomerang" line, point at upper right, across center of globe face. Large red "MILE-maker" across center of globe face.

(MARATHON) SUPER-M 6 1962-EXP Capco NL
Experimental logo. White globe face with large red "boomerang" line, point at upper right, across center of globe face. Large blue "SUPER-M" across center of globe face.

****TOWER OIL CO. — CINCINNATI, OHIO:** Tower was a Cincinnati based marketer that sold out to Marathon and was rebranded Marathon in 1958.

TOWER GASOLINE 4 1940s 13.5" Glass $1000-1600
Globe face has black outline. Upper half is yellow with large black/white "tower" logo. Lower half has green band above lower black area. White lettering "TOWER" with black dropshade on green band. Yellow script "Gasoline" in lower black area.

HI TOWER ETHYL 3 1940s 13.5" Glass $400-650
White globe face with black outline. Green band across center with black outline white tower logo above band. Yellow "-HI-" with black dropshade across tower, "TOWER" on green band. Small Ethyl(EC) logo in circle at bottom of globe face.

****AURORA GASOLINE CO. — DETROIT, MICHIGAN:** Aurora Gasoline Company was a Detroit refiner/marketer that introduced the Speedway brand in the early 1930s. The company's premium product, Speedway 79 Stratofuel, was introduced before World War II and the primary brand name used from the end of World War II until 1962 was "Speedway 79." In 1959 Marathon purchased the Aurora Gasoline Company and the Speedway 79 logo was redesigned. Gas pump globes were eliminated at this time. The Speedway 79 brand was replaced by Marathon in 1962; however, in 1975 the Speedway brand name was reintroduced at several company-owned retail "gas only" stations. Marathon began buying up regional private brands and consolidating them under Emro Marketing with the Speedway brand name. As of this writing, the Speedway name appears on petroleum products retailed at hundreds of Emro "Starvin Marvin" c-stores.

SPEEDWAY GASOLINE 6 1930s 15" Metal NL
Blue globe face with yellow ring around face near edge. Large yellow "Speedway" with speed lines across globe with yellow "GASOLINE" below.

SPEEDWAY OIL CO. W/FLAG 5 1938-1946 13.5" Glass $900-1300
White globe face with blue outline rings. Large blue and white checkered flag covering most of globe face. Red "SPEEDWAY" below flag, over smaller "OIL COMPANY".

SPEEDWAY 79

STRATOFUEL 5 1938-1946 13.5" Glass $750-1000
Red globe face with white "SPEEDWAY" with blue dropshade arched just above center. Checkered flag and speed lines above "SPEED-WAY". Large white "79" with blue dropshade in center above white "STRATOFUEL" with blue dropshade.

SPEEDWAY 79
RED/BLUE 4 1946-1955 Capco $250-400
Globe face split by white line across center. Upper half blue,lower half red. Large white "SPEEDWAY" with distorted lettering in upper blue area, white "79" in lower red area.

SPEEDWAY 79
(red background,
blue lettering) 3 1955-1962 13.5" Glass $400-650
Red globe face with white band across above center. Blue "SPEED-WAY" on white band with white "79" in lower red area offset to right.

SPEEDWAY 79
(blue background,
red lettering) 5 1955-1962 13.5" Glass $400-650
Blue globe face with white band across above center. Red "SPEED-WAY" on white band with white "79" in lower red area offset to right.

SPEEDWAY EXTRA4 1955-1962 13.5" Glass $300-500
Blue globe face with white band across above center. Red "SPEED-WAY" on white band with white "EXTRA" in lower blue area.

SPEEDWAY 79
FUEL OIL 6 1955-1962 13.5" Glass $400-650
Blue globe face with white band across above center. Red "SPEED-WAY" on white band with white "FUEL OIL" in lower blue area.

****OLD DUTCH REFINING CO. — MUSKEGON, MICHIGAN:** Founded in Muskegon about 1930, the Old Dutch brand appeared on a small number of stations, primarily jobber operated, in the Muskegon area. As best as can be determined, Aurora Gasoline Co. purchased Old Dutch Refining in the late 1950s, although jobbers kept the Old Dutch brand alive well into the 1970s. This brand's globes were sometimes found on clear ripple gills. For such examples, add $700 to the value.

OLD DUTCH
ECONOMY 6 1930-1935 15" Metal NL
White globe face with progressively fading blue band across center of globe face. Dark blue windmill on upper globe face with small dark blue "Old" to left of windmill and larger dark blue "Dutch" below windmill. Blue "ECONOMY" below "DUTCH".

OLD DUTCH
ECONOMY 4 1935-1940 Gill $1400-1800
White globe face with progressively fading blue band across center of globe face. Dark blue windmill on upper globe face with small dark blue "Old" to left of windmill and larger dark blue "Dutch" below windmill. Blue "ECONOMY" below "DUTCH".

OLD DUTCH
QUALITY 4 1935-1940 Gill $1400-1800
White globe face with progressively fading blue band across center of globe face. Dark blue windmill on upper globe face with small dark blue "Old" to left of windmill and larger dark blue "Dutch" below windmill. Red "QUALITY" below "DUTCH".

AROSPEED 4 1935-1940 Gill $400-650
White globe face with red outlined bar and circle area across center of globe face. Blue arrow "through" bar and circle. Blue "ARO-SPEED" with large "S" in bar and circle.

OLD DUTCH 4 1940-1946 Gill $1400-1800
White globe face with progressively fading blue band across center of globe face. Dark blue windmill on upper globe face with small dark blue "Old" to left of windmill and larger dark blue "Dutch" below windmill.

AROSPEED 5 1940-1946 Glass Oval $750-1000
White globe face with blue/red outlined bar and circle area across center of globe face. Blue arrow "through" bar and circle. Blue "ARO-SPEED" with large "S" in bar and circle.

OLD DUTCH 4 1946-1974 Capco $900-1300
White globe face with red outline ring. Blue line across center of globe face with progressively thinner red lines above. Blue windmill on upper globe face with blue "Old" to left and blue "Dutch" below.

****REPUBLIC OIL CO. — PITTSBURGH, PENNSYLVANIA:** Dating from the early 1920s, and believed to be a successor to Riverside Oil Company (originator of the Marathon trademark), Republic supplied branded and unbranded jobbers throughout the southeast. Republic marketing was concentrated in small areas and no attempt was made at complete statewide coverage in the states where they marketed. By the 1950s, the corporate name was Plymouth Oil Company and Marathon purchased the operation in 1962. Concentrating on the extensive unbranded market, the Republic brand name was eventually phased out, although some jobbers continue to use the brand name today.

REPUBLIC 5 1922-1946 OPC Shield $2000-2500
One piece cast shield shaped globe with three faces. Upper 1/3 of shield is blue with three white stars and white "REPUBLIC". Lower 2/3 is red/white verticle stripes.

REPUBLIC ETHYL 5 1928-1946 OPC Shield $2100-2600
One piece cast shield shaped globe with three faces. Upper 1/3 of shield is blue with three white stars and white "REPUBLIC". Lower 2/3 is red/white verticle stripes. Blue script "Ethyl" across red and white stripes.

REPUBLIC MOTOR
GASOLINE 5 1930s 15" Metal $500-750
White globe face with large black outlined red "MOTOR" across center of globe face. Black "REPUBLIC" arched around top, "GASOLINE" around bottom.

REPUBLIC Shield 3 1946-1958 13.5" Glass $400-650
White globe face with red/white/blue Republic shield covering most of globe face. Blue lettering "REPUBLIC" extends past sides of shield.

REPUBLIC SUPERIOR
GASOLINE 5 1946-1958 13.5" Glass $300-550
White globe face with blue letters, "SUPERIOR" arched around top "GASOLINE" around bottom. Double red diamond in center with "REPUBLIC" in red across diamond. Small "REPUBLIC PRODUCT" around bottom edge of diamond in blue letters.

REPUBLIC POWER
PAK' D 2 1958-1970 Capco $175-275
Red globe face with large blue/white outlined red and white shield covering most of globe face. Red "POW'R/PAK' D" with blue dropshade and red speed lines on white band on shield.

REPUBLIC HI-T 2 1958-1970 Capco $150-250
White globe face wtih large red/white/blue "HI/T" shield covering most of globe face.

REPUBLIC HI-WAY 4 1946-1950 Capco $175-275
White globe face with blue outer ring. Blue dot in center with white "HI-WAY" diagonal across dot. White "REPUBLIC" arched around top in blue area, "GASOLINE" around bottom.

HI-WAY 3 1950-1958 Capco $150-250
White globe face with blue dot in center. White "HI-WAY" with blue outline diagonal across dot.

HI-WAY 3 1958-1970 Capco $150-250
Blue globe face with white band horizontal across below center. Red lettering "HI-WAY" with blue/white lines extending up to point.

SUPER HI-WAY 6 1958-1970 Capco NL
Blue globe face with white band horizontal across below center. Red lettering "HI-WAY" with blue/white lines extending up to point. "SUPER" across top above "HI-WAY".

****BONDED — SPRINGFIELD, OHIO:** In the years following World War II, Paul Deer, a close associate of Russell Williams (who founded Gaseteria - Bonded), began operation of a chain of discount service stations in Ohio, Kentucky and West Virginia, using the Bonded brand name with friend William's blessings. Marathon purchased Bonded in January 1975 and, although many locations were converted to Speedway/Starvin-Marvin, the Bonded brand has remained in use at many locations.

BONDED 98 in
laurel wreath 5 1955-1968 13.5" Glass W $750-1000
White globe face with laurel wreath around outer edge of globe face. Blue Old English "Bonded" over large red "98" across center of globe face.

BONDED ETHYL in
laurel wreath 5 1955-1968 13.5" Glass W $600-850
White globe face with laurel wreath around outer edge of globe face. Blue Old English "Bonded" over script "Ethyl" and small Ethyl (EC) logo center of globe face.

****CONSOLIDATED — OSHKOSH, WISCONSIN:** Consolidated was a Wisconsin discount operation from the postwar era that operated about 30 stations throughout the state. The firm was purchased by Marathon in

July 1972 and rebranded Speedway about 1982.

CONSOLIDATED 5 1946-1960 Oval Glass $600-850
White globe face with black outlined green and black "Consolidated" oval across globe face.

CONSOLIDATED ETHYL 5 1946-1960 Oval Glass $600-850
Same as above with small Ethyl (EC) logo in circle at lower center of globe face.

****CHEKER OIL CO. — CHICAGO, ILLINOIS:** Cheker was a suburban Chicago collection of private brands, the Cheker name originally replacing "Chief" and "Road Chief" in about 1968. Cheker purchased the Sun-Glo operation in Indiana around 1970 and converted those stations to the Cheker brand. In 1976 Cheker, then 50% owned by Exxon, purchased the Imperial Refineries operation. Imperial, founded in St. Louis in 1915, was in its earliest days an integrated oil company involved with production, refining and marketing. After World War II, Imperial resumed marketing as a discount chain operation, eventually operating stations in nearly 20 states. Cheker operated Imperial Refineries as a separate brand until the September 1983 merger with Marathon/Emro. Emro replaced most of the Imperial brand with Ecol after the 1984 purchase of this Jackson, Mississippi, independent.
>>No Cheker globes known at this time.

****ROAD CHIEF — CHICAGO, ILLINOIS:** Predecessor to Cheker. See Cheker introduction for explanation.

ROAD CHIEF "RULES THE ROAD" 5 1960s Capco $175-275
Red outlined yellow globe face with red outlined white center circle. Red band across center with white script "Rules the Road" across band. Blue "ROAD" arched around top, "CHIEF" around bottom.

****SUN GLO — INDIANAPOLIS, INDIANA:** Sun Glo was a regional independent that was purchased by Cheker in about 1970.

SUN GLO 5 1950-1970 Capco $200-300
White globe face with large red circle in center. Blue "SUN-GLO" arched around top, "GASOLINE" around bottom.

****IMPERIAL REFINERIES — CLAYTON, MISSOURI:** Founded in Tulsa, Oklahoma, in 1915 as a refiner/marketer, Imperial was originally affiliated with Wickett Refining Co., Kettle Creek Refining, and San Angelo Refining, soliciting jobbers throughout the central United States, beginning in 1929. Beginning around 1940, Imperial Refineries withdrew from jobber marketing to concentrate on developing a chain of discount stations throughout the Midwest. Imperial affiliates were also active in Florida and in Michigan.

IMPERIAL W/LARGE OLD Shield 5 1920-1929 16.5" Metal $1250-1750
White globe face with large Imperial tan and maroon shield in center. Tan lions to either side with banners arched above and below. Unique typestyle "IMPERIAL" in upper banner, "REFINERIES" in lower banner.

IMPERIAL REFINERIES SMALL Shield 5 1929-1940 16.5" Metal $800-1250
White globe face with small tan/maroon Imperial shield at bottom of globe face and distinctive "IMPERIAL" across globe face.

IMPERIAL 2 1940-1982 13.5" Glass W $200-300

IMPERIAL 1 1946-1982 Capco $150-250
White globe face with black outline ring. Yellow and red Imperial shield covers most of globe face with black "IMPERIAL" across shield and small yellow "Refineries" below.

IMPERIAL ETHYL (EGC) 2 1940-1946 13.5" Glass W $200-350
Yellow globe face with black outline ring and black outline around white center circle. Red "IMPERIAL" arched around top, "Refineries" around bottom with small Ethyl (EGC) logo in center.

IMPERIAL ETHYL (EC) 2 1946-1965 13.5" Glass W $200-350

IMPERIAL ETHYL (EC) 3 1946-1965 Capco $175-275
Yellow globe face with black outline ring and black outline around white center circle. Red "IMPERIAL" arched around top, "Refineries" around bottom with small Ethyl (EC) logo in center.

IMPERIAL PREMIUM 5 1965-1982 Capco $175-275
White globe face with small red/yellow/black Imperial shield at top of globe face and italic red "PREMIUM" across lower globe face.

IMPERIAL NO LEAD 2 1975-1982 Capco $125-225
Same as IMPERIAL ETHYL, above, with large black "NO LEAD" replacing "IMPERIAL", black crosses in white circles around bottom and Imperial shield logo in center.

IMPERIAL DIESEL 5 1950-1965 Capco $200-300
White globe face with black outline ring. Yellow and red Imperial shield covers most of globe face with black "IMPERIAL" across shield and small yellow "Refineries" below, with "DIESEL" below "Refineries".

IMPERIAL STOVE OIL 4 1950-1965 Capco $200-350
Same as IMPERIAL ETHYL, above, with red "STOVE OIL" arched around bottom and large blue "1" in center circle. Black circle/red cross logo beside "1".

IMPERIAL 50 YEARS 5 1965 13.5" Glass W $500-750
Unique in globes, this globe was used at Imperial stations for one year, 1965, to commemerate their 50th anniversary. These are very rare. White globe face with black outline ring. Gold metal and ribbon design in center with black "50th" in center and black "1915" and "1965" to either side. Black "OUR" arched around top, "ANNIVERSARY" around bottom.

****JOHNSON OIL CO.—MT. PLEASANT, MICHIGAN:** Founded in 1926 in Lakeview, Michigan, and currently the Imperial Oil Company of Mt. Pleasant, Michigan—Johnson Oil was apparently an Imperial jobber during the time Imperial Oil had refineries and a jobber marketing network. Johnson continues to market in Michigan and Indiana under the Imperial name. This company is not related to Marathon marketing and is listed here only because of the ties to Imperial.

IMPERIAL OF MICHIGAN 4 1946-1970 Capco $150-250
White globe face with large, flat top yellow and red Imperial shield on globe face. Black "GASOLINE" across lower shield.

****STEPHEN:** An Imperial Refineries affiliate.

STEPHEN 5 1940s 13.5" Glass $200-350
Red/white/blue globe face with multiple outline rings. Large "S" in center in blue diamond box. Red "STEPHEN" across center.

STEPHEN 5 1940s 13.5" Glass $200-350
Red/white/blue globe face with multiple outline rings. Large "S" in center in red diamond box. Blue "STEPHEN" across center.

****ROCK ISLAND REFINING — INDIANAPOLIS, INDIANA:** Founded in Duncan, Oklahoma, in the 1920s, Rock Island was a mid-continent area marketer until World War II. Having shifted to wartime refining, virtually abandoning retail gasoline, the company was purchased by the Koch family in 1946 and moved to Indianapolis. There Rock Island became a major source of unbranded gasoline for independent marketers. Over the years the company acquired several Indiana and Michigan marketers including Colonial/Progressive of Ft.Wayne, Tulsa of Detroit, United of Indianapolis, Naph-Sol of Muskegon and Golden Imperial of Indianapolis. By the mid-1980s "United" had become the primary brand, replacing the others where convenient. In 1989 Marathon's Emro Marketing purchased Rock Island and many United locations have converted to the Speedway brand.

ROCK ISLAND GASOLINE 5 1930-1940 15" Metal $2800-3500
Red/white outlines around blue outer ring. Red/white rings around white center circle. White "ROCK ISLAND" arched around top, "GASOLINE" around bottom with red line drawing of the Rock of Gibraltar in the center.

ROCK ISLAND NUNBETTER 5 1930-1940 15" Metal $600-900
Red/white outlines around blue outer ring. Red/white rings around white center circle. White "ROCK ISLAND" arched around top, "GASOLINE" around bottom . Red outlined white band di-

agonal across plain white center (no rock). Red "NUNBETTER" in center band.

ROCK ISLAND ROCKILENE 4 1930-1940 15" Metal $600-900
Red/white outlines around blue outer ring. Red/white rings around white center circle. White "ROCK ISLAND" arched around top, "GASOLINE" around bottom . Red outlined white band diagonal across plain white center (no rock). Red "ROCKILENE" in center band.

ROCK ISLAND ANTI-KNOCK 5 1930-1940 15" Metal $600-950
Red/white outlines around blue outer ring. Red/white rings around white center circle. White "ROCK ISLAND" arched around top, "GASOLINE" around bottom . Red outlined white band diagonal across plain white center (no rock). Red "ANTI-KNOCK" in center band.

ROCK ISLAND ETHYL 5 1930-1940 15" Metal $600-850
Red/white outlines around blue outer ring. Red/white rings around white center circle. White "ROCK ISLAND" arched around top, "GASOLINE" around bottom. Large Ethyl (EGC) logo with yellow rays in center circle.

ROCK ISLAND GASOLINE (early rock) 5 1940-1946 Red Capco $1000-1400
Red/white outlines around blue outer ring. Red/white rings around white center circle. White "ROCK ISLAND" arched around top, "GASOLINE" around bottom with red line drawing of the Rock of Gibraltar in the center.

ROCK ISLAND GASOLINE (early rock) 5 1940-1946 Red Capco $1000-1400
Blue/white outlines around white outer ring. Blue/white rings around white center circle. Red "ROCK ISLAND" arched around top, "GASOLINE" around bottom with blue line drawing of the Rock of Gibraltar with green water in the center.

ROCK ISLAND ROCKILENE 4 1940-1946 Capco $200-300
Red/white outlines around blue outer ring. Red/white rings around white center circle. White "ROCK ISLAND" arched around top, "GASOLINE" around bottom . Red outlined white band diagonal across plain white center (no rock). Red "ROCKILENE" in center band.

ROCKILENE ETHYL 4 1940-1946 Capco $200-300
Red/white outlines around blue outer ring. Red/white rings around white center circle. White "ROCKILENE" arched around top, "GASOLINE" around bottom . Large Ethyl (EGC) logo with yellow rays in center.

ROCK ISLAND W/ ROCK (red/white) 4 1946-1960 Capco $600-900
White globe face with maroon "ROCK" across top, "ISLAND" across bottom and maroon rock and water picture in center.

ROCK ISLAND W/ROCK (red/white/blue) 4 1946-1960 Capco $600-900
White globe face with red "ROCK" across top, "ISLAND" across bottom and blue rock and water picture in center.

ROCK ISLAND ETHYL (EGC) (red/white/ blue) 4 1946-1960 Capco $600-900
White globe face with red "ROCK" across top, "ISLAND" across bottom and blue rock and water picture in center. Small Ethyl (EGC) logo in white circle superimposed over rock in center.

ROCK ISLAND W/ ROCK (red/yellow) 4 1946-1960 Capco $750-1100
Bright yellow globe face with maroon "ROCK" across top, "ISLAND" across bottom and maroon rock and water picture in center.

ROCK ISLAND ETHYL (EC) (red/yellow) 4 1946-1960 Capco $750-1150
Bright yellow globe face with maroon "ROCK" across top, "ISLAND" across bottom and maroon rock and water picture in center. Small Ethyl (EC) logo in white circle superimposed over rock in center.

****COLONIAL/PROGRESSIVE — FT. WAYNE, INDIANA:** Independent marketer Colonial/Progressive was founded in the 1930s and operated Colonial stations in Indiana and Progressive stations in Michigan. The company sold out to Rock Island in the 1960s.

COLONIAL 3 1950-1965 13.5" Glass $200-350
Orange upper and lower halves with black outlined white band across center. White band widens left-to-right. Black speed lines above and below white band. Black "COLONIAL" in progressively larger type on center band.

PROGRESSIVE 3 1950-1965 13.5" Glass $200-350
Orange upper and lower halves with black outlined white band across center. White band widens left-to-right. Black speed lines above and below white band. Black "PROGRESSIVE" in progressively larger type on center band.

O-70 6 1950s 13.5" Glass $200-400
Orange upper and lower halves with black outlined white band across center. White band widens left-to-right. Black speed lines above and below white band. Black "O-70" in progressively larger type on center band.

****GOLDEN IMPERIAL —INDIANAPOLIS, INDIANA:** This company was an independent marketer that operated "Golden Imperial" stations in the Indianapolis area in the 1950s and 1960s. Golden Imperial sold out to Rock Island in the 1960s.

GOLDEN IMPERIAL HI-TEST 9 1950s Oval Capco NL
From a description in corporate literature. White globe face with large blue "IMPERIAL" across center of globe face. Gold crown logo center at top above "IMPERIAL" with blue "HI-TEST" across crown.

GOLDEN IMPERIAL ETHYL 9 1950s Oval Capco NL
From a description in corporate literature. White globe face with large blue "IMPERIAL" across center of globe face. Gold crown logo center at top above "IMPERIAL" with red "ETHYL" across crown.

****TULSA OIL CO. — DETROIT, MICHIGAN:** Tulsa is an old Detroit independent brand that sold out to Crystal Oil in July 1971 and was in turn purchased by Rock Island around 1978. The United brand name replaces Tulsa in 1985.

TULSA 6 1940s 15" Metal NL
Identical in design to early "Tydol" globes. Orange ring around outer edge with black line around white center circle. Black "TULSA" (oversize "T") across white center.

TULSA 95 4 1950-1970 Capco $450-700
White globe face with yellow outline ring. Red "TULSA" with black dropshade across top of globe face. Red "95" (in quotes) in center above yellow and black stylized cow head and horns.

TULSA 100 3 1950-1970 Capco $450-700
White globe face with yellow outline ring. Red "TULSA" with black dropshade across top of globe face. Red "100" (in quotes) in center above yellow and black stylized cow head and horns.

TULSA HI-TEST 4 1950-1970 Capco $450-700
White globe face with yellow outline ring. Red "TULSA" with black dropshade across top of globe face. Red "HI-TEST" (in quotes) in center above yellow and black stylized cow head and horns.

TULSA DIESEL 5 1950-1970 Capco $475-750
White globe face with yellow outline ring. Red "TULSA" with black dropshade across top of globe face. Red "DIESEL" (in quotes) in center above yellow and black stylized cow head and horns.

TULSA FUEL OIL 5 1950-1970 Capco $475-750
White globe face with yellow outline ring. Red "TULSA" with black dropshade across top of globe face. Red "FUEL OIL" (in quotes) in center above yellow and black stylized cow head and horns.

****DURO OIL CO. — DETROIT, MICHIGAN:** Duro was a Detroit based independent that merged with Tulsa in the 1970s.

DURO "D" GASOLINE 4 1950s 13.5" Glass $200-400
Red/white outline ring around green globe face. White circle in center. White "DURO" arched around top, "GASOLINE" around bottom with large red "D" with green dropshade in center.

****UNITED — INDIANAPOLIS, INDIANA:** United was an Indianapolis based discount marketer with stations located throughout Indiana and a few in Michigan. In 1985, United was purchased by its primary supplier, Rock

Island, and became the brand name used to consolidate all of Rock Island's marketing. Most of the United stations were rebranded Speedway in 1990, but some locations remain United.

UNITED HI-TEST W/ARROW 5 1946-1965 Capco $200-350
Blue outlined red globe face with blue outlined white bar and circle across center of globe. White "UNITED" arched in upper red area, "GASOLINE" below with blue "HI TEST" in white center area. Red arrow passing through "HI TEST".

UNITED ETHYL W/ARROW 5 1946-1960 13.5" Glass $350-600
Black outlined red globe face with black outlined white bar and circle across center of globe. White "UNITED" arched in upper red area, "GASOLINE" below with Ethyl (EC) logo in white center area. Red arrow pointing left through center.

UNITED J-J W/ARROW 6 1946-1955 13.5" Glass $400-650
Black outlined red globe face with black outlined yellow center circle and band across center. Yellow "UNITED" arched around top, "GASOLINE" around bottom with black outlined red arrow pointing left through center. Red outlined black "J/J" superimposed over arrow.

UNITED FUEL OIL W/FLAME 5 1946-1965 13.5" Glass $600-850
Light blue outline around red outer ring. Light blue outline around white center circle. White "UNITED" arched around top, "FUEL OIL" around bottom with red/white/tan detailed flames in center and black/blue band across flames. White script on black band "GUARANTEED QUALITY".

UNITED ETHYL RED/BLACK 5 1960-1965 13.5" Glass $200-400
White globe face with black outlined white circle in center. Red "UNITED" arched around top, "GASOLINE" around bottom with black triangles to either side. Red and black stylized Ethyl logo in center.

UNITED 3 1965-1980 Capco $125-200
White globe face with thick red band arched across center. White "UNITED" on red band.

UNITED REGULAR 3 1965-1980 Capco $125-225
White globe face with thick red band arched across center. White "UNITED" on red band. Blue "REGULAR" below.

UNITED ETHYL 3 1965-1980 Capco $150-250
White globe face with thick blue band arched across center. White "UNITED" on blue band. Red "ETHYL" below.

UNITED SUPER ETHYL 3 1965-1980 Capco $150-275
White globe face with thick blue band arched across center. White "UNITED" on blue band. Red "SUPER/ETHYL" below.

UNITED UNLEADED 4 1975-1980 Capco $125-250
White globe face with thick red band arched across center. White "UNITED" on red band. Blue "UNLEADED" below.

****NAPH-SOL REFINING CO. — MUSKEGON, MICHIGAN:** Naph-Sol entered the gasoline market with the introduction of Zephyr gasoline in 1932. Marketing continued through jobbers and independent dealers in Michigan and Wisconsin until 1989 when Naph-Sol was bought out by Rock Island Refining. Several Michigan jobbers continue to use the Zephyr brand.

INDEPENDENT ZEPHYR GASOLINE 4 1936-1940 13.5" Glass $200-400
White globe face with thick blue outline ring and blue lines forming white band across center. Blue "INDEPENDENT" arched around top, "GASOLINE" around bottom with red "ZEPHYR" in center white band.

ZEPHYR 4 1940-1957 13.5" Glass $200-400
White globe face with thick blue outline ring and blue lines forming white band across center. Red "ZEPHYR" in center white band.

ZEPHYR ETHYL 4 1940-1957 13.5" Glass $200-425
White globe face with thick blue outline ring and blue lines forming white band across center. Red "ETHYL" at top, "ZEPHYR" in band and small Ethyl (EC) logo below.

ZEPHYR ETHYL 4 1940-1957 13.5" Glass $200-425
White globe face with thick blue outline ring and black lines forming white band across center. Red "ETHYL" at top, "ZEPHYR" in band and small Ethyl (EC) logo below.

ZEPHYR W/Shield 2 1957-1970 Capco $275-475
White globe face with blue outlined red rounded triangle Zephyr logo covering most of globe face. White "ZEPHYR" over blue "GASOLINE" in logo.

(ZEPHYR) BLU-FLAME FUEL OIL 3 1960s Capco $200-350

(ZEPHYR) BLU-FLAME FUEL OIL 3 1950s 13.5" Glass $300-500
Yellow globe face with black outline ring. Light blue/white flame design covering most of globe face with blue "BLU-FLAME" across flames.

****OSCEOLA REFINING CO. — REED CITY, MICHIGAN:** Osceola was an independent refiner/marketer purchased by Zephyr in the 1970s.

OSCEOLA REGULAR 4 1960s Plastic $250-400
White plastic formed trapezoidal halves assembled back to back. Blue "OSCEOLA" logotype across top with red Indian head in blue area below. Red "REGULAR" to right of Indian head.

OSCEOLA PREMIUM 4 1960s Plastic $250-400
White plastic formed trapezoidal halves assembled back to back. Blue "OSCEOLA" logotype across top with red Indian head in blue area below. Red "PREMIUM" to right of Indian head.

****MARTIN OIL SERVICE — ALSIP / BLUE ISLAND, ILLINOIS:** Separated from Martin Oil Company in 1957, Martin Oil Service was marketing through over 200 stations coast to coast by 1965. Following a significant downsizing during the gas shortages of the 1970s, Martin Oil Service was continuing to operate nearly 50 stations in Illinois and Indiana when the firm was purchased by Emro in 1994. The Martin name remains in use at many of the stations.

MARTIN SUPER REGULAR 3 1955-1962 13.5" Glass W $250-450
Red/white/blue repeating outline rings around outer red area. Outline rings around white center circle. Red "MARTIN across upper globe face, extending into outer red area. Blue "Super" diagonal over red "Regular" on lower globe face.

MARTIN XTRA SPECIAL 4 1955-1962 13.5" Glass W $250-450
Red/white/blue repeating outline rings around white globe face. Blue "MARTIN" arched around top, "SPECIAL" horizontal across bottom with blue outlined red "XTRA" diagonal across globe face. Outline rings around "X" in "XTRA".

MARTIN PURPLE MARTIN 6 1955-1962 13.5" Glass W NL
White globe face with blue "PURPLE" arched around top. Red/yellow/red line around lower globe face. Small blue bird under arch of "PURPLE" above red "MARTIN" over yellow script "The Finest" over red "ETHYL" over yellow script "of all". Small Ethyl(EC) logo in circle at end of "of all".

MARTIN #1 DIESEL 6 1955-1962 13.5" Glass W NL
Colors cannot be determined. Dark outline ring around pale colored globe face and around center circle. Large "#1" in center circle with dark colored "MARTIN" arched around top, "DIESEL" around bottom. Small "NO" to left of center circle, "FUEL" to right.

MARTIN #2 DIESEL 6 1955-1962 13.5" Glass W NL
Colors cannot be determined. Dark outline ring around pale colored globe face and around center circle. Large "#2" in center circle with dark colored "MARTIN" arched around top, "DIESEL" around bottom. Small "NO" to left of center circle, "FUEL" to right.

MARTIN SUPER REGULAR 3 1962-1970 Capco $200-300
White globe face with parallel red lines across upper globe face. Red outlined Martin blue "M" logo superimposed over lines. Large blue "SUPER" over smaller red "REGULAR" on lower globe face.

MARTIN EXTRA SPECIAL ETHYL 3 1962-1970 Capco $250-350
White globe face with parallel blue lines across upper globe face. Blue outlined Martin blue "M" logo superimposed over lines. Large blue outlined red "XTRA SPECIAL" arched below logo over red "ETHYL" and small Ethyl (EC) logo.

MARTIN PURPLE MARTIN ETHYL 4 1950s 13.5" Glass W $400-750

MARTIN PURPLE

MARTIN ETHYL 3 1962-1970 Capco $350-600
White globe face with parallel purple lines across upper globe face. Purple outlined Martin blue "M" logo superimposed over lines. Small purple bird just below logo with purple "PURPLE" to left and "MARTIN" to right of bird. Large red "ETHYL" and small Ethyl (EC) logo on lower globe face.

MARTIN DIESEL #2 4 1962-1970 Capco $200-300
White globe face with parallel green lines across upper globe face. Green outlined Martin blue "M" logo superimposed over lines. Green "DIESEL" over large red "2" with green "NO." to left and "H.I." to right on lower globe face.

MARTIN DIESEL #1 4 1962-1970 Capco $200-300
White globe face with parallel green lines across upper globe face. Green outlined Martin blue "M" logo superimposed over lines. Green "DIESEL" over large red "1" with green "NO." to left and "H.I." to right on lower globe face.

MARTIN RANGE OIL 5 1962-1970 13.5" Glass W $300-550
White globe face with parallel green lines across upper globe face. Green outlined Martin blue "M" logo superimposed over lines. Green "RANGE/OIL" on lower globe face.

****J.A. HOGSHIRE and SON — INDIANAPOLIS, INDIANA**: Founded in Lebanon, Indiana, in 1923, J.A.Hogshire was one of the original members of the IOMA organization. Originally marketing under the brand name "Criteria," Hogshire introduced the WAKE UP brand in 1937, operating discount stations throughout Indiana under the Wake Up name. The company was purchased by Emro marketing in 1994 and the brand name continues to appear on several dozen stations today.

CRITERIA GASOLINE 7 1930-1937 13.5" Glass NL
From an old photo. Dark colored outline ring around white globe face. Double outline rings around solid color center circle. Dark color "CRITERIA" arched around top, "GASOLINE" around bottom with "THE BEST" arched around top of center circle, "FOR LESS" below and white "100%/PENNSYLVANIA" in center circle.

CRITERIA ANTI-KNOCK ETHYL GASOLINE7 1930-1937 13.5" Glass NL
From an old photo. Dark colored outline ring around white globe face. Double outline rings around solid color center circle. Dark color "CRITERIA" arched around top, "GASOLINE" around bottom with "THE BEST" arched around top of center circle, "FOR LESS" below and white "100%/PENNSYLVANIA" in center circle. Small "HIGH TEST" arched just below "CRITERIA" at top, "ANTI-KNOCK" just above "GASOLINE" around bottom.

WAKE UP (cube) 7 1937-1940 OPB NL
From an old photo. White cube shaped globe with peaked top. Large red rooster of four sides with blue "WAKE UP" superimposed across rooster.

WAKE UP ANTI-KNOCK ETHYL (cube) 7 1937-1940 OPB NL
From an old photo. White cube shaped globe with peaked top. Small "WAKE UP" arched across top, larger "ETHYL below, on all four sides.

WAKE UP "YOU'RE BUYING THE BEST" 7 1940-1955 Gill NL
White globe face with large red rooster covering most of globe face. Blue "WAKE UP" across rooster. Small "You're Buying The Best" below "WAKE UP".

WAKE UP ETHYL "YOU'RE BUYING THE BEST" 7 1940-1955 Gill NL
White globe with small red rooster at top of globe face. Blue "WAKE UP" across rooster. Ethyl (EGC) logo in circle on lower globe face. Small "You're Buying The Best" "WAKE UP", above Ethyl logo.

WAKE UP ETHYL (no slogan) 7 1940-1955 Gill NL
White globe with small red rooster at top of globe face. Blue "WAKE UP" across rooster. Ethyl (EC) logo in circle on lower globe face.

WAKE UP ETHYL 99 7 1950-1955 Gill NL
White globe face red rooster in center of globe face. Blue "WAKE UP" across rooster, blue "99" at bottom of globe face. Ethyl (EC) logo in circle on lower globe face.

WAKE UP 99 7 1955-1960 Gill $900-1300
White globe face red rooster in center of globe face. Blue "WAKE UP" across rooster, blue "99" at bottom of globe face.

MARTIN & AFFILIATES

MARTIN OIL COMPANY — CARBONDALE, ILLINOIS and MARTIN OIL SERVICE — BLUE ISLAND, ILLINOIS: Martin, an Illinois based trackside discounter dating from the early 1930s, was actually a collection of several family owned operations within the Martin chain. In the earliest days dual headquarters in Chicago and Carbondale, Illinois, were listed on products. Later the Chicago division was moved to Alsip and later still to Blue Island, Illinois. About 1956 the chain separated into two distinctly separate operations, the Chicago (Alsip) division becoming Martin Oil Service while the Carbondale division remained Martin Oil Company.

Martin Oil Service expanded coast-to-coast, operating over 200 gas stations by the 1970s, while Martin Oil Company concentrated marketing through a small number of stations in the south central states. Cut off from supplies during the 1973-74 gas shortage, Martin Oil Service retreated to its home base. Martin Oil Service currently operates less than 50 stations in Illinois and Indiana. Martin Oil Company sold out to J. D. Street of Maryland Heights, Missouri, in 1977 and the stations were rebranded with J. D. Street's Zephyr brand. In the 1980s, J. D. Street became a branded jobber for several majors and the Zephyr brand remains in use on only a small number of stations today.

****MARTIN OIL COMPANY — CARBONDALE, ILLINOIS**: Martin Oil Company operated stations in the south central states from the 1930s until it was purchased in 1977 by J.D. Street.

MARTIN'S ALLWEATHER GASOLINE 6 1930s 15" Metal LP RED $600-900
White globe face with blue "MARTIN'S" arched around top, "GASOLINE" around bottom with red "ALLWEATHER" across center.

MARTIN'S SUPER REGULAR w/ tank car 5 1940-1955 13.5" Glass N $1600-2400
White globe face with repeating red/white/blue outline rings. Large red/white outlined red railroad tank car across center of globe face with blue outlined white "SUPER" on tank car. Old English red outlined blue "Martin's" arched around top, red outlined blue block "REGULAR" around bottom.

MARTIN'S PREMIUM ETHYL (EC) 4 1942-1955 13.5" Glass N $250-450
Red/white/blue outline rings around blue outer area of globe. Outline rings around white center circle. Large Ethyl (EC) logo in center with white "MARTIN" arched around top, "PREMIUM" around bottom.

MARTIN'S SAFETY KEROSENE 5 1940-1955 13.5" Glass N $400-650
White globe face with repeating red/white/blue outline rings. Old English blue "Martin's" arched around top, "Kerosene" around bottom with blue outlined script red "Safety" diagonal across center of globe face.

MARTIN SUPER REGULAR 5 1955-1970 13.5" Glass N $250-450
White globe face with gray lines, forming a large "M", screened in the background. Series of alternating blue and red lines across lower globe face bent in a right angle to a point at the bottom of the globe face. Large blue "MARTIN" over red script "Super" over blue "REGULAR" superimposed over screened "M" on globe face.

MARTIN PREMIUM ETHYL (EC) 5 1955-1970 13.5" Glass N $250-450
Similar to MARTIN SUPER REGULAR, above, with lines at bottom notched for black outlined circle and Ethyl (EC) logo. Blue "MARTIN" over red script "Premium" on upper globe face.

****J.D. STREET and CO. — MARYLAND HEIGHTS, MISSOURI**: Founded in 1884 as a lubricants distributor, J.D. Street entered the retail petroleum market with the introduction of the Zephyr gasoline brand in 1939. J.D. Street purchased Martin Oil Company in 1977 and converted its stations to the Zephyr brand. Currently J.D. Street is serving as a jobber for several major brands as well as operating a few remaining Zephyr brand stations.

ZEPHYR 2 1939-1954 Gill $250-400
White globe face with red lines forming white diagonal band across globe face. Red increasing width spiral around globe face. Red "ZEPHYR" with speed lines and blue dropshade on diagonal band.

ZEPHYR ETHYL 3 1939-1954 Gill $250-425
White globe face with red lines forming white diagonal band across globe face. Red increasing width spiral around globe face. Red "ZEPHYR" with speed lines and blue dropshade on diagonal band. Black outlined white circle with Ethyl (EGC) logo offset to right below band.

ZEPHYR HIGH OCTANE 3 1954-1960 Oval Capco $225-325
Blue globe face with blue/white outline ring. Red outlined white diagonal band across center. Red "ZEPHYR" with speed lines and blue dropshade on diagonal band. White "HIGH OCTANE" in upper blue area, "GASOLINE" below.

ZEPHYR ETHYL 3 1954-1960 Oval Capco $225-325
Blue globe face with blue/white outline ring. Red outlined white diagonal band across center. Red "ZEPHYR" with speed lines and blue dropshade on diagonal band. White "SUPREME" in upper blue area, white circle with Ethyl (EC) logo below.

ZEPHYR 2 1960-1975 Oval Capco $200-300
Much lighter blue oval with red "ZEPHYR" on white diagonal band. No lettering in blue area.

****MARTIN OIL SERVICE — ALSIP / BLUE ISLAND, ILLINOIS:** Separated from Martin Oil Company in 1957, Martin Oil Service was marketing through over 200 stations coast-to-coast by 1965 and, following a significant downsizing during the gas shortages of the 1970s, was continuing to operate nearly 50 stations in Illinois and Indiana when it was purchased by Emro in 1994. The Martin name remains in use at many of the stations.

MARTIN SUPER REGULAR 3 1955-1962 13.5" Glass W $250-450
Red/white/blue repeating outline rings around outer red area. Outline rings around white center circle. Red "MARTIN across upper globe face, extending into outer red area. Blue "Super" diagonal over red "Regular" on lower globe face.

MARTIN XTRA SPECIAL 4 1955-1962 13.5" Glass W $250-450
Red/white/blue repeating outline rings around white globe face. Blue "MARTIN" arched around top, "SPECIAL" horizontal across bottom with blue outlined red "XTRA" diagonal across globe face. Outline rings around "X" in "XTRA".

MARTIN PURPLE MARTIN 6 1955-1962 13.5" Glass W NL
White globe face with blue "PURPLE" arched around top. Red/yellow/red line around lower globe face. Small blue bird under arch of "PURPLE" above red "MARTIN" over yellow script "The Finest" over red "ETHYL" over yellow script "of all". Small Ethyl(EC) logo in circle at end of "of all".

MARTIN #1 DIESEL 6 1955-1962 13.5" Glass W NL
Colors cannot be determined. Dark outline ring around pale colored globe face and around center circle. Large "#1" in center circle with dark colored "MARTIN" arched around top, "DIESEL" around bottom. Small "NO" to left of center circle, "FUEL" to right.

MARTIN #2 DIESEL 6 1955-1962 13.5" Glass W NL
Colors cannot be determined. Dark outline ring around pale colored globe face and around center circle. Large "#2" in center circle with dark colored "MARTIN" arched around top, "DIESEL" around bottom. Small "NO" to left of center circle, "FUEL" to right.

MARTIN SUPER REGULAR 3 1962-1970 Capco $200-300
White globe face with parallel red lines across upper globe face. Red outlined Martin blue "M" logo superimposed over lines. Large blue "SUPER" over smaller red "REGULAR" on lower globe face.

MARTIN EXTRA SPECIAL ETHYL 3 1962-1970 Capco $250-350
White globe face with parallel blue lines across upper globe face. Blue outlined Martin blue "M" logo superimposed over lines. Large blue outlined red "XTRA SPECIAL" arched below logo over red "ETHYL" and small Ethyl (EC) logo.

MARTIN PURPLE MARTIN ETHYL 4 1950s 13.5" Glass W $400-750

MARTIN PURPLE MARTIN ETHYL 3 1962-1970 Capco $350-600
White globe face with parallel purple lines across upper globe face. Purple outlined Martin blue "M" logo superimposed over lines. Small purple bird just below logo with purple "PURPLE" to left and "MARTIN" to right of bird. Large red "ETHYL" and small Ethyl (EC) logo on lower globe face.

MARTIN DIESEL #2 4 1962-1970 Capco $200-300
White globe face with parallel green lines across upper globe face. Green outlined Martin blue "M" logo superimposed over lines. Green "DIESEL" over large red "2" with green "NO." to left and "H.I." to right on lower globe face.

MARTIN DIESEL #1 4 1962-1970 Capco $200-300
White globe face with parallel green lines across upper globe face. Green outlined Martin blue "M" logo superimposed over lines. Green "DIESEL" over large red "1" with green "NO." to left and "H.I." to right on lower globe face.

MARTIN RANGE OIL 5 1962-1970 13.5" Glass $300-550
White globe face with parallel green lines across upper globe face. Green outlined Martin blue "M" logo superimposed over lines. Green "RANGE/OIL" on lower globe face.

MIDWEST & AFFILIATES

****MIDWEST OIL CO. — MINNEAPOLIS, MINNESOTA:** Dating from 1919, Midwest was a large upper Midwest independent with offices in Minneapolis, Sioux Falls and Fargo. The company operated under the Midwest, Ace High, Ace, and Avio brands. In the postwar era Midwest Oil concentrated its brand image behind the trademarks "Ace" and "Ace High." The company purchased the retail service stations of Minneapolis based H.K. Stahl & Company in 1958. Around that time, the company rebranded from Ace High to Midwest. Midwest Oil operated stations under the Midwest brand as late as the 1980s and then became a branded Sinclair jobber. Today the firm operates a number of Sinclair stations from a base in Sioux Falls, South Dakota.

MIDWEST ACE HIGH (w/airplane) 5 1925-1930 15" Metal $4500-7000
White globe face with red "MIDWEST OIL COMPANY" arched around top, "GASOLINE" around bottom and blue/gray biplane in center. Blue "ACE" above plane, "HIGH" below.

MIDWEST GASOLINE WITH ETHYL 5 1925-1930 15" Metal $600-900
White globe face with large Ethyl (EGC) logo in center. Dark color "MIDWEST" arched around top, "GASOLINE" around bottom. From an old photo.

MIDWEST AVIO GASOLINE 5 1920s 15" Metal $800-1250
White globe face with red "MIDWEST OIL COMPANY" arched around top, "GASOLINE" around bottom and blue "AVIO" over blue outlined flat triangle in center. Blue "QUALITY" inside triangle outline.

ACE HIGH GASOLINE 5 1940s 13.5" Glass NL
Blue outlined white globe face with blue outlined red triangles projecting in at upper right and lower left. Blue "ACE/HIGH" offset around triangle points. Red "GASOLINE" arched around bottom of globe face with blue drawing of airplane at top.

ACE HIGH COMPRESSION GASOLINE 6 1930s 13.5" Glass $450-700
White globe face with blue outline ring and blue/white outlined blue triangle offset to right of globe face. Large red "ACE" above triangle, "GASOLINE" arched below with white "HIGH/COMPRESSION" on triangle.

AVIO QUALITY GASOLINE 4 1946-1958 13.5" Glass $300-500
White globe face with blue outline ring. Large blue "AVIO" over blue outlined flat triangle in center of globe face. Blue "QUALITY" inside triangle. Red "GASOLINE" arched around bottom.

MOCO 6 1940s 13.5" Glass $200-350
White globe face with unusual "MOCO" logotype across globe face.

ACE HIGH GASOLINE 5 1940s 13.5" Glass $400-650

ACE HIGH GASOLINE 5 1940s Gill $400-650
Red globe face with black outline ring and black ring around white center circle. Black "ACE" over "HIGH" in center circle. White "GASOLINE" arched around bottom.

ACE HIGH ETHYL w/attendant 6 1940s 13.5" Glass NL
White globe face with black outline ring. Large red outlined white circle with "ACE/HIGH" in circle. This circle logo forms the face of an attendant with a large red attendants hat with an ace logo on hat, red arm and hand saluting and red legs. Small half circle at bottom of globe face with white "ETHYL" lettering on half circle.

MIDWEST (points left) 4 1958-1970 Oval Capco $200-350
White globe with large red/white/blue "MIDWEST" arrow logo covering globe face, pointing to the left.

MIDWEST (points right) 5 1958-1970 Oval Capco $225-375
White globe with large red/white/blue "MIDWEST" arrow logo covering globe face, pointing to the right.

****H.K. STAHL & CO.—ST. PAUL, MINNESOTA:** Founded in Minneapolis in 1888, H.K. Stahl sold motor fuels under the Powerflash and Trophy trademarks for many years. In 1958, Midwest Oil Company purchased the retail marketing of H.K. Stahl, and it continues to produce lubricants well into the 1990s.

POWERFLASH INDEPENDENT GASOLINE 5 1930s 15" Metal HP Red $1600-2200
White globe face with blue band across center. Blue "INDEPENDENT" arched around top, "GASOLINE" around bottom with white script "Powerflash" in center band. Blue piston below band with red flash extending above blue band.

POWERFLASH INDEPENDENT GASOLINE 5 1930s Gill $700-1000
White globe face with red outline runing and red band across center. Blue "INDEPENDENT" arched around top, "GASOLINE" around bottom with white script "Powerflash" in center band.

TROPHY - OUR PREMIUM 4 1935-1958 Gill $450-750
White globe face with blue outline ring. Blue outlined white trophy in center with blue "TROPHY" in bow-tie lettering across center. Blue "OUR PREMIUM" below "TROPHY".

TROPHY POWER FLASH 5 1950-1958 Gill $500-850
White globe face with blue outline ring and blue band across bottom. Red "POWERFLASH" with speed lines across upper globe face with small red trophy just below center. Red outlined white "TROPHY" in bow-tie lettering in lower blue area.

MOBIL & AFFILIATES

MOBIL OIL CORPORATION — FAIRFAX, VIRGINIA:

****STANDARD OIL COMPANY of NEW YORK/SOCONY — NEW YORK, NEW YORK:** Founded in 1882 as the administrative division of the Standard Oil Trust, Socony eventually became involved in refining, and later marketing, in New York and New England. With the breakup of Standard in 1911, Socony was assigned its refining and marketing properties but left without crude oil production. Exploration began, but most of the crude oil production came through acquisition of Magnolia Petroleum in 1918 and General Petroleum in 1926. Socony expanded its marketing through Magnolia and General (and White Eagle) as well, marketing from coast-to-coast by 1930.

In 1931 Socony merged with another Standard affiliate, Vacuum Oil. The new corporate entity was named Socony-Vacuum Oil Company. Socony-Vacuum mated Socony's shield and flying horse with Vacuum' s "MOBILGAS" trademark and expanded the Mobilgas brand to all of the affiliated companies by the end of World War II. In 1955 the company was renamed Socony-Mobil and in 1966 Mobil Oil Corporation. See the SOCONY-VACUUM section below for more information.

>>NOTE: The Socony "SONY & Shield" logo will be referred to in the following descriptions as a blue circle with a large white/red outlined five-point shield positioned to lower edge of circle. A red line crosses the shield about 1/3 of the way down with red "SOCONY" above the line and red "MOTOR GASOLINE" below the line. Small blue "STANDARD OIL CO. OF NEW YORK" around the bottom edge of the shield. Distorted white "S" and "O" in the blue area above the shield and elongated "N" to left and "Y" to right of shield. Refer to this description from the globe text below.

SOCONY (sphere) 7 1915-1918 OPB NL
White spherical globe with blue "SOCONY" across globe face.

SOCONY MOTOR GASOLINE 5 1915-1920 OPE W $2200-3500
White globe with blue/white outlined SONY & Shield logo as described above on globe face.

SOCONY MOTOR GASOLINE (milk glass lens) 4 1915-1926 16.5" Metal HP Red $1200-1800

SOCONY MOTOR GASOLINE (milk glass lens) 4 1915-1926 15" Metal HP Red $1200-1800

SOCONY MOTOR GASOLINE (milk glass lens) 5 1915-1926 14" Metal HP Red $1700-2200

SOCONY MOTOR GASOLINE (milk glass lens) 5 1915-1926 13.675" Metal HP Red $1800-2400
Milk glass globe face with white outline ring around SONY & Shield logo as described above.

SOCONY MOTOR GASOLINE 5 1915-1926 16.5" Metal HP Red $1800-2400

SOCONY MOTOR GASOLINE 5 1915-1926 15" Metal HP Red $1800-2400
Globe has white outline ring around SONY & Shield as described above.

SOCONY MOTOR GASOLINE (flat porcelain lens) 5 1915-192615" Metal HP Red $2000-2600

SOCONY MOTOR GASOLINE (curved porcelain lens) 6 1915-1926 15" Metal HP Red $2000-2600

SOCONY MOTOR GASOLINE (porcelain lens) 5 1915-1926 14" Metal HP Red $2000-2500

SOCONY MOTOR GASOLINE (cast lens) 6 1915-1926 16.5" Metal HP Red NL

SOCONY MOTOR GASOLINE (cast lens) 5 1915-1926 15" Metal HP Red $2000-2600
White globe face with blue/white outlined SONY & Shield logo as described above on globe face.

SOCONY 3 1926-1934 16.5" Metal HP Red $300-550

SOCONY 3 1926-1934 15" Metal HP Red $300-550
White globe face with red "SOCONY" with blue dropshade across center.

SOCONY red lettering w/ white lines 3 1926-1934 16.5" Metal HP Red $300-550

SOCONY red lettering w/ white lines 3 1926-1934 15" Metal HP Red $300-550
White globe face with red "SOCONY" with blue dropshade across center - red lettering with white lines.

SOCONY SPECIAL milk glass lens 4 1922-1926 16.5" Metal HP Red $900-1300

SOCONY SPECIAL milk glass lens 4 1922-1926 15" Metal HP Red $900-1400
White milk glass globe face with red "SOCONY" arched around top, "GASOLINE" around bottom with blue "SPECIAL" across center.

SOCONY SPECIAL (normal painted

lens) 5 1922-1926 15" Metal HP Red $700-1000
White milk glass globe face with red "SOCONY" arched around top, "GASOLINE" around bottom with blue "SPECIAL" across center.

SOCONY ETHYL (EGC) 3 1926-1934 16.5" Metal HP Red $325-575

SOCONY ETHYL (EGC) 3 1926-1934 15" Metal HP Red $325-575
White globe face with red "SOCONY" with blue dropshade across center. Black outlined white circle at top of globe face with small Ethyl (EGC) logo. Red "ETHYL" with blue dropshade below "SOCONY".

SOCONY ETHYL (w/lines) 3 1926-1934 16.5" Metal HP Red $325-575

SOCONY ETHYL (w/lines) 3 1926-1934 15" Metal HP Red $325-575
White globe face with blue outline ring and red "SOCONY" with blue dropshade across center. Black outlined white circle at top of globe face with small Ethyl (EGC) logo. Red "ETHYL" with blue dropshade below "SOCONY". Blue line under "ETHYL".

SOCONY AVIATION GASOLINE 6 1920s 16.5" Metal HP Red $1600-2200
White milk glass globe face with red "SOCONY" arched around top, "GASOLINE" around bottom with blue "AVIATION" across center.

SOCONY MOTOR OILS sm w/4" base 6 1915-1920 OPE N $1300-1800
Small one piece globe with 4" base and blue "SOCONY" over "MOTOR OILS" on face.

SOCONY MOTOR OILS 5 1915-1920 OPE RW $1200-1750

SOCONY MOTOR OILS 5 1915-1920 OPB W $1000-1400
One piece globe with blue "SOCONY" over "MOTOR OILS" on face.

SOCONY BURNING OIL 4 1934-1940 16.5" Metal HP Red $1250-1750

SOCONY BURNING OIL 4 1934-1940 15" Metal HP Red $1250-1750
White globe face with small flying red horse at top. Blue "SOCONY BURNING OIL" on lower globe face.

STANDARD KEROSENE 4 1934-1940 15" Metal HP Red $750-1000
White globe face with small red flying horse at top. Blue "STANDARD" over "KEROSENE" on lower globe face.

****MAGNOLIA PETROLEUM — DALLAS, TEXAS:** A Texas oil producer/refiner/marketer dating from 1911, Magnolia sold a 45% interest to Socony in 1918. Socony completed the purchase in 1925. The Magnolia brand name was replaced by "MOBILGAS" about 1934, after the Socony-Vacuum merger. Magnolia remained a separate, named subsidiary primarily involved in pipeline transportation until 1960.

MAGNOLIA GASOLINE 7 1915-1920 OPE NL
White globe with red "MAGNOLIA" arched around top, "GASOLINE" around bottom of globe face.

MAGNOLIA GASOLINE 6 1920-1922 16.5" Metal HP Red NL
White globe face with NO outline ring. Black outlined red "MAGNOLIA" arched around top, "GASOLINE" around bottom with detailed brown and gold flower magnolia blossom in center.

MAGNOLIA GASOLINE w/ seven-leaf flower 4 1920-1926 16.5" Metal HP Red $1700-2200
White globe face with red outline ring. Blue outlined red "MAGNOLIA" arched around top, "GASOLINE" around bottom with detailed SEVEN green leaf / white flower magnolia blossom in center.

MAGNOLIA GASOLINE w/ five-leaf flower 5 1920-1926 16.5" Metal HP Red $1700-2200
White globe face with red outline ring. Blue outlined red "MAGNOLIA" arched around top, "GASOLINE" around bottom with detailed FIVE green leaf / white flower magnolia blossom in center.

MAGNOLIA MAXIMUM MILEAGE 3 1926-1934 16.5" Metal HP Red $1700-2200
White globe face with blue outline ring. Blue outlined red "MAGNOLIA" arched around top, "GASOLINE" around bottom with small detailed green leaf/white flower magnolia blossom in center. Blue "MAXIMUM" to left of flower, "MILEAGE" to right.

MAGNOLIA ANTI-KNOCK 5 1922-1926 16.5" Metal HP Red $1600-2100
White globe face with blue outline ring. Wide dark blue band across center with large white "ANTI-KNOCK" across center band. Red "MAGNOLIA" arched around top, "GASOLINE" around bottom with detailed flower above band on upper globe face.

MAGNOLIA ETHYL (EGC) 4 1926-1934 16.5" Metal HP Red $750-1000
Blue outline ring around red globe face. White/blue outline around white center circle. White "MAGNOLIA" arched around top, "GASOLINE" around bottom with small Ethyl (EGC) logo in center.

****GENERAL PETROLEUM CORPORATION — VERNON, CALIFORNIA:** A pioneer California oil producer and refiner, General was founded in 1910. In 1926 Socony purchased General Petroleum and maintained the General brands until the end of World War II, gradually replacing them with Mobilgas. As with Magnolia, General remained a separate, named subsidiary until 1960.

GENERAL GASOLINE 4 1928-1934 15" Metal $700-1000
White globe face with red line above and below center. Script blue "GENERAL" above blue "GASOLINE" between red lines.

GENERAL VIOLET RAY GASOLINE 5 1928-1932 15" Metal $3000-4000
White globe face with large green diamond covering most of globe face. White band across center of diamond with red outlined purple "VIOLET RAY" on band. White "GENERAL" and purple lightning bolt in upper area of green diamond. Red "ANTI-KNOCK" in lower area of green diamond above wide white band. Red "GASOLINE" extending beyond edge of diamond on lower band.

GENERAL ETHYL 3 1926-1934 15" Metal $2000-2500
Red upper globe face with green lower face. Large red outlined white triangular shield covers most of globe face with red/white/green checkerboard pattern at top of shield, small Ethyl (EGC) logo on lower shield.

GENERAL PETROLEUM ETHYL GASOLINE6 1942 13.5" Glass NL
White globe face with blue outline ring. Red "GENERAL" arched around top, "PETROLEUM" around bottom with small Ethyl (EC) logo in center and small blue "GASOLINE" below logo.

****WHITE EAGLE OIL and REFINING — KANSAS CITY, MISSOURI:** White Eagle was a mid-western producer/refiner/marketer with gas stations in 11 midwestern states. Chief White Eagle was a leader of the Ponca Indians in Oklahoma. White Eagle was first used as a corporate name in 1916 when L.L. Marcell founded the White Eagle Petroleum Company. That same year the new company built a refinery at Augusta, Kansas, having a capacity of 2000 barrels per day.

In 1919 the White Eagle Petroleum Company was reorganized and the name changed to the White Eagle Oil and Refining Company. This reorganization was in reality a merger of six independent oil companies. Within a few years, the new company enjoyed second position in the distribution of petroleum products in the Midwest. A second refinery was built at Casper, Wyoming, and another was purchased at Fort Worth, Texas.

The poised White Eagle was adopted as an emblem because of its dignity and general significance, particularly in their territory. Not to mention that the White Eagle was also in keeping with the company name. As a service station identification, this emblem—the big iron eagle in front of the station and the illuminated glass eagles on the pumps—was very effective.

In January 1930, the White Eagle Oil and Refining Company was purchased by the Standard Oil Company of New York (Socony) at which time the name was changed to the White Eagle Oil Corporation. When Socony and Vacuum merged in 1931, White Eagle Oil Corporation became a subsidiary company. In 1935 the White Eagle Oil Corporation was dissolved and the operation became the White Eagle division of Socony-Vacuum Oil Company, Inc.

WHITE EAGLE GASOLINE 5 1918-1920 OPE W $7500-10000

White globe with black and white perched eagle in center of globe face. Red "WHITE EAGLE" arched around top, "GASOLINE" around bottom.

WHITE EAGLE
GASOLINE & OILS5 1920-1925 15" Metal $1700-2200
White globe face with black octagon outline. Black outlined red circle in center of black/white outlined black band across center. White "WHITE EAGLE" in black band, white "GASOLINE" arched around top of red circle, "AND OILS" arched around bottom.

WHITE EAGLE
blunt nose 3 1924-1932 Eagle $1300-1600
Cast white eagle with very little detail and blunt nose.

WHITE EAGLE
slit throat 3 1924-1932 Eagle $1300-1600
Same as above with casting line around throat.

WHITE EAGLE
pointed nose,
some detail 4 1932-1937 Eagle $1600-2100
Cast white eagle with more feather detail and pointed nose.

WHITE EAGLE
pointed nose,
good detail 4 1925-1927 Eagle $1800-2400
Cast white eagle with very good feather detail and pointed nose.

****MANITOU OIL COMPANY — MANITOU SPRINGS, COLORADO**: Apparently a White Eagle jobber. No other information available.

MANITOU WHITE
HI TEST 6 1920-1925 15" Metal $1700-2200
White globe face with black octagon outline. Black outlined red circle in center of black/white outlined black band across center. White "wHITE EAGLE" in black band, white "MANITOU" arched around top of red circle, "HIGH TEST" arched around bottom.

****VACUUM OIL COMPANY — ROCHESTER, NEW YORK**: Founded as a lubricants manufacturer in 1866, Vacuum had developed an extensive line of specialty lubricants when Standard Oil purchased the company in 1879. During the Standard era, Vacuum's lubricants, under the Gargoyle and later Gargoyle Mobiloil brands, were sold by the various Standard marketing firms. With the breakup of Standard in 1911, Vacuum was left to market its oils through independent channels worldwide. They became involved with gasoline marketing, under the "MOBILGAS" brand, in the late 1920s and expanded the gasoline venture with the purchase of Lubrite Refining Co. of St. Louis, Missouri, in 1929. The following year Vacuum purchased Wadhams Oil Corporation of Milwaukee and the Vacuum "MOBILGAS" name was added to Wadhams extensive product line. Also in 1930 Vacuum purchased White Star Refining of Detroit and added over 1500 outlets for Mobilgas. In 1931 Socony and Vacuum merged to form Socony Vacuum Oil Company. After a two year reimaging program, the Mobilgas brand name became the company's primary brand, although several of the regional marketers retained their own identities until after World War II.

MOBILGAS -
(progressive red &
white lines) 4 1928-1930 16.5" Metal HP Red $800-1200

MOBILGAS -
(progressive red &
white lines) 4 1928-1930 15" Metal HP Red $800-1200

MOBILGAS -
(progressive red &
white lines) 6 1928-1930 13.5" Glass W NL
Red globe face with series of increasing width white horizontal lines above and below white center band. Black "MOBILGAS" across white center band.

MOBILGAS -
(milk glass) 6 1928-1930 15" Metal HP Red NL
Milk glass red globe face with series of increasing width white horizontal lines above and below white center band. Black "MOBILGAS" across white center band.

MOBILGAS -
(progressive red &
white lines & horse) 6 1931-1934 15" Metal HP Red NL
Red globe face with series of increasing width white horizontal lines above and below white center band. Black "MOBILGAS" across white center band. Small red pegasus logo at top of globe face, possibly a decal.

MOBILGAS - blue
lettering w/
red line 2 1932-1933 16.5" Metal L/H Red $300-550

MOBILGAS - blue
lettering w/
red line 2 1932-1933 15" Metal L/H Red $300-550

MOBILGAS - blue
lettering w/
red line 3 1932-1933 13.5" Glass W $250-375
White globe face with thin red line positioned above and below center so as to form a white band across the center of globe face. Dark blue "MOBILGAS" in center band.

MOBILGAS -(milk
glass)blue lettering
w/ red line 6 1932-1933 15" Metal L/H Red NL
Cast face with white globe face with thin red line positioned above and below center so as to form a white band across the center of globe face. Dark blue "MOBILGAS" in center band.

MOBILGAS - green
lettering w/red line 5 1932-1933 15" Metal L/H Red
$500-800
White globe face with thin red line positioned above and below center so as to form a white band across the center of globe face. Bright green "MOBILGAS" in center band.

MOBILGAS - blue
lettering w/ red line
& horse 6 1933-1934 15" Metal L/H Red NL
White globe face with thin red line positioned above and below center so as to form a white band across the center of globe face. Dark blue "MOBILGAS" in center band. Small red flying horse center above upper red line.

MOBILGAS ETHYL 3 1928-1933 16.5" Metal L/H Red $325-575

MOBILGAS ETHYL 3 1928-1933 15" Metal L/H Red $325-575
White globe face with small Ethyl (EGC) logo at top of globe face. Black "MOBILGAS" across center above black "ETHYL".

MOBILGAS ETHYL 3 1930-1933 16.5" Metal L/H Red $325-575

MOBILGAS ETHYL 3 1930-1933 15" Metal L/H Red $325-575

MOBILGAS ETHYL
(red ethyl) 3 1930-1933 13.5" Glass W $250-375
White globe face with small Ethyl (EGC) logo at top of globe face. Dark blue "MOBILGAS" across center above red "ETHYL".

MOBILGAS ETHYL
(black ethyl) 3 1930-1933 13.5" Glass W $250-375
White globe face with small Ethyl (EGC) logo at top of globe face. Dark blue "MOBILGAS" across center above black "ETHYL".

GARGOYLE MOBIL
OIL (large) 3 1920-1935 OPC Oval $1500-2000
Oval one piece cast globe with red outline ring. Detailed red and black gargoyle logo with black "GARGOYLE" arched over top of logo. Black "MOBILOIL" across lower globe face below logo.

GARGOYLE MOBILOIL
(small; raised
border) 3 1920-1935 OPC Oval $1500-2000
Smaller globe similar to above with raised red outline ring.

GARGOYLE MOBILOIL
(small; recessed
border) 3 1920-1935 OPC Oval $1500-2000
Smaller globe similar to above with red recessed outline ring.

GARGOYLE
MOBILOIL
PORCELAIN
SIGN GLOBE 3 1920-1935 OPC Oval NL
Oval porcelain sign globe with Gargoyle Mobiloil logo similar to that used on glass globes. Two sided porcelain, with globe type base.

GARGOYLE
MOBILOIL 6 1930s 16.5" Metal HP Red $1800-2500
White globe face with black "Mobiloil" logotype across lower center of globe face. Red and black detailed gargoyle logo on upper center of globe face with black "GARGOYLE" arched across top. Lettering of "GARGOYLE" has red line arching through center of lettering.

****LUBRITE REFINING CO.—ST. LOUIS, MISSOURI**: Lubrite was an early refiner/marketer, a pioneer in the development of service stations in the

Midwest. Lubrite was purchased by Vacuum Oil Company in 1929 to greatly expand their growing presence with the Mobilgas brand.

LUBRITE 6 1925-1929 15" Metal HP Red $600-900
White globe face with blue outlined red "Lubrite" logotype across center of globe face.

LUBRITE SKY HIGH 6 1925-1929 15" Metal HP Red $650-950
White globe face with red "Lubrite" logotype across upper center of globe face. Large tapering blue "SKY-HY" with speed lines across lower globe face.

****WADHAMS OIL CORPORATION—MILWAUKEE, WISCONSIN:** Founded in the 1870s as a lubricants manufacturer, Wadhams was a pioneer in gasoline marketing, operating Wadhams stations throughout Wisconsin prior to 1920. Wadhams was briefly affiliated with the Independent Oil Men of America in about 1925 and expanded with the 1929 purchase of Milwaukee competitor Bartles-Maguire Oil Company. Noted for its Chinese pagoda style stations, Wadhams had developed a first-class gasoline marketing organization when the company was purchased by Vacuum in 1930. Vacuum's Mobilgas trademark was added to the Wadhams brands and finally replaced the Wadhams name in about 1939.

"W" LAMP POST GLOBE 5 1910s OPC $900-1200
Unusual vase shaped white glass lamp post globe with a large red "W" on two opposite sides of view.

WADHAMS TRUE GASOLINE CHIMNEY CAPPED 6 1910s OPE NL
White chimney capped globe with red "WADHAMS" (large "W") across center of globe face. Small black "TRUE GASOLINE" below "WADHAMS".

WADHAMS w/can 6 1910s OPE NL
White globe with black "WADHAMS" and red gasoline can on globe face.

WADHAMS TRUE GASOLINE 6 1910s OPE NL
White globe with red "WADHAMS (large "W") across center of globe face. Small black "TRUE GASOLINE" below "WADHAMS".

WADHAMS 370 w/can 5 1920-1922 15" Metal HP NL
White globe face with black "WADHAMS" across top. Red "370" with black dropshade arched across center. Detailed red and black gasoline carry can at bottom of globe face.

WADHAMS SPECIALO 5 1920-1922 15" Metal HP $1700-2250
White globe face with black "WADHAMS" across top. Red "SpeciaLo" with black dropshade arched across center. Detailed red and black gasoline carry can at bottom of globe face.

WADHAMS 370 STEEL FACE 5 1922-1925 15" Metal HP $900-1250
White reflective steel face with black "WADHAMS" across top. Red "370" with black dropshade across center over red circle. Black "W" on red band below.

WADHAMS 370 5 1922-1925 15" Metal HP $800-1100
White globe face with black "WADHAMS" across top. Red "370" with black dropshade across center over red circle. Black "W" on red band below.

WADHAMS ANTI-KNOCK STEEL FACE 4 1922-1925 15" Metal HP $900-1250
White reflective steel face with black "WADHAMS" across top. Red "ANTI-KNOCK" with black dropshade arched across center over red bar. Black "W" superimposed over red bar.

WADHAMS RED HAT 9 1925-1926 15" Metal HP NL
White globe with black outline ring around globe face. Large red top hat with black hatband in center of globe. Three white stars on black hatband. Small black triangle/red thunderbird IOMA logo below hat. Black "WADHAMS" arched around top, "GASOLINE" around bottom with "RED" to left of hat and "HAT" to right.

WADHAMS "INDEPENDENT" IOMA 5 1926-1929 15" Metal HP $1200-1600
White globe face with red outline ring. Red semi-circle at top of globe face. Black "Wadhams" across center with black triangle/red thunderbird IOMA logo at bottom of globe face.

WADHAMS ETHYL (EGC) 4 1926-1932 15" Metal HP $750-950
White globe face with red outline ring. Yellow "bar" with black "W" at top of globe face. Black "Wadhams " across center with large black and red outlined white circle on lower globe face. Large Ethyl (EGC) logo in circle.

WADHAMS MOBILGAS 5 1929-1932 15" Metal HP $900-1300
Red globe face with series of increasing width white horizontal lines above and below white center band. Black "MOBILGAS" across white center band. Black "Wadhams" in upper red area above "Mobilgas".

WADAMS METRO "W" 4 1929-1932 15" Metal HP $600-900
White globe face with yellow outline ring. Black "Wadhams " across upper globe face above large black "METRO". Red "bar" with black "W" on lower globe face.

WADHAMS MOBILGAS w/shield 7 1932-1935 13.5" Glass W NL
Red globe face with series of increasing width white horizontal lines above and below white center band. Black "MOBILGAS" across white center band. Black "Wadhams" in upper red area above "Mobilgas". Small black outlined white shield emblem just below "Mobilgas" with small flying red horse in shield.

WADAMS METRO "W" 3 1932-1940 13.5" Glass W $300-500
White globe face with yellow outline ring. Black "Wadhams" across upper globe face above large black "METRO". Red "bar" with black "W" on lower globe face.

WADHAMS GIANT 5 1932-1940 13.5" Glass W $300-550
White globe face with red outline ring. Black "Wadhams" across upper globe face with red "GIANT" with black dropshade arched across center. Red "bar" with black "W" on lower globe face.

WADHAMS ETHYL 3 1932-1940 13.5" Glass W $300-550
White globe face with red outline ring. Red "bar" with black "W" at top of globe face over black "Wadhams". Black/yellow/black outlined white circle on lower globe face with large Ethyl (EGC) logo on circle.

****BARTLES-McGUIRE OIL CO. — MILWAUKEE, WISCONSIN:** Bartles-McGuire was a large Wisconsin marketer that merged with Wadhams in 1929, shortly before Vacuum purchased Wadhams.

BARTLES WHITE EAGLE GASOLINE 7 1920s OPE Oval NL
White oval one piece etched globe with small eagle design in center. Red band across center with white "WHITE" to left of eagle and "EAGLE" to right. Black "BARTLES" arched around top, "GASOLINE" around bottom.

BARTLES BONDED GASOLINE 5 1921-1929 15" Metal $650-950
White globe face with large red "B" covering most of globe face. Red arrow diagonal through "B". White "BARTLES" in upper area of "B", "BONDED" in lower area with white "GASOLINE" on diagonal arrow.

****WHITE STAR REFINING CO. — DETROIT, MICHIGAN:** Also added to Vacuum's Mobilgas marketing was Detroit's White Star Refining. White Star operated over 1500 stations in Ohio, Indiana and Michigan when purchased by Vacuum in 1930. Mobilgas replaced the White Star brand almost immediately, but former White Star outlets used a special Mobilgas shield sign with a small blue circle and white star at the bottom until after World War II.

WHITE STAR GASOLINE 6 1918-1920 OPE W $2200-2800
White globe with blue globe face. Large white star in center with blue band across center of star. White "WHITE STAR/GASOLINE" on center band with white "WHITE STAR REFINING CO." arched around top and "A/QUALITY/PRODUCT" below star.

WHITE STAR ETCHED MILK GLASS FACES 5 1918-1920 15" Metal L/H $1800-2500

Blue globe face with large white star in center. Blue band across star with white "WHITE STAR/GASOLINE" on band. White "WHITE STAR REFINING CO." arched around top, "A/QUALITY/PRODUCT" below star.

WHITE STAR GASOLINE 6 1918-1920 15" Metal L/H $900-1250
Blue globe face. Large white star in center with blue band across center of star. White "WHITE STAR/GASOLINE" on center band with white "WHITE STAR REFINING CO." arched around top and "A/QUALITY/PRODUCT" below star.

STAROLINE GASOLINE IS BETTER 6 1918-1920 15" Metal L/H $900-1250
Blue globe face. Large white star in center with blue band across center of star. White script "Staroline" on center band with white "WHITE STAR REFINING CO." arched around top and "GASOLINE/IS BETTER" below star.

WHITE STAR GASOLINE 5 1918-1920 15" Metal L/H $900-1250
White globe face with blue circle in center. Blue "WHITE STAR" arched around top. "GASOLINE" around bottom. Large white star in blue center circle with blue logotype "Staroline" above "MOTOR/OILS" on star.

STAROLINE BETTER GASOLINE 4 1920-1922 15" Metal L/H $900-1250
Blue globe face with white outline ring. White script logotype "Staroline" arched across top above small "BETTER". White "GASOLINE" arched around bottom.

STAROLINE GASOLINE IS BETTER 4 1922-1926 15" Metal L/H $900-1250
Blue globe face with white outline ring. Small white outlined blue circle with white star at top. Script white "Staroline" across center with white "GASOLINE/IS BETTER" below.

WHITE STAR GASOLINE 4 1926-1930 15" Metal L/H $850-1200
Blue globe face with white outline around blue center circle. White star in center circle with white "WHITE STAR" arched around top and "GASOLINE" around bottom.

WHITE STAR ETHYL 4 1926-1930 15" Metal L/H $650-900
Blue globe face with white/blue outlined white center circle. Large Ethyl (EGC) logo in center circle. White "WHITE STAR" arched around top, "GASOLINE" around bottom.

STAROLINE AND WHITE STAR ETHYL 4 1926-1930 15" Metal L/H $750-1000
Blue globe face with small white circle at top. Small Ethyl (EGC) logo in circle. White script "Staroline" across center above small white "AND" above white "WHITE STAR". Yellow "ETHYL" below "WHITE STAR".

WHITE STAR GASOLINE 4 1930-1934 15" Metal L/H $900-1250
White globe face with red circle in center. White star in red center circle. Blue "WHITE STAR" arched around top, "GASOLINE" around bottom.

WHITE STAR GASOLINE 4 1930-1934 15" Metal L/H $900-1250
Red globe face with blue circle in center. White star in blue circle. White "WHITE STAR" arched around top, "GASOLINE" around bottom.

WHITE STAR STAROLINE MOTOR OILS (no star) 6 1930s 15" Metal L/H NL
Description not available.

****WHITE STAR OIL CO. of CANADA**: This firm was a Canadian affiliate of Detroit's White Star Refining Company.

STARLITE GASOLINE 6 1920s 15" Metal HP NL
Description not available.

WHITE STAR HI TEST 6 1920s 15" Metal HP $900-1250
Blue globe face with large white/red outlined white star in center. Red "HIGH/TEST" on star. White outlined black "WHITE STAR" arched around top of globe face, "OIL" to left of star, "CO." to right and "GAS" arched around bottom.

****INDEPENDENT OIL CO. of PENNSYLVANIA — ALTOONA, PENNSYLVANIA**: Purchased by Socony-Vacuum in 1934, Independent Oil Company was rebranded Mobilgas.

INDEPENDENT (W/3 MEN) 4 1925-1934 15" Metal HP $3500-5000
White globe face with black "INDEPENDENT" arched around top. Black outlined white circle on lower globe face with multi-color scene with colonial soldiers and drummer.

INDEPENDENT ETHYL (EGC) 6 1925-1934 15" Metal HP $400-600
White globe face with black "INDEPENDENT" arched around top. Black outlined white circle on lower globe face with Ethyl (EGC) logo with rays in circle.

INDEPENDENT GASOLINE (W/3 MEN) great detail 5 1925-1934 15" Metal HP $4000-6500
Yellow globe face with black outlined white circle in center. Multi-colored scene with colonial soldiers and drummer in center circle. Black "INDEPENDENT" arched around top, "GASOLINE" around bottom.

INDEPENDENT GASOLINE (W/3 MEN) less detail 5 1925-1934 15" Metal HP $3500-6000
Yellow globe face with black outlined white circle in center. Less detailed multi-colored scene with colonial soldiers and drummer in center circle. Black "INDEPENDENT" arched around top, "GASOLINE" around bottom.

****METRO — JAMESTOWN/OLEAN, NEW YORK**: Metro was a small New York marketer purchased by Socony-Vacuum in 1934. The Mobilgas brand was added at Metro stations (note the red/white/green "Mobilgas" globe in the Vacuum section) and Metro became a motor-grade product sold at Mobil stations in the various marketing territories until after World War II.

METRO w/diamond 6 1920s 15" Metal NL
Description not available.

METRO ETHYL 6 1920s 15" Metal NL
Description not available.

METRO 68-70 6 1920s 15" Metal NL
Description not available.

METRO GASOLINE/ green lines 4 1925-1933 15" Metal $400-650
White globe face with thin green line positioned above and below center. Large green "METRO" over smaller green "GASOLINE" between the lines across center of globe face.

METRO GAS GARGOYLE MOBILOIL 5 1933-1935 16.5" Metal $2200-2700

METRO GAS GARGOYLE MOBILOIL 5 1933-1935 15" Metal $2200-2700
White globe face with red/white/blue lines across center. Blue "Metro" arched over "Gas" on upper half, red and black gargoyle logo with black "GARGOYLE" with red stripe arched above over black "Mobiloil" on lower globe face.

METRO green/ green on white/ green 3 1935-1940 16.5" Metal $400-650

METRO green/ green on white/ green 3 1935-1940 15" Metal $400-650

METRO green/ green on white/ green 3 1935-1940 13.5" Glass W $250-400
Green globe face with white band across center. Large green "METRO" over smaller green "GASOLINE" on center band.

METRO red/green on white/red 2 1940-1955 15" Metal $400-650

METRO red/green

on white/red 2 1940-1955 13.5" Glass W NL

METRO red/green on white/red 2 1940-1955 Capco $175-275

Red globe face with white band across center. Red outlined green "Metro" across center on white band.

METRO red/black on white/red (GENERAL) 5 1930-1935 16.5" Metal $450-700

Red globe face with white band across center. Red outlined black "Metro" across center on white band.

****GILMORE OIL CO.—LOS ANGELES, CALIFORNIA:** Founded in 1900, Gilmore entered gasoline marketing in 1923. The company established an elaborate network of dealers in California, Oregon and Washington, totaling more than 3500 outlets at its peak in the late 1930s. In 1940 Socony-Vacuum purchased control of Gilmore and assigned Gilmore marketing to the General Petroleum subsidiary. Transition from the Gilmore brand to Mobilgas began in 1942 and was completed in 1945, just after World War II.

GILMORE "ROAR WITH GILMORE" 4 1925-1942 15" Metal HP Red $7000-9000

Yellow globe face with black outline ring. Large red and black jumping lion at top of face with black "ROAR" over black script "with" beside lion. Black outlined red "GILMORE" over black and white checkered flag below lion.

GILMORE BLU-GREEN GASOLINE 4 1925-1933 15" Metal HP Red $8000-11000

Yellow globe face with black outline ring and black outlines around white semi-circles in center. Large multi-color lion's head in center superimposed over semi-circles. Black outlined red "GILMORE" arched around top, "GASOLINE" around bottom with blue "BLU" to left of lion and "GREEN" to right.

GILMORE W/ETHYL (block letters) 5 1926-1942 15" Metal HP Red $2000-3500

Yellow globe face with black outline ring around white center circle. Small Ethyl (EGC) logo with red rays in center circle with black *BLOCK* "with" above. Black outlined red "GILMORE" arched around top, "GASOLINE" around bottom.

GILMORE W/ ETHYL (script letters) 5 1926-1942 15" Metal HP Red $2000-3500

Yellow globe face with black outline ring around white center circle. Small Ethyl (EGC) logo with red rays in center circle with black *SCRIPT* "with" above. Black outlined red "GILMORE" arched around top, "GASOLINE" around bottom.

GILMORE RED LION W/TETRAETHYL 4 1933-1942 15" Metal HP Red $8000-10000

Yellow globe face with black outline ring. Large red and black lion at top of globe face. Black outlined red "GILMORE/RED LION" over black script "plus" with black "TETRAETHYL" arched around bottom.

GILMORE FLEET W/FLAG 6 1933-1942 15" Metal HP Red NL

Yellow globe face with black outline ring. Black outlined red "GILMORE" across center over italic black "FLEET". Small black and white checkered flag at top of globe face.

****CLIPPER OIL COMPANY — SEATTLE, WASHINGTON:** Exact history unknown. Clipper Oil is believed to have been purchased by Mobil in the 1950s and continued as a secondary brand into the 1960s.

CLIPPER W/ AIRPLANE 4 1940s 15" Metal $3000-4500

White globe face with green/red airplane in center of globe face. Red "CLIPPER" with green dropshade & speed lines across lower globe face.

CLIPPER W/ AIRPLANE 4 1940s Oval Capco $1100-1600

White globe face with green/red airplane in center of globe face. Red "CLIPPER" with green dropshade & speed lines across lower globe face.

****MOBILGAS/SOCONY-VACUUM:** As is noted in the introduction, Socony and Vacuum merged in 1931. By 1934 it was decided to join the Socony shield and flying horse with Vacuum's Mobilgas to create a single brand name for all the affiliates. Several, including Magnolia and White Eagle, readily converted to the new Mobilgas identity. Others, including General and Wadhams, combined the Mobilgas brand with their own brand names, at least for the first few years. Conversion was nearly complete when World War II interrupted, and was completed as soon as possible after the war.

Please note however, that many manufacturers were involved in making the following globes and color variations—including some certain to exist but not listed—exist as well as typestyle and size variations. A complete listing would be nearly impossible, but two years of research has compiled the following list. Note also that there are no differences in value between black and blue lettering, so if your globe is not listed, the value can be determined by comparing with the same globe with the other colored lettering. The "Mobilgas" designations for products was changed to Mobil in 1962 and all of the older globes were replaced at that time.

MOBILGAS w/ horse black letters 4 1934-1962 16.5" Metal L/H Red $500-750

MOBILGAS w/ horse black letters 4 1934-1962 15" Metal L/H Red $500-750

MOBILGAS w/ horse black letters 4 1934-1962 13.5" Glass WN $300-500

White globe face with flying red horse at top of globe face. Black "Mobilgas" across lower globe face.

MOBILGAS w/ horse blue letters 2 1934-1962 16.5" Metal L/H Red $500-750

MOBILGAS w/ horse blue letters 2 1934-1962 15" Metal L/H Red $500-750

MOBILGAS w/ horse blue letters 2 1934-1962 13.5" Glass WN $300-500

MOBILGAS w/ horse blue letters 4 1934-1962 Gill $350-550

MOBILGAS w/ horse blue letters 5 1934-1962 Balcrank $400-650

MOBILGAS w/ horse blue letters 2 1955-1962 Capco $200-350

White globe face with flying red horse at top of globe face. Blue "Mobilgas" across lower globe face.

MOBILGAS ETHYL w/ horse 4 1934-1936 Gill $375-575

MOBILGAS ETHYL w/ horse 4 1934-1936 15" Metal L/H Red $550-800

MOBILGAS ETHYL w/ horse 4 1934-1936 13.5" Glass WN $325-525

White globe face with flying red horse at top of globe face. Black "Mobilgas" over red "ETHYL" on lower globe face.

MOBILGAS SPECIAL black letters 4 1936-1962 16.5" Metal L/H Red $500-750

MOBILGAS SPECIAL black letters 4 1936-1962 15" Metal L/H Red $500-750

MOBILGAS SPECIAL black letters 4 1936-1962 13.5" Glass WN $300-500

White globe face with flying red horse at top of globe face. Black "Mobilgas" over red "SPECIAL" on lower globe face.

MOBILGAS SPECIAL blue letters 2 1936-1962 16.5" Metal L/H Red $500-750

MOBILGAS SPECIAL blue letters 2 1936-1962 15" Metal L/H Red $500-750

MOBILGAS SPECIAL blue letters 2 1936-1962 13.5" Glass WN $300-500

MOBILGAS SPECIAL blue letters 4 1936-1962 Gill $350-550

MOBILGAS SPECIAL blue letters 5 1936-1962 Balcrank $400-650

MOBILGAS SPECIAL blue letters 2 1955-1962 Capco $200-350

White globe face with flying red horse at top of globe face. Blue "Mobilgas" over red "SPECIAL" on lower globe face.

MOBILGAS SPECIAL large "SPECIAL" 4 1946-1950 13.5" Glass WN $325-525

MOBILGAS SPECIAL large "SPECIAL" 4 1946-1950 Gill $375-575

White globe face with small flying red horse at top of globe face. Small blue "Mobilgas" over large red "SPECIAL" on lower globe face.

MOBILGAS/MOBIL GAS SPECIAL dual 6 1955-1962 Capco NL

White globe face with small flying red horse at top of globe face. Thin blue line across center of globe with blue vertical line down from midpoint of horizontal line. Blue "Mobilgas" to one side of vertical line with red arrow pointing down diagonal with blue "Mobilgas" over red "SPECIAL" with red arrow on other side of vertical line. Note that faces would be opposite for opposite sides of the pump. Very rare globe for 1950s dual pump.

MOBILFUEL DIESEL
black letters 4 1935-1962 15" Metal L/H Red $800-1100

MOBILFUEL DIESEL
black letters 4 1935-1962 13.5" Glass WN $500-800
White globe face with flying red horse at top. Black "Mobilfuel" over red "DIESEL" on lower globe face.

MOBILFUEL DIESEL
blue letters 4 1935-1962 16.5" Metal L/H Red $800-1100

MOBILFUEL DIESEL
blue letters 4 1935-1962 15" Metal L/H Red $800-1100

MOBILFUEL DIESEL
blue letters 4 1935-1962 13.5" Glass WN $500-800
White globe face with flying red horse at top. Blue "Mobilfuel" over red "DIESEL" on lower globe face.

MOBILFUEL DIESEL
SPECIAL
blue letters 4 1935-1962 16.5" Metal L/H Red $850-1200

MOBILFUEL DIESEL
SPECIAL
blue letters 4 1935-1962 15" Metal L/H Red $850-1200
White globe face with flying red horse at top. Blue "Mobilfuel" over red "DIESEL SPECIAL" on lower globe face.

MOBILGAS MARINE w/
non-outlined horse 6 1930s 16.5" Metal L/H Red
$1000-1500
White globe face with flying red horse at top. Early, local made globe with poorly drawn horse. Black "Mobilgas" over red "MARINE" on lower globe face.

MOBILGAS MARINE
w/ horse 4 1950-1962 15" Metal L/H Red $1000-1500

MOBILGAS MARINE
w/ horse 5 1950-1962 13.5" Glass WN $750-1000
White globe face with flying red horse at top. Black "Mobilgas" over red "MARINE" on lower globe face.

MOBILGAS MARINE
WHITE w/ horse 6 1950-1962 15" Metal L/H Red $1200-1750
White globe face with flying red horse at top. Blue "Mobilgas" over red "MARINE WHITE" on lower globe face.

AERO MOBILGAS
w/ horse 5 1935-1950 15" Metal L/H Red $1800-2400
White globe face with flying red horse at top. Blue "AERO" offset to left over blue "Mobilgas".

MOBILGAS AIRCRAFT
w/ horse 4 1935-1962 16.5" Metal L/H Red $1500-2000

MOBILGAS AIRCRAFT
w/ horse 4 1935-1962 15" Metal L/H Red $1500-2000

MOBILGAS AIRCRAFT
w/ horse 4 1935-1962 13.5" Glass WN $900-1300
White globe face with flying red horse at top. Black "Mobilgas" over red "AIRCRAFT" on lower globe face.

MOBILGAS AIRCRAFT
italic letters 5 1935-1962 13.5" Glass WN $950-1350
White globe face with flying red horse at top. Blue "Mobilgas" over italic red "AIRCRAFT" on lower globe face.

STOVE/MOBILHEAT/
OILS 5 1940-1955 16.5" Metal L/H Red $900-1400

STOVE/MOBILHEAT/
OILS 5 1940-1955 15" Metal L/H Red $900-1400
White globe face with black "Mobilheat" across center of globe face. Red "STOVE" above "Mobilheat", "OILS" below.

MOBILHEAT STOVE
OIL w/ no outline
horse 5 1955-1962 13.5" Glass WN $750-1000
White globe face with flying red horse at top. NO outline around horse. Blue "Mobilheat" above red "STOVE OIL" on lower globe face.

MOBILHEAT STOVE
OIL with horse 5 1955-1962 Capco $450-700
White globe face with flying red horse at top. Horse is outlined. Blue "Mobilheat" above red "STOVE OIL" on lower globe face.

KEROSENE
w/ horse 5 1940s 15" Metal L/H Red NL
White globe face with flying red horse at top. Blue "KEROSENE" on lower globe face.

MOBIL/KEROSENE
with horse 5 1950-1966 15" Metal L/H Red $750-1000

MOBIL/KEROSENE
with horse 4 1950-1966 13.5" Glass WN $500-800

MOBIL/KEROSENE
with horse 5 1950-1966 Capco $300-500
White globe face with blue outlined flying red horse positioned upper right of center. Blue "Mobil/Kerosene" positioned lower right.

MOBILOIL (can
shaped globe) 6 1947-1955 OPP NL
White glass globe cylindrically shaped. Upper area had flying red horse above blue "Mobiloil" lower third is covered by red band. Shaped and imaged in the likeness of the 1947 Mobiloil motor oil can.

****MOBIL**: Although the Mobil name replaced Mobilgas as the primary station identification on a gradual basis between 1955 and 1958, gasoline brands remained Mobilgas and Mobilgas Special until 1962 when they were replaced with Mobil Regular and Mobil Premium. Mobil again reimaged in 1966, replacing the elongated Mobil shield with the modern "red O" logo. No US globes are known with the 1966 logo.

MOBIL REGULAR 2 1962-1966 Capco $200-300
White globe face with red/white/blue "1958" flat "Mobil" shield above "Regular" in blue.

MOBIL PREMIUM 2 1962-1966 Capco $200-300
White globe face with red/white/blue "1958" flat "Mobil" shield above "Premium" in red.

MOBILFUEL
DIESEL 4 1962-1966 Capco $250-350
White globe face with red/white/blue "1958" flat "Mobil" shield above "Mobilfuel/Diesel" in blue.

Glossary

Glossary of Terms

Term	Definition
4" base	Some early one-piece globes had smaller 4 inch bases.
6" base	Ninety-eight percent of all globes come with a standard 6 inch diameter base.
7" base	Some companies used larger bases on globes, including Standard Oil Company and Eldred Oil.
13.5"	Diameter of inserts that typically fit plastic or glass bodies.
15"	A metal-frame globe—its size, or its inserts—that measures 15 inches.
16.5"	A metal-frame globe—its size, or its inserts—that measures 16.5 inches.
Air globe	Globe that sat on an air pump, usually saying "Air." This is not a gas globe.
Aircraft	A 1920s super premium grade. Later, a fuel for airplanes.
All-glass	Can mean glass body and glass inserts. Sometimes used interchangeably with the term "one-piece."
Aviation	See "Aircraft."
Balcrank	Any glass-body globe that uses three screws, positioned at twelve o'clock, four o'clock, and eight o'clock, to hold the insert in place. Manufactured by Balcrank.
Band	Globe body, as in "metal band", or a retaining ring designed to hold a glass insert to a body.
Banded glass	Glass-body globe that uses an internal ring to hold the insert in place.
Blunt-nose Eagle	White Eagle gasoline globe with a large, rounded beak.
Body	Glass or metal holder that holds inserts and lenses.
Branded	Any item with the oil company logo or brand name on it.
CAPCOLITE/ CAPCO	Plastic body for 13.5 in inserts made by the Cincinnati Advertising Products Company (CAPCO) in Cincinnati, Ohio.
Cast face	An insert, usually 15 in, for a metal-frame body. It has its letter or design cast deep into the insert, similar to an etched one-piece globe.
Chimney top	One-piece etched globe that has a smaller opening at the top which looks similar to a base. This opening should then have a metal cap that covers it—hence the term "chimney cap."
Clam	See "Shell."
Clover body	Clover-shaped body usually holding clover-shaped inserts; used by Cities Service Oil Company.
Co-op/farm co-op	Gasoline marketer owned by an association of farmers to supply their farm needs. Most market to the general public as well, but profits are distributed on a cooperative basis.
Crimped on Metal Base	See Pressed on Metal Base.
Crown	Crown-shaped globe used by the Standard Oil Company of Indiana.
Cut-throat Eagle	White Eagle gasoline globe with a slit under its throat. The beak is similar to the blunt-nose Eagle.
Diesel	Kerosene-type fuel used primarily in trucks and farm equipment. This was a specialty fuel prior to the mid-1930s.
Discounter/ tracksider	Independent or chain-operated service stations often located along a rail line. They received products by rail car, and sold gasoline at prices noticeably lower than surrounding major brand stations.
Etched letters	Letters etched into an insert, which is very rare, or more typically etched into a one-piece globe.
Ethyl	Premium-grade gasoline named for the chemical additive tetraethyl lead. This chemical was distributed on an exclusive basis by the Ethyl Gasoline Corporation.
Fantasy Globe	a phony globe that never originally existed in the past. Someone created it.
Flame	One-piece flame that went on Standard Oil Company signs only. Sought after by most globe collectors, *although it was never actually used on gas pumps.*
Frame	another term for a holder or body for globes. Frames hold inserts.
Fuel oil	Heating oil used for oil-burning heating equipment.
Full Feather Eagle	a White Eagle globe with greater feather detail than the more common blunt nose eagle.
Gargoyle	Oval one-piece Mobil Oil globe used only on oil cabinets.
Generic globes	Describes the early globes before company logos were used. These globes said "Gasoline," "Filtered Gasoline," and so on.
Gill body	A hollow glass body similar in design to the hull body, but the 13.5 in inserts are held in with a narrow metal band. Original manufacturer was Gill Glass Company, Pennsylvania. Other sizes exist.
Gill lenses	The 13.5 in inserts or other similar sizes that have no notches at nine o'clock or three o'clock. Must go on a Gill body.
Glass	Glass-body globe with glass inserts which could be a Gill body, Hull body, regular glass body, or other style. Also refers to the body itself.
High profile	Metal body that holds 15 or 16.5 in inserts. It has one seam only in the middle. The metal above the insert is raised up and has a "high profile" when viewed from the front.
Holder	See "Body."
Hull body	Glass body made by the Hull Glass Company of Swissvale, Pennsylvania. It holds 13.5 in glass inserts and the body center is hollowed out, unlike the typical solid-glass bodies.

Independent	Single station or chain operation that is a regional marketer without production. May or may not be involved in refining.
Inserts	Any glass advertisements with an oil company name or generic logo that attach to a body.
Jewel body	Metal body with twenty-four colored, glass jewels around body.
Jobber	Petroleum wholesaler operating in a limited geographic area.
Jobber association	Group of petroleum wholesalers operating under a common cooperative brand.
Jobber brand	Regional brand used by a petroleum wholesaler.
Kerosene	Fuel grade used for heating or light or as a tractor fuel.
Leaded glass	Assembled multi-piece stained-glass globe or window insert.
Lenses	See "Inserts."
Logotype	Distinctive lettering as it appeared in a logo, without the use of the entire logo design.
Lollipop-shaped	Typical round globe, distinguishing this style from an odd-shaped or globe sphere-shaped one.
Low profile	Metal body that holds 15 or 167.5 in inserts and has several ridges across the top. Overall it is very flat and when viewed from the front, the inserts are what is mostly seen.
Major	Integrated oil company; operates exploration and production, refining, transportation, and marketing over a regional or national area.
Metal	Metal-frame globe with glass inserts, or just the metal body itself.
Metal rim	Metal body.
Metal ring	Any metal ring that holds the inserts into a metal-frame globe.
Milk glass inserts	Inserts, usually for 15 in metal-frame bodies, made of milk glass. The company logo was then etched into these or fired onto the surface of the insert.
Milk glass	Certain globes that were made with actual milk-colored glass, sometimes giving off a pearlescent color effect. Term is also used by collectors to mean any glass globe, which is not always correct.
Motor	A 1930s economy grade of gasoline with no tetraethyl lead content.
Narrow body	Glass body that typically is only 5 in wide.
Neon or Neon Metal	Special globe body manufactured by Neon Globe Company of Norfolk, VA that featured clear rings around the outer edge of a lens and a neon tube inside the frame so as to light up the lens, in particular the outer clear rings.
Notched inserts	The 13.5 in inserts with notches usually at nine o'clock and three o'clock.
Oil globes	Globes typically used on oil cabinets or such places. Some globes were used on both oil cabinets and gasoline pumps.
One-piece	Globe that has no inserts and is a single piece of glass.
One-Piece Baked	a one-piece globe where the logo is baked onto the surface. This is not similar to an etched globe. The surface is very smooth overall.
One-Piece Cast	same as a raised letter one-piece. The one-piece globe was cast from a mold and part or all of the logo is typically raised up from the surface.
One-Piece Etched	The logo or print on the globe is etched down into the glass. It feels very rough or scratchy as you run your fingers over it.
One-Piece Painted	a one-piece globe wit painted, non baked details. Rarely seen.
One-piece plastic	Formed or assembled one-piece plastic globe. Does not have removable inserts.
Picker	Person who picks up collectibles to resell them.
Plastic	Usually meaning a plastic body with glass inserts, or the body itself.
Premium	Top grade of gasoline—high tetraethyl lead content.
Pressed on Metal Base	Some rare early Hull bodies had a copper base that was pressed on around the glass bottom of the globe. This base is not removable. Texaco and Indian were the biggest users of these.
Raised-letter Crown	Red Crown, Red Crown Ethyl, Solite, or Crown are the raised-letter crowns used by Standard Oil. These names are on both sides of the globe.
Raised letters	One-piece cast globe; where the letters are raised up off the surface.
Regular	Primary grade of gasoline—minimal tetraethyl lead content.
Repro	Reproduction globe.
Rim	See "Metal."
Ring	See "Metal ring" or "Gill body."
Ripple	Gill body globe, often colored, with a ripple texture to it.
Round globe	A few companies used globes shaped like a ball or sphere. Examples include many sphere-shaped, early generic "gasoline" globes.
Screw-on base	Any globe with a copper, aluminum, or metal base that screws onto the bottom of the globe.
Screw Base	See threaded base.
Shell	Shell-shaped globe used by the Shell Oil Company.
Sphere globe	See "Round globe."
Standard flame	See "Flame."
Threaded Base	Many globes have a threaded glass bottom as part of the glass globe itself. Metal, aluminum or copper bases can be screwed onto the bottom of these globes.
Three-piece globe	Any globe with two inserts and a body. More often used to reference a glass-body globe with two inserts.
Torch	See "Flame."
Two-piece	A few companies used a globe bodystyle that had two glass halves which attached in the middle by a piece of metal with screws.
White glass	White-colored glass. Unlike milk glass, it has a somewhat duller off-white appearance.
Wide body	Glass body that is wide—usually 7 in.
Window insert	Globe insert intended for use as a lighted insert in a building or other structure. Not used as a pump-top globe.

Index

**=affiliates to the major gas companies.

C

Cans, 29
Collecting globes, 12
Cross Reference to Affiliated Gas Companies, 5-6

G

Gas Companies & Their Affiliates:
Amoco, Pan-Am & Affiliates, 31-33, 129-130
  **American Oil Company, 31, 129
  **Pan-Am Southern Corporation, 32-33, 129-130
  Capitol Oil Company, 130
Asamera & Affiliates, 33, 130
  **Oriental Refining Company, 130
  **Andy-Dandy Oil Co., 33, 130
  **Gasamat, 33, 130
Ashland & Affiliates, 33-45, 130-137
  **Pepper, 33-34, 130-131
  **Ashland, 34-35, 131
  **Tri-State, 35-36, 131-132
  **National Refining, 36-37, 132-133
  **White Rose, 37-38, 133
  **White Rose/Mid-West Refining, 38, 133
  **Aetna Oil, 38-39, 133
  **Frontier Refining, 39, 133
  **Valvoline Oil Company, 39-40, 133-134
  **Freedom Oil Works, 40, 134-135
  **Independent Quality Service, 40-41
  **Dance, 41, 135
  **Payless, 41-42, 135
  **SuperAmerica (SA), 42, 135
  **Oskey Bros. Petroleum, 43, 135
  **Red Head Oil Co., 136
  **Southern Oil of NY, 43-44, 136
  **Hood Oil, 44, 136
  **Tresler Oil, 44-45, 136-137
  **Cincinnati Oil Works Co., 45, 137
  **Union Oil Co., 45, 137
  **Pyroil Chemical Co., 45, 137
Atlantic & Affiliates, 45-47, 137-139
  **The Atlantic Refining Co., 45-47, 137-139
  **Red "C" Oil Company, 139
Champlin & Affiliates, 48, 139-140
  **Champlin Oil & Refining, 48, 139
  Champlin Oil & Refining/Harbor, 48, 139
  **Eason Oil Co., 139
  **Golden Rule Oil Company, 139-140
Chevron & Affiliates, 49-50, 140-141
  **Standard Stations, Inc. & Standard Oil/Chevron Dealers, 49, 140
  **The California Company, 49, 140
  **California Oil Company-Northeast Marketing, 50, 141
  **Stanavo, 141
  **PDQ-Chevron Secondary Brand, 141
  **Signal, 141
Cities Service & Affiliates, 50-57, 141-145
  **Cities Service Oil Company, 50-54, 141-143
  **Citgo, 143
  **Lincoln Oil Co., 143
  **Elwood Oil Co., 143
  **Carter Hall Co., 143
  **Louisiana Oil Refining Co., 143
  **Crew-Levick Co., 143
  **Warner-Quinlan Company, 54, 143
  **Shoemaker Coal Co., 54, 144
  **A.R. Newcombe Oil Co., 54, 144
  **Jenney Manufacturing Co., 55, 144
  **Lindsay-McMillan Co., 144
  **Winona Oil Company, 55, 144
  **Milton Oil Co., 55-56, 144-145
  **Pioneer, 56, 145
  **Milton-Savex, 56, 145
  **Valley Oil Co., 57, 145
Clark & Affiliates, 57, 145-146
  **Clark Oil & Refining Corp., 57, 145
  **Target-Western Oil Co., 145
  **Owens Oil Co., 57, 146
Colonial & Affiliates, 57-58, 146
  **Colonial Oil Industries, 57-58, 146
Conoco & Affiliates, 58-60, 146-148
  **Continental Oil Company, 58, 146
  **Marland Oil Company, 59, 147
  **National Oil Co., 147
  **Western, 59, 147
  **Zenith Gas Pool, 59, 147-148
  **Direct Service Oil Co., 60, 148
  **Malco Refining, 60, 148
  **Valley Oil Company, 60, 148
  **New Mexico Oil Co., 148
  **Douglas Oil Co., 148
  **Kriegler, 60, 148
  **F.P. Kendall & Co., 148
Co-Operative Associations, 60-66, 148-152
  **Farmland Industries, 60-61, 148-149

**Falcon, 61, 149
**Union Oil Company, 61, 149
**Farmers Union Central Exchange, 62, 149
**Midland Cooperatives, 62-63, 149-150
**Farmers Educational & Co-Operative State Union, 63, 150
**Missouri Farm Bureau, 150
**Southern States Co-Operatives, Inc., 64, 150
**Tennessee Co-Operatives, 64, 150
**Growmark Industries, 64, 150-151
**Missouri Farmers Association, 64, 151
**Farmers Gas & Oil Company of Michigan, 64-65, 151
**Central Co-Operatives Wholesale, 65, 151
**Co-Operative GLF Farm Supplies, Inc., 65, 151
**Landmark Co-Operatives, 65-66, 151-152
**Farmers Union Co-Operative Organization of Iowa, 66, 152
**Iowa Farm Service Company, 66, 152
**Minnesota Farm Bureau, 152
**Maryland Co-Operatives, 152
**Kansas Farmers Union, 66
**UFA Central Co-Operatives Association, Ltd., 66, 152
Crown & Affiliates, 67, 152-153
**Crown Central Petroleum, 67, 152-153
**Thrift, 67, 153
**Energy Oil Co., 153
Crystal & Affiliates, 67-68, 153-154
**Crystal Oil & Land Co., 153
**Mercury Discount Stations, Inc., 153
**Berry Asphalt Co., 153
**Tulsa Oil Co., 153-154
**Crystal Petroleum Co., 154
**Eastern Petroleum Corp., 154
**Stone Oil Co., 67, 154
**High Octane Terminal Co., 154
**Platolene, 68, 154
Derby & Affiliates, 68-69, 154-155
**Derby Oil Co., 68-69, 154-155
**Soverign Service, 69, 155
Dixie & Affiliates, 69-71, 155-156
**Dixie Distributors, Inc., 69-70, 155-156
**Burford Oil Company, 70-71, 156
DX & Affiliates, 71-75, 156-159
Mid-Continent Petroleum, 71-73, 156-157
**Sunray Oil, 73, 157
**Premier, 74, 158
**Barnsdall Refining Co., 74, 158
**Bareco-Barnsdall, 74, 158
**Monarch Manufacturing Co., 74, 159
**Hawkeye Oil Co., 159
**Sweney Gasoline & Oil Co., 75, 159
Erickson & Affiliates, 75, 159-160
**Erickson Oil Co., 75, 159
**Life Oil Co., 159
**Erickson Oil Products, 75, 159-160
**Northwestern Refining Co., 159
Exxon & Affiliates, 75-83, 160-166
Standard Oil Company of New Jersey, 75-78, 160-161
**Exxon, 78, 161
**Standard Oil Co. of Louisiana, 161
**Humble Oil & Refining, 78, 162
**Texas Pacific Coal & Oil, 162
**Colonial Beacon Oil Company, 79, 162
**Stanavo, 162
**Powerine Oil Co., 79, 162
**Northwest Refining Co., 79, 163
**Yale Oil Co., 79, 163
**Oval E, 80, 163
**Carter Oil Co., 80, 163
**Enco, 163
**Pate Oil Co., 80, 163
**Oklahoma Oil Company, 80, 163
**Perfect Power Oil Co., 164
**Gaseteria Oil Co., 164
**Quality Gasoline Company, 81, 164
**Johnson Oil Refining Co., 81-82, 164
**Brilliant Bronze, 164
**Hoosier Petroleum Corp., 82, 164
**Vortex Oil Co., 165
**Ryan's, 82, 165
**Senco, Inc., 165
**Southern Oil Co., 82, 165
**Olentangy Oil Co., 165
**Wood Oil Company, 82, 165
**Alert Oil Co., 165
**Imperial Oil Limited, 82-83, 165-166
**Home Oil Co. Ltd., 166
**Champlain Oil Co., Ltd., 166
Fina & Affiliates, 84-86, 166-168
**Fina, 166
**Panhandle Oil Co., 84, 166
**Guthrie Refining Co., 166
**American Liberty, 84, 167
**Petro-Atlas, 84-85, 167
**Petroleum Marketing Co., 167
**Colonial Oil Co., 85, 167
**Cosden Petroleum, 85-86, 167-168
**Col-Tex, 86, 168
**Onyx, 86, 168
**Wides Oil Company, 86, 168
Flying A/Getty & Affiliates, 86-91, 168-171
**Tidewater Oil Company, 86-89, 168-170
**Associated Oil Company, 89-90, 170
**Tidewater Associated Oil Co.-Flying A Marketing, 91, 170-171
**Seaside Oil Company, 91, 171
**Pathfinder Petroleum, 171
**Mohawk Oil Co., 91, 171
Gulf & Affiliates, 92-96, 171-175
**Gulf Oil Co., 92-93, 171-172
**Wilshire Oil Co., 93-94, 172-173
**Sunset Oil Company, 94, 173
**Carpenter Oil Co., 173
**L.L. Coryell & Sons, 94, 173
**Superior Oil Co., 95, 173-174
**Bulk Petroleum Co., 174
**Smith Oil Co., 95, 174
**British American Oil Co., Ltd., 95, 174
**Royalite Oil Co., Ltd., 175
**Hi-Way Refineries, Ltd., 95, 175
**Puritan Oil Company, Ltd., 95, 175
**Gas & Oil Products, Ltd., 96, 175

**Anglo-American Oil Co., 175
**Artic Oil Co., 175
Hess & Affiliates, 96, 176
**Hess Oil & Chemical Co., 176
**Olixir, 96, 176
**Billups Eastern Petroleum, 176
**Meadville Corp., 96, 176
Hudson & Affiliates, 96-97, 176-177
**Silver King, 96, 176
**Hudson Oil Co., 96-97, 177
**Hornet Oil Company, 97, 177
Husky & Affiliates, 97-99, 177-178
**Western Oil & Fuel, 97, 178
**H. Earl Clack, Inc., 98, 178
**Husky-Hi Power, Inc., 98, 178
**Frontier Oil, 98-99, 178
**Western States Petroleum, 99, 178
IOMA (Independent Oil Men of America), 99-103, 179-182
Generic Red Hat Globes, 99, 179
High Hat/Jones Brothers Oil Co./Weeks Oil Co., 99, 179
Blake Oil Company, 99, 179
Consumers Oil Co., 99, 179
Gerkins Oil Company, 100, 179
Gibson Brothers Oil Co., 100, 179
H.S. Goodell & Sons, 100, 179-180
Home Oil Company, 100, 180
Independent Oil Company, 100, 180
Matheson Oil Company, 100, 180
Musco Oil Company, 100, 180
Northland Gasoline, 101, 180-181
O'Neils Paint & Hardware Co., 101, 181
Pendergast Oil Company, 101, 181
Petersons Oil Company, 101, 181
Quincy, 101-102, 181
Royal & Royal 400, 102, 181
Swanson Oil Corporation, 102, 182
Westland Oil Co., 102-103, 182
Kendall & Affiliates, 103-104, 182-183
**Kendall Oil, 103, 182-183
**Waverly Oil Works, 103-104, 183
Kerr McGee & Affiliates, 104-107, 183-186
**Deep Rock Oil Co., 104-105, 183-184
**Kermac, 105, 184
**Triangle Refineries, 105, 184
**Southport Oil Corporation, 106, 184-185
**Mileage Mart, 106, 185
**Trackside Service Stations, 106, 185
**Peoples Oil Co., 106, 185
**Wallace & Walco, 106, 185
**Mutual Oil Co., 106-107, 185
**Golden Rule, 107, 185-186
**Sparky, 107, 186
**Vaught, 107, 186
Lion & Affiliates, 107-108, 186
Mapco & Affiliates, 108, 186-187
**Red Ace Petroleum, 108, 186
**Capitol Oil Company, 108, 187
Marathon & Affiliates, 108-116, 187-193
**Lincoln Oil Refining Co., 108, 187
**Red Fox Oil Company, 108, 187
**Transcontinental Oil Co., 109, 187-188
**The Ohio Oil Company, 109-110, 188
**Tower Oil Co., 110, 188
**Aurora Gasoline Co., 110-111, 188-189
**Old Dutch Refining Co., 111, 189
**Republic Oil Co., 111-112, 189
**Bonded, 112, 189
**Consolidated, 112, 190
**Road Chief, 112, 190
**Imperial Refineries, 113, 190
**Stephen, 113, 190
**Rock Island Refining, 113-114, 190-191
**Colonial/Progressive, 114, 191
**Tulsa Oil Co., 114-115, 191
**United, 115-116, 192
**Naph-Sol Refining Co., 116, 192
**Osceola Refining Co., 116, 192
Martin & Affiliates, 116-118, 193-194
**Martin Oil Company, 116, 193
**J.D. Street & Co., 117, 194
**Martin Oil Service, 118, 194
Midwest & Affiliates, 118-120, 194-195
**Midwest Oil Co., 118-119, 194-195
**H.K. Stahl & Co., 119-120, 195
Mobil & Affiliates, 120-128, 195-201
**Standard Oil Company of New York/SOCONY, 120, 195-196
**Magnolia Petroleum, 121, 196
**General Petroleum Corporation, 121, 196
**White Eagle Oil & Refining, 121-122, 196-197
**Vacuum Oil Company, 123, 197
**Lubrite Refining Co., 123, 198
**Wadhams Oil Corporation, 124-125, 198
**White Star Refining Co., 125, 198-199
**Independent Oil Co. of Pennsylvania, 125, 199
**Metro, 125-126, 199-200
**Gilmore Oil Co., 126, 200
**Clipper Oil Company, 126-127, 200
**Mobilgas/SOCONY-Vacuum, 127-128, 200-201

Globe body styles, pictorial overview with values, 8-11
Globe categories or types, 13-17
Globe investing, 19
Globe listings, examples, 7
Globe pricing, 18
Globe reproductions, 20-25
Globe restoration, 27-28
Globe types, definitions, 8

I

Internal corporate items, 30

L

Lens removal & installation, 17

M

Maps & Paper Items, 30

P

Pumps, 29-30

R

Rarity guide, 7-8

S

Signs, 29

T

Toys, 30

Notes